WORLD ATLAS

KIMBERLEE PRESS MIAMI, FLA.

1982

FOREWORD

The gratitude of the authors goes to those who have contributed to and advised on the preparation of this Atlas, as well as to the authors of the numerous books which were used as reference material in its preparation. A special mention goes to Tirso Echeandia of AGUILAR S.A. de Ediciones, Madrid, executive vice-president and editor of Atlas Mundial Grafico from which the maps contained in this atlas are derived. The collaboration of the following has also been essential to the realization of this work:

Albertine Sirois Coulombe, translation
Maria Fernanda Enriquez de Salamanca, cartographic coordination
Gaetano Martinez, M.P. Photos Reproductions Ltd.
Montreal Lithographing Ltd., printing
Harpell's Press Cooperative, binding
Kimberlee Press, publisher

INTRODUCTION TO THE WORLD OF TODAY

If Herodotus had been commissioned to produce an Atlas in the year 420 BC, of necessity, the end result would have been a somewhat modest effort, its scope being limited by the paucity of information available to him.

Since those early days, however, our knowledge of the Universe and of the environment in which we live has increased considerably as a result of centuries of travel and exploration Science is forever probing new, challenging areas, and thus, man is now in a better position to collate all this information, condense it and pass it on to Posterity in the form of treatises, scientific manuals and atlases.

Whereas Herodotus may have been somewhat short of material, the problem of modern historians and geographers is precisely the opposite. Indeed, it would require an army of researchers, historians and writers to compile all available data on the World and transpose it into conventional reading material - but then again, the product would be so voluminous as to be totally impractical for most readers, from the point of view of time available to read it. It would therefore seem that the Atlas format is the most practical method of bringing a maximum amount of condensed information to the average modern man whose leisure time is limited by his life-style.

Following logically on our assessment of the current situation, we have attempted to give the reader in these pages, an opportunity to take a quick "trip" around the Globe, gathering information on each continent and each country's history, size, population, its main cities, ethnic components, languages, geographical features, temperature, etc. The summary which ensues is not, of course, meant to be exhaustive; it merely supplements a study of the related maps and the annotations found thereon.

For ease of reference, continents and land masses have been listed in alphabetical order, as have the various countries within same.

André D. Coulombe,
Editor

CONTENTS

STATISTICAL REFERENCE GUIDE (UNITED NATIONS STATISTICS 1980)

	Capital	Area in km²	Population	Pop. density (km²)	Annual Growth rate	Currency and value Sept. 21, 1981 *= Latest avail. info.	Annual income per capita	Inflation rate (1970 = 100)	Unemployment rate	Exports in million US$	Imports in million US$
AFGHANISTAN	Kabul	647 497	15 490.0	31	2.5	Afghani(.0234)*	241	122	8.0	429	328
ALBANIA	Tirana	28 748	2 670.0	89	2.5	Lek(.1824)*	540				
ALGERIA	Algiers	2 381 741	17 300.0	7	3.2	Dinar(.2164)	1 253	184	25.0	8 198	10 882
American Samoa(US)	Pago Pago	197	33.0	167	2.5	Dollar(1.00)	5 380	–	–	120	100
Andorra(Fr.& Sp.)	Andorra	453	28.4	62	6.5	Peseta,Frank					
ANGOLA	Luanda	1 246 700	6 900.0	4	-0.5	Kwanza(.0275)*	407	–	–	907	778
Anguilla(U.K.)	Basseterre	357	66.0	185	0.5	Dollar(.3876)*	–	185	–	9	31
Antarctica		13 200 000									
Antigua(U.K.)	St.John's	442	71.0	161	1.4	Dollar(.3717)*	509	205	–	41	87
Arctica		14 200 000									
ARGENTINA	Buenos Aires	2 776 889	26 730.0	9	1.3	Peso(.0002)	2 590	129M	2.3	8 503	5 723
AUSTRALIA	Canberra	7 686 848	14 420.0	2	1.1	Dollar(1.1629)	8 126	244	5.6	189 743	16 432
AUSTRIA	Vienna	83 849	7 510.0	90	0.2	Schilling(.0623)	7 730	173	1.8	15 483	20 254
Azores(Portugal)		2 313	330.0			Escudo(.0155)					
BAHAMAS	Nassau	13 935	220.0	15	3.6	Dollar(1.0077)	3 310	164	11.0	2 256	2 614
BAHREIN	Manama	622	350.0	416	3.2	Dinar(2.6683)	5 138	–	–	2 416	2 286
BANGLADESH	Dacca	143 998	85 560.0	546	2.4	Taka(.0654)*	118	355	–	632	1 368
BARBADOS	Bridgetown	431	250.0	580	0.7	Dollar(.5049)	1 931	331	8.0	130	312
BELGIUM	Brussels	33 099	9 890.0	137	0.1	Frank(.0270)	9 848	191	9.9	56 258	60 410
BELIZE	Belmopan	22 965	144.0	6	3.1	Dollar(.5455)*	757	–	–	50	86
BENIN	Porto-Novo	112 622	3 470.0	28	2.8	Frank(.0045)*	224	–	–	26	267
Bermuda (U.K.)	Hamilton	53	57.0	1075	1.7	Dollar(1.0077)	6 770	170	–	46	260
BHUTAN	Thimbu	647 497	1 240.0	26	2.3	Ngultrum(.1211)*	103	74	–	1	51
BOLIVIA	La Paz,Sucre	1 098 581	5 430.0	5	2.7	Peso(.0503)*	790	381	3.2	777	1 011
BOTSWANA	Gaberone	600 372	790.0	1	1.7	Pula(1.2644)*	632	173	–	2 200	220
BRAZIL	Brasilia	8 511 965	108 750.0	13	2.8	Cruzeiros(.0117)	1 635	742	2.4	15 244	19 804
Brunei(U.K.)	Bandar	5 765	177.0	31	5.2	Dollar(.4314)*	3 540	146	–	1 613	277
BULGARIA	Sofia	110 912	8 810.0	79	0.4	Lev(1.0498)	2 413	103	–	8 879	8 447
BURMA	Rangoon	676 552	32 910.0	43	2.2	Kyats(.1476)*	133	258	–	363	319
BURUNDI	Bujumbura	27 834	4 380.0	139	2.7	Frank(.0116)*	146	164	–	105	153
Caiman Is.(U.K.)	Georgetown	259	11.0	42	0.8	Dollar(1.2001)					
CAMBODIA	Phnom-Penh	181 305	8 570.0								
CAMEROON	Yaounde	475 442	8 060.0	14	2.9	Frank(.0045)*	595	235	–	1 129	1 271
CANADA	Ottawa	9 976 139	24 088.0	2	1.4	Dollar(.8524)*	8 735	197	8.1	46 065	43 434
Canary Is.(Port.)		7 273	1 139.0			Peseta(.0151)					
CAPE VERDE	Praia	4 033	320.0	74	1.9	Escudo(.0155)	230	437	–	2	44
CENTRO AFRICAN REP	Bangui	622 984	1 830.0	3	1.6	Frank(.0045)*	248	210	–	72	57
CHAD	N'Djamena	1 284 000	4 420.0	3	2.3	Frank(.0045)*	2 188	145	–	59	118
CHILE	Santiago	756 945	10 920.0	14	1.8	Peso(.0259)	1 289	310M	–	3 763	4 218
CHINA	Peking	9 596 961	1014 000.0	89	1.4	Yuan(.5940)	410	–	2.0		
COLOMBIA	Bogota	1 138 914	25 640.0	21	2.7	Peso(.0182)	890	428	7.0	3 381	4 437
COMOROS	Moroni	2 171	310.0	143	2.5	Frank(.0045)*	248	–	–	9	18
CONGO	Brazzaville	342 000	1 500.0	4	2.6	Frank(.0045)*	689	198	16.0	139	261
Cook Islands(N.Zel)	Avarua	234	18.0	7.7	-2.6	Dollar(.9880)					
COSTA RICA	San Jose	50 700	2 090.0	40	2.6	Colon(.1171)	1 646	2365	–	923	1 409
CUBA	La Habana	114 524	9 850.0	83	1.4	Peso(1.3779)	860	–	–	4 456	4 687
CYPRUS	Nicosia	9 251	620.0	69	0.0	Pound(2.6425)*	2 224	118	3.0	456	1 009
CZECHOSLOVAKIA	Prague	127 869	15 250.0	117	0.8	Koruna(.1406)	3 837	109	–	13 792	14 262
DEMOCRAT.KAMPUCHEA	Phnom-Penh	181 035	8 350.0	46	2.8	Rial(.0088)*	147	–	–	15	35
DENMARK	Copenhagen	43 079	5 120.0	118	0.3	Krone(.1431)	10 958	278	7.7	18 450	18 450
DJIBOUTI	Djibouti	22 000	106.0	5	2.2	Frank(.0056)*	1 047	–	–	36	134
DOMINICAN REP.	Santo Domingo	48 734	5 280.0	99	3.0	Peso(1.0068)	917	221	14.0	822	1 062
DOMINIQUE	Roseau	751	76.0	101	1.1	Dollar(.3876)*	419	254	–	16	28
ECUADOR	Quito	283 561	8 080.0	26	3.4	Sucre(.0413)*	958	290	5.0	1 494	1 627
EGYPT	Cairo	1 001 449	40 980.0	38	2.2	Pound(1.4599)	557	184	2.5	1 840	3 837
EL SALVADOR	San Salvador	21 041	4 660.0	196	3.2	Colon(.3971)*	679	205	10.0	1 118	1 028
EQUATORIAL GUINEA	Malabo	28 051	360.0	11	1.3	Eqpwele(.0130)*	355				
ETHIOPIA	Addis Abeba	1 221 900	30 420.0	23	2.6	Birr(.5049)*	143	204	–	423	576
Falkland,Is.(U.K.)	Stanley	12 173	2.0	0	0.0	Pound(1.8414)	–	163	–	4	7
FAROES,ISLANDS	Thorshavn	1 399	40.0	29	0.7	Pound	5 030	106	–	144	152
FIJI	Suva	18 272	610.0	32	1.8	Dollar(1.1441)	1 535	228	6.0	249	470
FINLAND	Helsinki	337 009	4 750.0	14	0.3	Markka(.2276)	7 132	262	6.1	11 175	11 400
FRANCE	Paris	547 026	53 480.0	97	0.7	Frank(.1891)	8 851	221	2.2	98 059	106 994
French Guyana	(Fr.)Cayenne	91 000	60.0	1	3.5	Frank(.1891)	1 680	176	–	7	144
French Polynesia	(Fr.)Papeetee	4 000	132.0	33	3.3	Frank(0.119)*	–	176	–	16	327
GABON	Libreville	267 667	540.0	2	1.7	Frank(.0045)*	4 926	240	–	1 307	589
GAMBIA	Banjul	11 295	580.0	48	2.8	Dalasi(.4729)*	244	250	–	58	141
GERMANY,DEM. REP.	Berlin	108 179	16 790.0	155	0.2	Mark(.4610)	4 222	97	–	13 267	14 572
GERMANY,FED. REP.	Bonn	248 577	61 340.0	247	0.3	Mark(.4488)	10 419	156	4.6	171 540	157 747
GHANA	Accra	238 537	10 320.0	43	3.6	Cedi(.3669)*	900	687	–	965	1 143
Gibraltar(U.K.)	Gibraltar	6	30.0	6	2.1	Pound(1.8414)	2 990	-211	34.5 –	4	40
Gilbert Is.(U.K.)	Tarawa	886	68.0	77	3.2	Dollar(1.11817)*	1 070	–	–	20	13
GREECE	Athens	131 944	9 440.0	69	1.1	Drachma(.0180)	3 375	304	15.6	3 855	7 648
Greenland(Denmark)	Godthaab	2 175 600	47.0	0	1.1	Krone(.1825)*	–	177	–	83	155
GRENADA	St.George's	344	96.0	279	0.4	Dollar(.3706)	484	–	–	22	45
Guadeloupe(France)	Basse-Terre	1 779	360.0	202	1.6	Frank(.1891)	–	181	–	79	376
Guam(U.S.)	Agana	549	102.0	186	3.0	Dollar(1.00)	5 620	-129	9.7	28	290
GUATEMALA	Guatemala	108 889	6 620.0	57	2.9	Quetzal(1.0068)	932	150	–	1 089	1 286
GUINEA	Conakry	245 947	4 890.0	18	2.6	Syli(.0500)*	262	–	–	258	173
GUINEA-BISSAU	Madina-do Boe	36 125	560.0	15	1.7	Peso(.0228)*	250	–	–	811	32
GUYANA	Georgetown	214 969	860.0	4	1.7	Dollar(.3376)	618	196	15.0	289	279
HAITI	Port-au-Prince	27 750	4 920.0	168	1.8	Gourde(.1988)*	278	235	–	155	212
Hawaii(U.S.)	Honolulu	16 215	770.0	46		Dollar(1.0000)					
HONDURAS	Tegucigalpa	112 088	3 560.0	25	3.6	Lempira(.4981)*	529	178	–	596	693
Hong-Kong(U.K.)	Victoria	1 045	4 078.0	4249	1.9	Dollar(.1690)	2 099	–	–	9 626	10 407
HUNGARY	Budapest	93 030	10 700.0	114	0.5	Florint(.0300)	2 277	142	–	7 938	8 674
ICELAND	Reyjavik	103 000	230.0	2	0.8	Krone(.0026)*	9 904	759	0.5	791	838
INDIA	New Delhi	3 287 590	650 980.0	186	2.1	Roupee(.1180)	184	190	8.0	6 398	8 150
INDONESIA	Jakarta	2 027 087	148 470.0	69	2.6	Rupiah(.0017)	340	351	10.0	15 578	7 225
IRAN	Tehran	1 648 000	33 510.0	20	2.2	Rial(.0116)*	2 335	222	10.9	19 307	7 261
IRAQ	Baghdad	434 924	12 770.0	26	3.5	Dinar(3.3534)	1 876	152	4.0	21 449	4 213
IRELAND	Dublin	70 283	3 370.0	45	1.2	Pound(1.6147)	3 756	304	11.8	73 180	9 837
ISRAEL	Jerusalem	20 770	3 780.0	170	2.2	Pound(.0792)	3 913	1041	3.9	1 417	7 398
ITALY	Rome	301 225	56 910.0	186	0.5	Lira(.0009)	4 587	304	7.2	72 242	77 970
IVORY COAST	Abidjan	322 463	7 920.0	16	4.3	Frank(.0045)*	1 014	272	9.0	2 516	2 488
JAMAICA	Kingston	10 991	2 160.0	188	1.5	Dollar(.5668)	1 266	321	24.2	769	1 010
JAPAN	Tokyo	372 313	115 870.0	303	1.0	Yen(.0044)	8 476	219	2.0	103 032	110 672
JORDAN	Amman	97 740	3 090.0	28	3.2	Dinar(3.3350)	842	156	8.0	402	1 949
KASHMIR	Jammu	222 237	5 700.0								
KENYA	Nairobi	582 646	15 320.0	24	3.6	Schilling(.1343)*	370	201	11.0	1 022	1 709
KIRIBATI	Bairiki	683	56.0			Dollar(1.1097)*	730		–	16	10
KUWAIT	Kuwait	17 818	1 270.0	58	6.1	Dinar(3.5623)	13 367	175	2.0	17 748	5 352
LAOS	Vientiane	236 800	3 630.0	14	2.5	Kip(.0052)*	83	457	–	11	65
LEBANON	Beirut	10 400	3 090.0	285	2.5	Pound(.2138)*	374	–	6.0	640	1 631
LESOTHO	Maseru	30 355	1 310.0	35	2.5	Maluti(1.2082)*	145	193	–	28	160
LIBERIA	Monrovia	111 369	1 800.0	16	3.4	Dollar(1.00)	427	204	20.0	486	481

STATISTICAL REFERENCE GUIDE (UNITED NATIONS STATISTICS 1980)

	Capital	Area in km²	Population	Pop. density (km²)	Annual Growth rate	Currency and value Sept. 21, 1981 * = Latest avail. info.	Annual income per capita	Inflation rate (1970 = 100)	Unemployment rate	Exports in million US$	Imports in million US$
LIBYA	Tripoli	1 759 540	2 860.0	1	4.2	Dinar(3.2915)	7 262	178	–	9 907	4 602
LIECHTENSTEIN	Vaduz	157	21.4			Frank(.6355)*	16 864				
LUXEMBOURG	Luxembourg	2 586	360.0	138	1.2	Frank(.0355)*	97 489	179			
Macao(Portugal)	Macao	16	280.0	17500	1.7	Pataca(.2094)*	780	–		241	443
MADAGASCAR	Tananarive	587 041	8 510.0	14	3.0	Frank(.0044)*	253	205	–	386	330
Madeira Is.(Port.)		789	270.0			Escudo(.0155)					
MALAYSIA	Kuala Lumpur	329 749	13 300	37	2.9	Ringgit(.4304)	1 194	166	8.0	11 079	7 844
MALAWI	Lilongwe	118 484	5 820.0	44	2.9	Kwacha(1.21)*	175	206	–	226	399
MALDIVES	Male	298	122.0	409	2.1	Rupee(.2663)*	78	–	–	5	4
MALI	Bamako	1 240 000	6 470.0	5	2.7	Frank(.0022)*	131	326	–	107	219
MALTA	Valletta	316	350.0		1.2	Pound(2.6085)*	2 121	165		424	753
Marianas Is.U.S.)	Saipan	404				Dollar(1.0000)					
Martinique	Fort-de-France	1 102	369.0	335	1.5	Frank(.1891)	2 089	184	–	128	428
MAURITANIA	Nouakchott	1 030 700	1 590.0	1	2.0	Ouguiya(.0223)*	353	135	–	119	181
MAURITIUS	Port-Louis	2 045	910.0	425	1.5	Rupee(.1643)*	1 002	274	17.0	326	501
Mayotte(France)	Dzaoudzi	374	40.0	107		Frank(.1891)					
MEXICO	Mexico	1 972 547	69 380.0	32	3.6	Peso(.0410)	1 413	280	4.0	8 768	12 004
Midway(U.S.)		5				Dollar(1.0000)					
MONACO	Monaco	1	23.0			Frank(.1891)					
MONGOLIA	Ulaan Bator	1 565 000	1 620.0	1	3.0	Tughrik(.3494)*	860			352	460
Montserrat(U.K.)	Plymouth	98	13.0	133	1.4	Dollar(.3876)*	1 036	–	–	1	7
MOROCCO	Rabat	446 550	19 470.0	40	3.0	Dirham(.2682)*	654	157	9.0	1 925	3 807
MOZAMBIQUE	Maputo	783 030	10 200.0	12	2.3	Escudo(.0315)*	197	172	–	129	278
Namibia(S-W.Africa)	Windhoek	824 292	900.0	1	2.6	Rand(1.21)*	929				
NAURU	Yangor	21	8.0	381	3.3	Dollar(1.1080)*	17 000			123	15
NEPAL	Kathmandu	140 797	13 710.0	91	3.3	Rupee(.0873)*	119	172		109	254
Nether. Antilles	Willemstad	961	241.0	251	1.4	Florin(.5568)	2 722	164		2 646	3 128
NETHERLANDS	Amsterdam	40 844	14 030.0	337	0.7	Guilder(.4058)	9 383	190	5.3	63 667	11 774
New Caledonia(Fr.)	Noumea	19 058	135.0	7	3.1	Frank(.0119)*	6 658	166	–	315	292
New Hebrides	U.K.& Fr.) Vila	14 763	97.0	7	2.8	Frank(.0141)*	480	103	–	28	35
NEW ZEALAND	Wellington	268 676	3 090.0	12	0.4	Dollar(.8404)	5 917	277	0.3	4 694	4 542
NICARAGUA	Managua	130 000	2 640.0	17	3.6	Cordoba(.0995)*	889	123	7.3	646	594
NIGER	Niamey	1 267 000	5 150.0	4	2.7	Frank(.0045)*	332	242	–	134	127
NIGERIA	Lagos	923 768	74 600.0	70	3.2	Naira(1.7655)*	730	167	8.0	9 483	12 857
Nioue Is.(N.Zel.)	Alofi	259	4.0	15	-4.5	Dollar(1.0550)					
NORTH KOREA	Pyongyang	120 538	16 246.0	135	2.6	Won(1.1204)*	470				
NORWAY	Oslo	324 219	4 070.0	12	0.4	Krone(.1723)	7 949	202	1.8	13 271	13 818
Occ.Irian(Indone.)	Jayapura	412 781	924.0			Rupiah(.0016)					
OMAN	Muscat	212 457	860.0	4	3.1	Riyal(3.0384)*	3 081	–	–	2 294	1 387
Pacific Islands	(U.S.)Saipan	1 779	125.0	70.0	4.6	Frank(.0135)*	990				
PAKISTAN	Islamabad	803 943	76 770.0	90	3.0	Roupee(.1088)	257	197	–	2 056	4 061
PANAMA	Panama	75 650	1 880.0	23	3.1	Balboa(1.00)	1 260	122	6.5	244	861
Panama, Canal Zone	Balboa Heights	1 676	50.0			Balboa(1.0000)					
PAPUA NEW GUINEA	Port Moresby	461 691	3 080.0	6	2.1	Kina(1.4518)*	628	197	–	883	808
PARAGUAY	Asuncion	406 752	2 970.0	7	3.0	Guarani(.0080)*	886	280	6.0	305	432
PERU	Lima	1 285 216	17 290.0	12	2.8	Sol(.0023)	665	884	5.2	3 533	2 022
PHILIPPINES	Manila	300 000	47 720.0	146	2.9	Peso(.1267)	506	250	3.9	3 425	5 143
POLAND	Warsaw	312 677	35 440.0	110	1.0	Zloty(.0325)	2 856	124	0.8	16 233	17 488
PORTUGAL	Lisbon	92 082	9 870.0	106	1.3	Escudo(.0155)	1 816	155	–	3 485	6 542
PUERTO RICO	San Juan	8 897	3 210.0	361	2.9	Dollar(1.00)	3 027	153	19.9		
QATAR	Ad Doha	11 000	210.0	9	5,7	Rial(.2709)*	14 856	–	–	3 598	1 425
Reunion(France)	Saint-Denis	2 510	510.0	203	2.3	Frank(.1891)	–	180	–	114	502
ROMANIA	Bucarest	237 500	21 850.0	90	1.0	Leu(.2250)	1 448	103	–	9 731	10 917
RWANDA	Kigali	26 338	4 650.0	163	2.4	Frank(.0112)*	188	199	–	70	179
SAMOA	Apia	2 842	151.0	7	1.0	Tala(1.3870)*	520	185	–	18	66
San Marino(Italy)	San Marino	61	19.0			Lira(.0009)					
SAO TOME &PRINCIPE	Sao Tome	964	81.0	84	1.6	Escudo(.0155)	407	–	–	23	14
SAUDI ARABIA	Riyadh	2 149 690	8 140.0	4	3.1	Riyal(.2945)	7 375	–	–	60 106	20 424
SENEGAL	Dakar	196 192	5 520.0	26	2.6	Frank(.0045)*	402	219	7.0	623	762
SEYCHELLES	Victoria	280	60.0	214	2.2	Rupee(.1425)*	1 378	360	–	3	46
SIERRA LEONE	Freetown	71 740	3 380.0	43	2.6	Leone(.9495)*	262	257	2.3	161	278
SIKKIM	Gangtok	7 107	205.0			Dollar(.4735)	3 316	104	10.0	14 233	17 635
SINGAPORE	Singapore	581	2 360.0	3924	1.2	Dollar(1.1817)*	433	115	–	36	35
Solomon Is.(U.K.)	Honiara	28 446	210.0	7	3.5	Schilling(.1663)*	220	205	20.0	107	241
SOMALIA	Mogadiscio	637 657	3 540.0	5	2.8	Rand(.0591)	1 594	243	–	9 956	8 352
SOUTH AFRICA	Cape Town	1 221 037	28 480.0		2.8	Won(.0015)	707	235	3.8	10 047	10 811
SOUTH KOREA	Seoul	98 484	35 860.0	364	1.8						
SOUTH YEMEN	Aden	332 968	1 850.0	5	2.7	Peseta(.0108)	3 999	359	9.8	17 903	25 432
SPAIN	Madrid	504 782	37 180.0	71	1.4	Rupee(.0520)	183	182	11.0	890	1 441
SRI LANKA	Colombo	65 610	14 740.0	209	2.0	Dollar(.3744)					
St.Helena(U.K.)	Jamestown	122	5.2	42		Dollar(.3744)*	598	230	–	23	59
St.Lucia(U.K.)	Castries	616	110.0	179	1.2	Frank(.1891)					
St.Pierre&Miquelon	(Fr.)St.Pierre	242	5.0	21	0.0	Dollar(.3744)*	339	–	–	10	30
St.Vincent(U.K.)	Kingston	388	100.0	258	1.2	Pound(2.0399)*	424	295	10.0	533	1 198
SUDAN	Khartoum	2 505 813	17 890.0	6	3.4	Florin(.5850)*	2 261	224	27.0	308	396
SURINAME	Paramaribo	163 285	380.0	3	0.8	Lilangeli(1.208)*	693	218	–	150	170
SWAZILAND	Mbabane	17 363	540.0	29	3.3	Krona(.1841)	10 543	212	1.2	27 240	28 488
SWEDEN	Stockholm	449 664	8 290.0	18	0.3	Frank(.5241)	13 335	156	1.5	26 507	29 354
SWITZERLAND	Bern	41 288	6 350.0	154	0.4	Pound(.2529)*	959	195	6.2	1 634	3 309
SYRIA	Damascus	185 180	8 090.0	41	3.3	Dollar(.0278)					
Taiwan(Nat. China)	Taipei	35 961	14 990.0			Schilling(.1293)	263	250	–	457	1 117
TANZANIA	Dar es Salaam	945 087	15 160.0	16	2.7	Baht(.0438)	484	207	–	5 308	7 156
THAILAND	Bangkok	514 000	46 140.0	84	2.5	Frank(.0045)*	327	218	10.0	159	284
TOGO	Lome	56 000	2 490.0	41	2.6	Paanga(1.3824)*	428	178	–	8	18
TONGA	Nukualofa	699	90.0	129	0.5	Dollar(.4208)	3 429	289	15.0	2 476	1 946
TRINIDAD & TOBAGO	Port of Spain	5 128	1 130.0	214	1.6	Dinar(2.5152)*	934	113	9.0	1 690	2 748
TUNISIA	Tunis	163 610	6 200.0	35	2.7	Lira(.0240)*	1 159	483	4.0	2 261	4 597
TURKEY	Ankara	780 576	44 310.0	53	2.3	Dollar(1.1097)*	50				
TUVALU	Fongafale	25	10.0			Rouble(1.2607)	2 759	100	–	64 762	57 773
U.S.S.R.	Moscow	22 402 000	264 110.0	11	0.9	Schilling(.1352)*	415	434	–	427	187
UGANDA	Kampala	236 036	13 220.0	51	3.4	Dirham(.2652)*	12 914	–	–	12 793	5 875
UNITED ARAB EMIR.	Abu Dhabi	83 600	750.0	3	8.4	Pound(1.8414)	5 545	306	6.2	91 030	102 969
UNITED KINGDOM	London	244 046	55 820.0	229	0.0	Dollar(1.00)	9 687	187	7.0	178 578	217 664
UNITED STATES	Washington, D.C.	9 363 123	220 099.0	23	0.7	Frank(.0045)*	126	–	3.0	42	191
UPPER VOLTA	Ouagadougou	274 200	6 730.0	23	2.6	Peso(.1329)*	1 713	4769	12.7	789	1 173
URUGUAY	Montevideo	176 215	2 800.0	16	0.6	Lira(.0009)					
Vatican City		44	1.0			Bolivar(.2344)	3 024	184	6.0	9 126	10 614
VENEZUELA	Caracas	912 050	13 520.0	14	3.0	Dong(.4387)*				59	618
VIET NAM	Hanoi	329 650	51 080.0	141	2.9	Dollar(1.00)	2 667				
Virgin Is.(U.K.)	Road Town	153	12.0	78	3.1	Dollar(1.00)	5 050	–	–	2 550	2 975
Virgin Is.(U.S.)		344	67.0	195	0.9	Rial(.2285)*	447			7	1 283
YEMEN,DEM.POP.REP.	San'a	195 000	5 790.0	35	2.3	Dinar(.0255)	1 721	432	16.5	6 491	12 862
YUGOSLAVIA	Belgrade	255 804	22 160.0	84	0.9	Zaire(1.2730)*	219	1132	10.0	925	589
ZAIRE	Kinshasa	2 345 409	27 940.0	11	3.7	Kwacha(1.3931)*	513	231	3.0	853	630
ZAMBIA	Lusaka	752 614	5 565.0	7	3.2	Dollar(1.51)*	589	199	–	1 154	937
ZIMBABWE	Salisbury	390 580	7 140.0	17	3.5						

THE COUNTRIES OF THE WORLD

AFRICA

Africa has frequently been referred to as the Black Continent, mainly due to the fact that most of its land still lay unexplored, deep, dark and mysterious, and also because civilization had made but few inroads into this large land - mass. However, for the last 35 years, a tremendous awakening has been taking place; a massive wave of nationalism has swept one country after another, until nearly all of Africa has shaken the bonds of colonialism. Some of this emancipation was achieved peacefully, with the cooperation of the European governments concerned; others were realised in bloodshed. Nonetheless, the light of modern day philosophy had been turned on; what it meant was that every man was entitled to influence his own destiny and that no nation had a divine right to rule others. For better or for worse, emancipation had come. The transition was, in most cases, difficult. The emerging nations, despite years of yearning, were simply not ready to govern themselves and had to accept the inevitable, i.e. the need to call back "advisers" from the ousted colonial powers to steady their hand at the helm of their newly-acquired ship. Internal strife also made it difficult for many new governments to stay on course in the pursuit of their goals, having to constantly look over their shoulders for possible coups aimed at over-throwing them, or having to take time out to quell a rebellion or turn back an invasion. Turmoil or not, Africa marches on with the definite intention of finding its rightful place in the sun of a world that appears to have long denied it this right.

Djibouti
Area: 23,000 km²
Pop.: 110,000
Cap.: Djibouti

A small enclave at the head of the Gulf of Aden, on the NE Coast of Ethiopia, Afar and Issas were formerly part of French Somaliland. The capital has a railhead linking it to Addis-Ababa. A strategic post on the "Petrol Route".

Algeria
Area: 2,381,741 km²
Pop. Grth.: 3.2
Pop.: 19,130,000
Cap.: Algiers

Algeria lies on the Mediterranean Coast. To the West is Morocco whilst in the East, its boundaries touch Tunisia and Libya. Most of the Country is covered by the hot and arid Sahara Desert, while the remaining portion of the land is occupied by the rugged Atlas Range whose corridors and plateaus are used for cattle-grazing.

Algeria produces tropical fruit, olives, grapes, cereal grain, tobacco and cotton. It is also a large producer of natural gas and petroleum.

The population is made up of Berbers, Moors, Arabs and Blacks, as well as a number of Europeans. Occupied initially by Berbers, the Country fell under the rule of ancient Carthage, and then, that of the Romans. It was later devastated by Vandals, passed under Byzantine domination, was taken over by Arabs and finally by Turks. In the 14th Century, the region of Algiers was a haven for pirates, the most famous of whom was Barbarossa. After unsuccessful attempts by France and England at ridding the seas of this scourge, the former power succeeded in conquering the Country in 1830. Algiers became French territory until 1962 when the Front de Libération National, after clashing with French forces from 1954 on, contributed to its emancipation. A one-party governement was set up and since then, the political situation has been fairly stable.

Angola
Area: 1,246,700 km²
Pop. Grth.: 2.5
Pop.: 6,900,000
Cap.: Luanda

Angola lies on the Atlantic Coast, SW of Zaire with which it has constant brushes due to the fact that Katangese rebels use the Angolan enclave as a base for their operations against the central government of Zaire. Angola's SE boundary looks upon Zambia, while to the North, it is bounded by the French Congo.

A former Portugese possession, Angola gained its independence in 1975. Coffee, diamonds, petroleum and iron ore are its main products.

Benin
Area: 112,622 km²
Pop. Grth.: 2.8
Pop.: 3,470,000
Cap.: Porto Novo

Formerly known as Dahomey, Benin is squeezed between Nigeria to the East and Togo to the West; it has a narrow strip of land on the Atlantic Ocean. After securing its independence from France in 1960, it suffered no fewer than five coups before settling down to farming and to palm oil exporting under its present military government.

Botswana
Area: 600,372 km²
Pop. Grth.: 1.7
Pop.: 790,000
Cap.: Gaborone

Botswana, formerly Bechuanaland, is bounded in the West by S.W. Africa, in the South by the Republic of South Africa and in the East by Rhodesia. Botswana contains the Kalahari Desert where Bushmen eke out an existence, much as their primitive ancestors did. Cattle-raising is the main economic activity.

Burundi
Area: 27,834 km²
Pop. Grth.: 2.7
Pop.: 4,380,000
Cap.: Bujumbura

Burundi is a small nation located at the north end of Lake Tanganyika, East of Zaire; it is densely populated and its people are essentially farmers. The country exports coffee. Formerly linked with Rwanda, it was first under German rule, then under Belgian administration until 1962.

Cameroun
Area: 475,442 km²
Pop.: 8,060,000
Cap.: Yaoundé

The Cameroun has the dubious honor of having the wettest place in Africa, Mt. Cameroun, with 10,160 mm of rain a year ! Located on the Atlantic between Nigeria in the North and Gabon in the South, it is a mountainous country which lives off its natural resources such as wood, rubber, coconut oil, cocoa, cotton, coffee, bananas, rice, maize and cattle-raising.

Formerly occupied by Germany (1884), it was shared by England and France after WW I. It is now an independent state.

Central African Republic
Area: 622,984 km²
Pop. Grth.: 1.6
Pop.: 1,830,000
Cap.: Bangui

Up to 1960, the Central African Empire was part of French Equatorial Africa; it then became an independent state and one of the poorest in Africa. Lack of maturity and political stability have led the Country to a shaky military government. Its population farms to live.

Chad
Area: 1,284,000 km²
Pop. Grth.: 2.3
Pop.: 4,420,000
Cap.: N'djamena

Like the Central African Empire the Congo and Gabon, Chad gained its independence from France in 1960. Situated between the Sudan in the East and Niger in the West, its northern boundary includes the Sahara.

Chad is essentially an agricultural nation and cattle-raising is its main source of revenue. However, it has uranium and petroleum deposits. Chad is a military republic.

Congo
Area: 342,000 km²
Pop. Grth.: 2.6
Pop.: 1,500,000
Cap.: Brazzaville

Like Chad, the Congo (not to be confused with the former Belgian Congo, now Zaire), gained its independence from France in 1960. Politically unstable, it was quickly taken over by the military.

Timber is its main export; it also produces peanuts, palm oil, maize, coffee and rice. It mines copper and stanum.

Egypt
Geography
Area: 1,001,449 km²
Pop. Grth.: 2.1
Pop.: 40,980,000
Cap.: Cairo

Situated on the Mediterranean, Egypt is separated from Israel, Jordan and Saudi Arabia by the Red Sea, while it is bounded in the West by Libya. To its South, lies the Sudan. Geographically and except for the Nile Valley where most people live, Egypt is a desert of sand and rocky outcrops. To the East, lie mountains that overlook the Red Sea. The mountainous Sinai Peninsula has frequently changed hands between Arabs and Israelis and now contains a United Nations buffer zone. The desert region to the West contains the Qattara Depression of WW II fame. The climate of Egypt is hot and dry; rainfall is practically non-existent in the South.

History and Culture

Egypt has produced one of the Ancient World's greatest civilizations. The feats of Egyptian astrologers, architects, physicians and embalmers are legendary and do not pale before modern day achievements. Conquered by Alexander the Great in 332 BC, the Country later became a Roman province, and later still, the centre of the Coptic Church. In AD 640 it fell under Arab domination, adopted Islam as a religion and Arabic as its language. In the 16th Century, it became a Turkish province and from 1798 to 1801, was ruled by France. Mohammed Ali Pasha, appointed governor by the Turks in 1805, defeated the Ottomans in 1841; he did a great deal to modernize Egypt, and his death was followed by a decline in the Nation's prosperity. The Suez Canal was open to navigation in 1869. During WW I and until 1922, Britain occupied and held the Country as a protectorate; it then became a monarchy. During WW II and though neutral, Egypt served as a battle-ground for Allied forces fighting German and Italian troops. The abuses of King Farouk's government and its inefficiency resulted in the King being overthrown by General Naguib who abolished the monarchy and instituted a republic in 1952-53. Naguib was followed by Colonel Gamal Abdul Nasser who made Egypt into a nationalist state and embraced the cause of Pan-Arabism; this resulted in the agreed departure of British troops in 1954 and the nationalization of the Suez Canal in 1956, despite a Franco-British landing aimed at holding the Canal. From 1948 to 1973, Egypt was involved in four wars with Israel over the partition of Palestine and the establishment of the State of Israel. Nasser's successor Anwar Sadar who is making history by taking positive steps towards settlement of the ARAB-ISRAELI problem. The signing of a peace treaty with ISRAEL, March 26, 1979 in Washington after months of negociations ended a state of war that had persisted over the last 30 years.

Economy:

WWII did not make Egypt prosperous nor have its constant wars with Israel which have resulted in the latter country occupying the Sinai Peninsula and in the Suez Canal being closed to navigation. Egypt has considerable petroleum deposits and an imposing number of thriving industries (textiles, cotton, tobacco, fertilizers, paper, footwear, sugar refining, steel, iron, and machinery). Most of the Country's hydro-electric power comes from the Aswan Dam on the Nile, just before the first cataract. The majority of the Egyptians farm the land of the Nile Valley, aided by large scale irrigation of their fields. Crops include cotton, beans, barley, maize, millet, rice, onions, wheat and sugar-cane. Cattle-raising is also of importance. Although Egypt has received military equipment and advisers from the USSR, it tries to "stay on the fence" between East and West, keeping its eyes focussed on its main objective, Arab unity, which it has not been able to achieve so far.

Equatorial Guinea

| Area: 28,051 km² |
| Pop. Grth.: 2.3 |
| Pop.: 360,000 |
| Cap.: Malabo |

Equatorial Guinea is made up of the small territory of Rio Muni, an enclave NW of Gabon, on the Atlantic coast of Africa, and of Macias Nguema Buyoga, formerly Fernando Po. Prior to 1968, the Country was under Spanish administration. Coffee and cocoa are its main exports.

Ethiopia

| Area: 1,221,900 km² |
| Pop. Grth.: 2.6 |
| Pop.: 30,420,000 |
| Cap.: Addis-Ababa |

On the southern shore of the Red Sea, Ethiopia is lodged between the Sudan to the NW and the Somali Republic to its SE. It consists of a rift valley enclosed by two mountainous, fertile plateaus. The northern plateau where the capital is located, is a range of volcanic peaks, while the southern plateau, not so lofty, descends toward arid, desert areas in the SE. The climate of the plateaus is cool.

Ethiopians are farmers who cultivate the land and raise cattle ; their main export is coffee, wax and ivory. The Blue Nile and the Atbara river drain this hot land, while its Red Sea coast is arid.

In the centuries BC, Ethiopia was under the rule of the Pharaohs of Egypt; its modern era started in 1855. The Country had to put up with Italian ambitions as far back as 1896, when its emperor, Menelik II defeated the Italians at Adoua. However, in 1935, the Italians who already ruled Erithrea and Somalia, came back to conquer Abyssinia which they held until 1941 when British troops liberated the Country. Haile Selassie was back on his throne in 1947. The Emperor was deposed by a coup in 1974 when Ethiopia became a republic.

Gabon

| Area: 267,667 km² |
| Pop.: 540,000 |
| Cap.: Libreville |

From 1910 to 1960 when it acquired its independence, Gabon was part of French Equatorial Africa. It has retained close ties with France and lives off the revenues derived from its exports (timber, rubber, wax, gold, iron, coffee, manganese diamonds, petroleum and uranium).

Gabon's political stability has contributed to make it one of the prosperous countries of Africa. It is situated on the southern portion of the Gulf of Guiana, between the French Congo and the Cameroun.

Gambia

| Area: 11,295 km² |
| Pop. Grth.: 2.8 |
| Pop.: 580,000 |
| Cap.: Banjul |

Gambia is a small enclave surrounded by Senegal; it has a narrow front on the Atlantic. The Country acquired its independence in 1965. Its chief exports are groundnuts.

Ghana

| Area: 238,537 km² |
| Pop. Grth.: 3.6 |
| Pop.: 11,320,000 |
| Cap.: Accra |

A former British possession known as the Gold Coast, Ghana gained its independence in 1957 and became a republic in 1960. Between two military coups, one in 1966, the other in 1972, Ghana had a brief period of civilian rule. Currently, a military government holds the power.

The World's largest producer of cocoa, Ghana also exports timber, gold, diamonds. Hydro-electric power provided by the Volta River permits aluminum smelting.

Geographically, Ghana is divided into a southern forested area and a northern region where savannas are used for cattle-grazing and cultivation. The Country is situated on the Bight of Benin; its neighbors are the Ivory Coast to its West and Togo to its East. Accra, the capital, lies on the Atlantic seaboard.

Guinea

| Area: 245,957 km² |
| Pop. Grth.: 2.6 |
| Pop.: 4,890,000 |
| Cap.: Conakry |

Prior to 1958 when it was a French colony, Guinea was an agricultural country. Today, the independent republic mines and exports bauxite, diamonds and iron ore. Its inhabitants are Muslims but the official language is French.

Guinea is situated on the NW coat of Africa, SW of Mali, between Guinea-Bissau to its NW, Sierra Leone, Liberia and The Ivory Coast to its SE and East.

Guinea-Bissau

| Area: 36,125 km² |
| Pop. Grth.: 1.7 |
| Pop.: 560,000 |
| Cap.: Bissau |

Guinea-Bissau was a former Portugese colony which gained its independence in 1974, after a long struggle. The economy of the Country is based on farming. Guinea-Bissau is immediately NW of Guinea, on the Atlantic.

Ivory Coast

| Area: 322,463 km² |
| Pop. Grth.: 4.3 |
| Pop.: 7,920,000 |
| Cap.: Abidjan |

This country on the Gulf of Guinea, lies between Ghana to the East and Liberia/Guinea to the West. Its southern forests supply timber for export and plantations are numerous in the region. The northern savannas are used for cattle-grazing and for growing cotton and millet. Coffee and cocoa are also on the export list. Processed farm products, paper and cement come out of the Ivory Coast's industrial pipeline.

Formerly a French possession, the Country has been an independent republic since 1960.

Kenya

| Area: 582,646 km² |
| Pop. Grth.: 3.5 |
| Pop.: 15,320,000 |
| Cap.: Nairobi |

More than half of Kenya is a high plateau flanked in places by precipitous walls and volcanic mountains; these are rich in scenery and wild life which attract tourists and safaris from all over the World. After tourism, coffee is the important source of revenue. The SW Plateau is ideal for farming, hence the influx of Europeans who settled the land from 1920 on. These, with Asian emigrants who joined them, ran afoul of the native Kikuyu who considered they were already short of land. There followed bloody uprisings with Mau-Mau massacres shaking the World. Nonetheless, Kenya gained its independence from Britain in 1963 and became a one-party state. Jomo Kenyatta, its leader, then proceeded to redistribute the land, africanize its people and develop the economy. Mombasa is the core of Kenya's industrial development.

Lesotho

| Area: 30,355 km² |
| Pop. Grth.: 2.5 |
| Pop.: 1,310,000 |
| Cap.: Maseru |

A land-locked country in the middle of South-Africa, somewhat toward its East Coast, Lesotho was formerly British and known as Basutoland; it became independent in 1966. Cattle-raising and the supplying of migrant labor to the Republic of South Africa are important to Lesotho.

Liberia

| Area: 111,369 km² |
| Pop. Grth.: 3.4 |
| Pop.: 1,800,000 |
| Cap.: Monrovia |

Between the Ivory Coast and Sierra Leone, Liberia rests on the Atlantic Ocean. Like that of Sierra Leone, the history of Liberia, has slave trade connotations. Indeed, Afro-American slaves freed from their bondage, came back to Africa to found Monrovia in 1822. Although independent since 1847, Liberia has remained in the orbit of the USA and its governent is modelled on that of America.

Iron ore and rubber are the Country's chief exports. Liberia has a large commercial fleet whose ships sail under different flags because of the low rates charged by Liberia. Numerous naval mishaps to such ships, resulting in extensive oil spilling with ensueing pollution, are being looked into and may result in a tightening of control over ship-leasing.

Libya

| Area: 1,759,540 km² |
| Pop. Grth.: 4.2 |
| Pop.: 2,860,000 |
| Cap.: Benghazi |

On the shores of the Mediterranean, between Egypt and Algeria, Libya's land is 9/10th desert. The remainder, which is fertile, is found on the coast and yields typical Mediterranean crops such as barley, nuts and olives. Libya's low, southern plateau runs into the Sahara. In this desert region, life is found only near oases and petroleum fields, the latter contributing to the Country having the highest per capita income of the Continent.

Formerly under Turkish, then Italian rule (1912) Libya, like Egypt, had to suffer the presence of WW II antagonists. In 1943, Britain took control of the northern part of the Country, while France looked after the South. In 1951, Libya became an independent monarchy, but political instability soon appeared with the military taking over in 1969. A Revolutionary Command Council then established a form of socialism acceptable to the Islamic mentality.

Malagasy Republic

| Area: 587,041 km² |
| Pop.: 8,510,000 |
| Cap.: Tananarive |

This group of islands includes Madagascar, the 4th largest island in the World; it lies off the East Coast of Africa, opposite Mozambique, Swaziland and Natal. The eastern part of the Island is forested, while the western coast is covered by plains. The S.W. is rather arid and not propitious to cultivation. Farming is the main activity, although less than 5% of the land is under cultivation.

Settled by Indonesians 1000 years ago, Madagascar itself was ruled by France from 1885 until it became an independent military republic in 1960.

Malawi

| Area: 118,484 km² |
| Pop. Grth.: 2.9 |
| Pop.: 5,820,000 |
| Cap.: Lilongwe |

Formerly the British Protectorate of Nyasaland, Malawi spreads along the western shore of Lake Nyasa, West of Mozambique and Tanzania. The Country has no opening on the Indian Ocean. Agriculture is the main occupation of a large population which must migrate to other countries to seek work. Main crops are tea and tobacco.

Malawi obtained its independence in 1963; it has a one-party republican form of government.

Mali

| Area: 1,240,000 km² |
| Pop. Grth.: 2.7 |
| Pop.: 6,470,000 |
| Cap.: Bamako |

Mali is land-locked nation of N.W. Africa; it is surrounded by Niger, Upper Volta, Senegal, Guinea, Mauritania and Algeria. Gaining its independence from France in 1960, Mali soon became a military republic in 1968. The northern part of the Country is covered by the Sahara, but the South is savanna. Agriculture is the mainstay of Mali.

Mauritania

| Area: 1,030, 700 km² |
| Pop. Grth.: 2.0 |
| Pop.: 1,590,000 |
| Cap.: Naouakch |

Like the Western Sahara, Mauritania is largely a desert country facing the Atlantic from the N.W. shoulder of Africa; it has been independent since 1960. The Country has important deposits of iron ore.

Mauritius

| Area: 2,045 km² |
| Pop. Grth.: 1.5 |
| Pop.: 910,000 |
| (mainly Indian) |

Mauritius is an island due East of Madagascar; it formerly belonged to France, but has been within British influence since 1810. Volcanic soil and a damp, warm climate combine to help the inhabitants cultivate sugar-cane.

Morocco

| Area: 446,550 km² |
| Pop. Grth.: 3.0 |
| Pop.: 19,470,000 |
| Cap.: Rabat-Sale |

Morocco is situated on the N.W. Coast of Africa, opposite Gibraltar. Its land is made up of the Atlas Mountain Range which overlooks low, fertile plateaus, separated from the higher ones by an arid strip of land. The temperature of the plateaus is fairly clement, but that of the mountains gets extremely hot, the thermometer frequently reaching 37°C (100°F) in the summer, only to dip to below freezing in the winter.

Agriculture and fishing are the main activities of the Moroccan population. Phosphates are produced in large quantities and manufacturing is gaining momentum. Main cities are Rabat, Casablanca and Marrakech.

From 1912 to 1956 when it became a constitutional monarchy, most of Morocco was a French protectorate. The northern part of the Country was ruled by the Spaniards. The presence of important deposits of phosphates in the Spanish Sahara was instrumental in Morocco claiming the territory in 1970. Spain agreed to leave in 1976, but Saharan nationalist elements opposed the decision. (see Western Sahara)

Mozambique

| Area: 801,590 km² |
| Pop. Grth.: 2.3 |
| Pop.: 10,200,000 |
| Cap.: Maputo |

Mozambique is a former Portugese colony situated on the Indian Ocean Coast of Africa; it gained its independence in 1975 after many years of guerilla activity. Farming is of great importance to this Country which produces and exports rice, sugar, tea, nuts, copra and cotton. Mozambique also benefits from its ports and resulting transport activities.

Namibia
(South-West Africa)

| Area: 824,295 km² |
| Pop.: 900,000 |
| Cap.: Windhoeck |

Situated on the Atlantic Coast between Angola and South-Afica, Namibia is a huge, arid, low-density populated country which has its own Namib desert and shares the Kalahari with Botswana to its East. Live-stock raising in the highlands and the mining of uranium, copper and lead are the two most important economic activities of this country.

The UN has had to exert considerable pressure on South-Africa (mandated to rule this former German possession), to free Namibia. The UN eventually declared the mandate to be illegal after South-Africa's refusal to abide by its decision.

Niger

| Area: 1,267,000 km² |
| Pop. Grth.: 2.8 |
| Pop.: 5,150,000 |
| Cap.: Niamey |

Like Mali, Chad and Upper Volta, Niger obtained its independence from France in 1960. Like the two first mentioned, its northern region is covered by the Sahara. Agriculture is the Country's mainstay.

Nigeria

| Area: 923,768 km² |
| Pop. Grth.: 3.2 |
| Pop.: 74,600,000 |
| Cap.: Lagos |

Nigeria (not to be confused with Niger), is a former British possession situated on the Gulf of Guinea; it has a large soft water lake, Lake Volta. The history of Nigeria goes back two centuries BC. From 1861 onwards, Britain's influence in Nigeria grew until in 1900, the southern part of the Country had become a colony, whilst by 1906, the northern sector had become a protectorate. Both regions merged in 1914. In 1960, Nigeria became independent and in 1963, it was constituted a republic.

The most densely populated country of Africa, Nigeria has no fewer than 250 tribal groups which has caused many problems, including military coups in 1966, and a civil war from 1967 to 1970 between the Ibos and Hausas. An attempt was made at easing tension by breaking up the Country into 12 regions, but this had the result of causing the eastern tribal groups to break away to form the Republic of Biaffra. More bloodshed followed which echoed around the World. Biaffra was eventually defeated in 1970, and the Supreme Military Council is trying to promote racial unity whilst endeavoring to rebuild the economy.

Low plateaus, plains, mangrove swamps off the coast, a rich river delta, (Niger), tropical forests which become savanna in the hinterland are characteristic of Nigeria's geography. The climate is hot and humid the year round, averaging 27°C (80°F) and reaching as much as 46°C or 115°F at the height of summer.

Economy:

Major petroleum deposits, coal and tin, engineering and metal industries combine with agriculture, food and timber-processing to provide the Country with revenue. Cattle-raising and fishing are also important activities in Nigeria. Britain heads the list of the Country's trading partners.

Zimbabwe

Area: 390,622 km²
Pop.: 7,140,000
Cap.: Salisbury

Rhodesia is a land-locked country surrounded by Mozambique, Zambia, Botswana and South-Africa.

In the 1880's, a High Commissioner based on South-Africa, ruled Rhodesia. The Country obtained self-government in 1923 and from 1953 to 1963, was part of a short-lived federation with Mali and Zambia. In 1965, Rhodesia declared itself independent, but Britain refused to ratify this decision and imposed economic sanctions which failed to alter the situation. In 1970, Rhodesia had become a republic with a government dominated by Europeans. Ethno-political problems continue to crop up, despite all efforts at finding a solution acceptable to the Black majority and the ruling White minority. The Matabele in the South and the Mashona in the North are the largest tribal groups. Sixty per cent of the natives live on reserves; others work in towns, mines or on White-owned farms. Rhodesia is divided into a high and a low veld, the latter including the Zambesi trough and the Limpopo Valley. Savanna covers a large portion of the Country. Due to the high altitude, the tropical climate is quite pleasant and has been a factor in attracting European settlers to the land. Rainfall is adequate.

Manufacturing with Salisbury and Bulawayo as the main centres accounts for 23% of the GNP. Hydro power comes from the Kariba Dam in the North. Agriculture runs close second to manufacturing, with tobacco, the main cash crop. Products of the land include maize, tea, sugar-cane and citrus fruit. Cattle raising is of importance. The mining industry is more modest; asbestos, coal, copper, gold and chrome are mined. The economic sanctions already mentioned, have had an impact on Rhodesia's economy; Mozambique having closed it's doors, Rhodesia was left with only two outlets to the Sea, the railway line through Botswana, and one through South-Africa, only recently put into operation (1974).

Rwanda

Area: 26,338 km²
Pop. Growth.: 2.4
Pop.: 4,650,000
Cap.: Kigali

Like Burundi, Rwanda is a small, agricultural, densely-populated nation located on the shores of Lake Tanganyika, on the eastern frontier of Zaire. As Rwanda-Urundi, the Country was colonized by Germany in the 1800s and then passed to Belgian administration after WWII. In 1962, the Country split into two independent nations. Coffee is the main export.

Senegal

Area: 196,192 km²
Pop. Grth.: 2.6
Pop.: 5,520,000
Cap.: Dakar

On the NW Coast of Africa, Senegal is surrounded by Mauritania, Mali, Guinea and Guinea-Bissau; it is also penetrated by the Gambian enclave on its West Coast. A former French colony, Senegal has been a one-party republic since 1960.

Groundnuts constitute its main export, but recent discoveries of iron ore look very promising.

Sierra-Leone

Area: 71,740 km²
Pop. Grth.: 2.6
Pop.: 3,380,000
Cap.: Freetown

Between Liberia and Guinea on the NW Coast of Africa, Sierra-Leone's history, like that of Liberia, is linked with slavery. Freetown, its capital, was founded in 1788 by freed American slaves. Under British rule from 1808 to 1961, the Country became a republic ten years later; its main exports are minerals including diamonds and iron ore. Farming occupies 80% of the population.

Somalia

Area: 637,657 km²
Pop. Grth.: 2.8
Pop.: 3,540,000
Cap.: Mogadiscio

The Somali Republic is situated on the NE Coast of Africa, on the Indian Ocean. It is a mountainous land in its northern reaches, while in the South, it is arid and desert-like. The only fertile regions are those along the Juba and the Shabelle rivers where livestock is raised.

Historically, the Republic stems from British and Italian Somaliland, united after 1960, when both colonies secured their independence. The government is pro communist.

South-Africa

Area: 1,221,037 km²
Pop. Grth.: 2.8
Pop.: 28,480,000
Caps: Cape Town & Pretoria

South-Africa which occupies the tip of the Continent, is essentially a high plateau, 1200 m above sea level, bordered by steep cliffs leading away towards the sea or neighboring lands. The Country is well drained by the Orange and Limpopo rivers. The altitude and the vicinity of the ocean soften the tropical climate; in fact, it gets quite cold in the winter and frosts occasionally damage crops. The Cape Province has possibly the most pleasant climate.

Dutch emigrants (Boers) first settled the land in 1652, on the site of present day Cape Town. Spreading steadily inland, they came to grips with the Bantu Nation in 1781. Meanwhile, the British ruled the Cape area from 1795 to 1806. The Afrikaners or White settlers opposed British rule, and at the end of a long march North (Great Trek), founded the Orange Free State and the South African Republic, now known as the Transvaal. This was eventually to lead to two wars with Britain (1880-81 and 1899-1902), both won by the latter. The Union of South Africa was born, with Natal, the Transvaal and the Cape Province joining the Orange Free State.

After WWI, South-Africa had been mandated to administer German SW Africa (present-day Namibia) but its post-WWII racial policies did not go down well with an evolving world opinion. Feeling the pressure and deciding to exercise discretion, South-Africa broke away from the British Commonwealth and declared itself independent in 1961.

Less than 20% of South-Africa's population consists of European Whites. Despite their numerical inferiority, the Whites control the government and dominate the economy. The largest groups of natives, the Zulus and the Xhosas are relegated to reserves in an effort to segregate the races in accordance with the government's Apartheid policy. However, the scheme is not very viable, as more than half the Blacks work in White-controlled areas. The Colored or people of mixed origin live in the Cape Province and provide the unskilled labor force. Asians are found mainly in Natal where they farm, trade or work in factories.

Despite racial frictions and their impact abroad, South-Africa continues to trade with the World at large. The year 1979 being the year when the Bantus are due to be freed from rule, the World is poised, waiting to see what in fact will happen; perhaps there will then be a better atmosphere, more conducive to freer trade and better relations.

Sudan

Area: 2,505,813 km²
Pop. Grth.: 3.4
Pop.: 17,890,000
Cap.: Khartoum

Located between Ethiopia to the East, Libya and Chad to the NW & W, and Egypt to the North, the Sudan has a relatively narrow coast on the Red Sea. It is the largest country of the Continent. The South is forested while the North is desert; in between, lie savannas.

The South is essentially Black and has over 100 dialects, while the North is predominantly Muslim and Arabic-speaking. This division of the Country led to a civil war from 1964 to 1972, out of which the South gained a large degree of autonomy.

Farming is the main activity and industrialization is making strides. The Sudan has been an independent military republic since 1956.

Swaziland

Area: 17,363 km²
Pop. Grth.: 3.3
Pop.: 540,000
Cap.: Mbabane

A former British possession, Swaziland is a small enclave on the SE tip of Africa, on the Indian Ocean, just South of Mozambique. Independent since 1968, the chief occupation of its population is cattle-raising. Mining is however, promising.

Tanzania

Area: 945,087 km²
Pop. Grth.: 2.7
Pop.: 15,160,000
Cap.: Dar-Es-Salaam

Situated on the shores of the Indian Ocean, between Kenya to its North and Mozambique to its South, Tanzania consists of mainland Tanganyika and the offshore Island of Zanzibar, a union which occured in 1964.

Two main features single out this country; one is famous Mount Kilimanjaro, immortalized by Hemingway, the other is the most important port of Dar-Es Salaam on the Indian Ocean, the heart of its industry.

German from 1880 to 1918, Tanganyika was taken over and administered by Britain until its independence in 1961 - that of Zanzibar came in 1963.

The Country has a hot coastal belt and a cooler inland plateau; its farming is hampered by irregular rainfall and poor communications. The completion of the Tanzania-Zambia railway has improved the situation. Tanzanians work in cooperatives not unlike those in Israel.

There are over 120 ethnic groups in the Country.

Togo

Area: 56,000 km²
Pop. Grth.: 2.6
Pop.: 2,470,000
Cap.: Lomé

Situated on the Bight of Benin, Togo has a narrow coast-line; it is flanked by Benin in the East and Ghana in the West. Its southern forests provide timber while its northern savannas are used for cattle-grazing as well as the growing of millet and cotton. The Country's main exports are phosphates, palm oil, cocoa and coffee.

What was left of the initial Togoland, after incorporation into Ghana, was ruled by France until 1960. Two coups turned Togo into a military republic.

Tunisia

Area: 163,610 km²
Pop. Grth.: 2.7
Pop.: 6,200,000
Cap.: Tunis

Lodged between Libya and Algeria on the Mediterranean Coast, East of Gibraltar, Tunisia evokes memories of the past when Carthage of Hannibalic fame, thrived close to the site of present-day Tunis.

Bright sunshine is characteristic of this Country which is getting increasingly popular with tourists. The South is desert while the North gets an adequate amount of rain. Mining is the main industry, but farming occupies the time of most inhabitants - Barley, grapes, dates, figs, olives, oranges, maize, oats and wheat are the main products of the land.

An independent monarchy in 1956, Tunisia became a one-party republic the following year.

Uganda

Area: 236,036 km²
Pop. Grth.: 3.4
Pop.: 13,220,000
Cap.: Kampala

North of Tanzania and Lake Victoria, Uganda lies between Zaire, Kenya and the Sudan. Fairly high above sea level, its temperature is milder than one usually expects of an equatorial region. The rich soil encourages farming which is the Country's major industry.

Ruled by Britain from 1893 to 1962, when it acquired its independence, Uganda has been the scene of considerable tribal unrest which has resulted in the banishment of 50,000 Uganda-born Asians, under the pretext of "Africanization". Uganda is presently invaded by Tanzanian forces.

Upper Volta

Area: 274,200 km²
Pop. Grth.: 2.6
Pop.: 6,730,000
Cap.: Ouagadougou

Situated between Mali and Ghana on the NW shoulder of Africa, Upper Volta has been a military republic since 1966; it had a brief period of civilian rule in 1971, but returned to the Military in 1974. The Country experiences considerable if irregular rainfall. Agriculture is its mainstay. A large portion of the work-force has to migrate South to find work.

Western Sahara

Area: 266,000 km²
Pop.: 100,000
Cap.: El Aaiun

On the NW Coast of Africa, the Western Sahara gets its only moisture from the Atlantic. It is situated between Morocco and Mauritania.

Formerly known as the Spanish Sahara, it took its present name when Spain pulled out in 1976. Morocco was to take over the North, but Saharan nationalists resisted and full independence resulted.

Phosphates are mined. Fishing and the growing of dates are the main industries.

Zaire

Area: 2,345,409 km²
Pop. Grth.: 3.7
Pop.: 27,940,000
Cap.: Kinshasa

Formerly known as the Belgian Congo, Zaire, the second largest country in Africa, is just short of being land-locked; it has a narrow corridor to the Atlantic through the Matadi-Boma-Cabinda strip. In the centre of the Continent, it is surrounded by the Central African Republic, the Sudan, Uganda, Rwanda, Burundi, Lake Tanganyika, Zambia, Angola and the Congo. Most of the Country lies in the basin of the River Zaire (Congo) and its tributaries; it is covered with hot, humid, tropical forests. Savannas in the South are drier. There are mountain ranges in the South and East. Its capital, Kinshasa, lies across the Equator; the climate there is hot and humid, while it is far more pleasant in the Lubumbashi region, to the SE, and in the area bordering Lake Kivu in the East. The Inga Dam being constructed near Matadi, will be the largest hydro-electric complex in the World.

Zaire is mainly an agricultural nation, but its wealth comes from the Shaba (Katanga) region in the form of cobalt, copper and industrial diamonds.

Henry Morton Stanley explored the Congo basin in the 1870s. The Country was the property of Leopold II of Belgium from 1884 and Belgium ruled the land until its independence in 1960. Like many other African nations, Zaire achieved its independence in bloodshed, a situation which necessitated the intervention of the UN. By 1965, order had been restored and an element of stability returned to the land. At the present time, the situation between the central government and the Katangese rebels, based on neighboring Angola, is almost as critical as it was in the 1960s; fighting and bloodshed have returned to the point where France and Belgium accepted to send troops to protect the lives and property of their nationals in the Kolwezi region.

Zambia

Area: 752,614 km²
Pop. Grth.: 3.2
Pop.: 5,565,000
Cap.: Lusaka

Zambia is a land-locked plateau East of Angola and South of Zaire. Its 70 ethnic groups are mainly composed of farmers who eke out an existence from the land. The copper belt which runs SE from Zaire extends into Zambia and provides it with its real wealth and most of its exports. Kitwe and N'Dola are in the centre of this densely populated region on the southern border of Zaire. The Kariba Dam on the Rhodesian border provides the hydro power required for copper smelting. Before independence in 1964, Zambia was the British Protectorate of Northern Rhodesia.

ANTARCTICA

Area: 13,200,000 km²

Antarctica, which is the name given to the ice mass covering the South Pole, is one of the largest, coldest and bleakest land masses in the World. A mountain range which extends right through the Peninsula divides the Continent into an eastern and a western region (the former faces Africa while the latter looks at the Pacific); both regions are ice-covered, more so the latter where ice thickness reaches 4270 M. The main geographic features of Antarctica are the Vinson Massif; two bays, large enough to be named the Ross Sea and the Weddell Sea; McMurdo Sound which boasts an active volcano (Mt. Erebus), and numerous glaciers which grind their way to the sea. Basically, the land is desolate and swept by cold winds.

Needless to say, climatic conditions are the worst; the lowest temperature ever known, was recorded near the South Pole at (-) 88°C or (-) 127°F!

Sea-life is plentiful along the coasts of Antarctica and attracts the fishing fleets of many nations. Coal deposits are known to exist on the Continent and oil and gas are suspected.

Although whalers are thought to have visited Antarctica in the early 1820's, exploration did not start until the late 1830's. Roald Amundsen, a Norwegian, was the first to reach the South Pole in 1911, followed, barely a month later, by the ill-fated British expedition led by Capt. Scott. U.S. Admiral Richard Byrd also carried out extensive explorations of the region from 1920 on.

THE ARCTIC

The "Roof of the World" is covered by a thick crust of ice which reaches the northern edges of America, Europe and Asia; it accounts for an area of 14.2 M km². The forbidding cold of the Arctic is no longer an absolute barrier to man's progress, for he has found means of reaching the mineral wealth which lies beneath the pack ice and has learned to use its weather system for his meteorological work.

Geographers place the Arctic's southern limits at what is called the northern tree line, a line beyond which trees do not grow and where midsummer temperatures do not normally exceed 10°C (50°F). Greenland lies North of this line, as do parts of Iceland, Alaska, Canada, Finland, the USSR, Sweden and Norway. Numerous islands are also found in this bleak region.

The few inhabitants of this land where winter temperatures average - 34°C or -30°F, only see the sun between March and September. Although the land is free of ice during the short summer and the top soil thaws, perma-frost is found at a depth of 18 inches. In summer, the plain or tundra "explodes" in a multitude of colorful arctic flowers, grasses and low shrubs. Moss and lichens feed large herds of reindeer and wandering caribou. Apart from an abundant bird life, the fauna includes polar bears, foxes, lemmings, squirrels, hares and voles. Whales and seals in large numbers gorge themselves on abundant sea life. Minerals exist in great quantity and variety in the arctic soil, to include uranium, thorium and gold.

The Arctic is sparsely populated, but humans do manage to live and thrive in what most would consider a totally inhospitable land. Apart from 100,000 Eskimos believed to have been in the land for 12,000 years, Lapps, Finns and the Yakut of Siberia live comfortably within the Arctic Circle and several cities of the USSR are to be found therein.

The search for a short-cut to the Pacific through the roof of the world, was the reason why exploration of the Arctic began circa 1500. Robert Peary, an American, was the first to reach the North Pole in 1909. In 1958, America's atomic-powered submarine, the Nautilus, proved the feasibility of travelling under the pack-ice. Major air-lines now use the northern route as a matter of course. American and Russian strategic bases dot the Arctic, as do several important weather stations.

Greenland

Area: 2,175,600 km²
Pop.: 47,000
Cap.: Godthaab

16 kms at its nearest point from the N.E. coast of Canada, Greenland, ice covered the year round, is the largest island in the World. Its permanent ice cap, fed by average temperatures of -47°C (-53°F) in winter, and -11°C (12°F) in summer, is 1.6 kilometers thick.

Imposing mountains in the North, West and East dominate a small plateau in the Southern part of the Island. The land is cut by deep fjords and the ice shelf on its coasts breaks off into glaciers - one of them, the Humboldt, results in cold currents which cool off the West Coast of South America.

Greenland is a Danish possession; its population is a mixture of Danes and Eskimos. Its strategic value in the North, warranted the establishment of an American airbase at Thule on the northwest coast. Cryolite, a substance used for making aluminum and ceramics, is found only in Greenland.

ASIA

The largest, most populous and certainly the most diverse continent in the World, is the best way to describe Asia. It is bounded in the North by the Arctic, in the South by the Indian Ocean where it claims a great deal of territory, and in the East, by the Sea of Japan and the Pacific. In the West, Europe accounts for a portion of the USSR, whilst the Middle-Eastern countries of Turkey, Iran and Afghanistan complete the circle. The main countries and land masses of Asia are: the USSR, the Peoples' Republic of China, Mongolia, Pakistan, India, Nepal, Bangladesh, Burma, Thailand, Vietnam, Laos, Cambodia, Malaysia, Indonesia, the Philippines and Taiwan, as well as Polynesia, Melanasia and Micronesia which are made up of numerous islands in the Western Pacific, North of Australia and New-Zealand.

Bangladesh

Area: 143,998 km²
Pop. Grth.: 2.4
Pop.: 85,650,000
Cap.: Dacca

A very young nation born in 1971, Bangladesh is a small, very densely populated country (530 persons per sq. km.), squeezed between India to its West and Burma to its East. The Country is a vast, fertile plain in the delta formed by the Ganges, the Brahmaputra and the Meghna rivers. Its low land, coupled with the incidence of monsoons which occur between April and October, make the region susceptible to disastrous flooding. The frequency of tidal waves, stirred up by cyclones, further adds to the woes of an over-crowded country whose people constantly face starvation, disease and death. The climate is hot and humid.

Bangladesh came into being when the Bengali majority of East Pakistan opted in favor of a costly and bloody civil war to secure independence from the rest of the Country. The already poor and under-nourished population had to pay a high price for its freedom! Agriculture is of great importance to the Country, but the large population and the acts of God already mentioned, which can wipe out entire crops in a matter of hours, make this important activity very undependable, Bangladesh is therefore frequently obliged to import grain and other basic staples as well as fuel and fertilizers to avert prolonged famines. Antiquated agricultural methods produce rice, jute and tea. Fishing is also of importance. The Bay of Bengal has an oil potential, but funds lack for its development.

The population is 80% Muslin. The government is of a military nature.

Bhutan

Area: 47,000 km²
Pop. Grth.: 2.3
Pop.: 1,240,000
Cap.: Trimphu

A small, independent, Buddhist constitutional monarchy, Bhutan is located between Tibet and India, in the Himalayas. The land varies from snow-capped peaks to near tropical valleys where farmers and herdsmen eke out an existence grazing cattle or growing crops of rice and barley on terraced mountain slopes.

Brunei

Area: 5,765 km²
Pop.: 177,000
Cap.: Banda Seri Begawan

Brunei is a British protectorate situated on the NW Coast of the Island of Borneo - It is encircled by Sarawak. This oil-rich country has a Chinese-Malay population.

Burma

Area: 676,552 km²
Pop. Grth.: 2.2
Pop.: 32,910,000
Cap.: Rangoon

Burma is a socialist republic whose philosophy is permeated by Buddhism. A British province since 1885, Burma secured its independence in 1948. The land is mountainous with thick forests and many rivers which form a delta and empty in the Gulf of Bengal. The Irrawady of WWII fame, is one of these rivers. The land stretches into a narrow coastal strip which penetrates half-way down the Malay Peninsula. The climate, hot and humid, is propitious to rice growing; its forests produce teak and bamboo. The ground yields jade and other precious and semi-precious stones.

Cambodia Kampuchea

Area: 181,305 km²
Pop.: 8,570,000
Cap.: Phnom-Penh

Cambodia is the land of the great Khmer Civilization of the 11th and 12th centuries. Under French administration from 1863 to 1953, Cambodia, by virtue of its geographical situation, had to get involved in the Vietnamese War (1970). Civil war ensued when the monarchy fell to the Khmer Rouge who founded a communist state in 1975 and proceeded to reorganize an economy which had been all but obliterated by the war. (see Vietnam)

The Country is mostly flat and dominated by the Mekong River, the level of which varies with the monsoons; the flooding which occurs then, is vital to rice-growing. Fishing is also an important subsistence activity. Cambodia's rubber plantations and its embryonic industries were dealt a telling blow by the civil war.

Hong-Kong

Area: 1,032 km²
Pop.: 4,078,000
Cap.: Victoria

The British Crown-Colony of Hong-Kong is made up of the main island and 240 smaller ones lying off the South Coast of China. The last outpost of Western influence in Asia, Hong-Kong is important to the Peoples' Republic of China as a contact point with the West. The Kowloon Peninsula and the New Territories to the North, are part of the Island complex. Steep hills and consequently limited arable land make mandatory the importing of food to supplement fishing, fruit and vegetable-farming.

Hong-Kong is largely autonomous; it has an important ship-building industry as well as textiles, plastics and electronics. The colony's skilled, but cheap labor has resulted in many of its products flooding world markets to the detriment of many national industries. Its strategic location enables it to handle 50% of China's external trade which ensures its survival in the midst of a world somewhat hostile to Western life-styles and philosophy.

The climate of the Island is tropical; hot, wet and subjected to annual monsoon activity.

The British acquisition of Hong Kong goes back to 1843, that of Kowloon to 1860 and finally, that of the New Territories, to 1898.

India

Area: 3,287,590 km²
Pop. Grth.: 2.1
Pop.: 650,980,000
Cap.: New Delhi

India's independence secured in 1947, has not yet managed to eradicate the effects of 200 years of British occupation. Administration at government and other levels, land communications, military organization, language and traditions still reflect British ways of life.

Geographically, India is a large peninsula in the southern part of the Asian Continent, with an almost inpenetrable barrier to its North, the Himalayas; these mountains make communications with the rest of Asia, anything but easy. South of the "Top of the World", lies an alluvial plain irrigated by the Ganges, the Brahmaputra and the Indus. Most of India's population is found in this plain, though coastal areas are densely populated.

India's climate is hot and humid, more so in the Ganges valley; monsoons occur from June to September and usually take place after a dry period.

Crops are vital to the survival of India's masses; monsoons arriving too early can wash crops out entirely - too late, they will be seared by the implaccable sun. In either case, famines ensue. One can readily appreciate the importance of Nature's timing.

India is poor and undernourishment commonplace. Domestic cattle and buffalo abound and could alleviate the situation, were it not for Hinduism which forbids their slaughter for food. Agricultural methods are antiquated and yield relatively little. The need to farm as a means of survival has meant the clearing of wooded areas to the point where forests are scarce. As a result of this state of affairs, India, like Bangladesh, must import basic staples such as rice and wheat. Protracted famines have often resulted in the World at large sending huge supplies of wheat to alleviate suffering; unfortunately, primitive storage facilities, looting and generally bad administration have resulted in much of this salvation food being lost or wasted. On the positive side, sugar cane, tea, peanuts and pepper are produced in large quantities. The soil of India contains coal and high-grade iron ore. Oil production is increasing and minerals such as bauxite, mica and manganese are mined.

Indonesia

Area: 1,904,345 km²
Pop. Grth.: 2.4
Pop.: 148,470,000
Cap.: Djarkarta

Between SE Asia and Australia, over 13,600 islands of various sizes dot the Pacific and the Indian oceans, forming the Republic of Indonesia. The main islands are Kalimantan (formerly Borneo), Sumatra, Irian Barat (formerly Western New-Guinea), and Java, the most fertile and most populated. The climate is hot and wet.

Indonesia is rich in oil, coal, timber, nickel, bauxite and tin; its rubber industry is staging a come-back.

For 300 years prior to 1949, Indonesia was under Dutch control. A Communist attempt to take over in 1965 failed, and the Country grew gradually closer to the West. Main trading is with the USA and Japan, with a good portion going through Singapore. The Republic is ruled by an army-backed president.

Japan

Area: 372,313 km²
Pop. Grth.: 1.0
Pop.: 115,870,000
Cap.: Tokyo

For centuries, Japan seemed content to hide behind its cherry blossoms and the fans of its geishas; its people were aloof and inscrutable. However, the defeat which Japan took at the hands of the USA in WWII, was to alter its way of life. Starting with the occupation by US troops after 1945, the Empire of the Rising Sun quickly began to adopt a democratic way of life. Its energetic and industrious people rebounded to pick up the economy and rebuild industries, transforming the Country into one of the World's most industrialized and progressive nations.

It does not much matter where one travels abroad, one meets Japanese, cameras slung over their shoulders, intelligent and alert, ready to learn and even readier to use that knowledge. Japanese cars, motor-cycles, TVs, cameras and electronic equipment are flooding world markets and attest to a most efficient industrialization. Ship-building, textiles and chemicals are active industries.

Geographically, Japan lies off the East Coast of China and is made up of the islands of Honshu, Hokkaido, Kyushu, and Shikoku. It is a rugged and generally mountainous country, being in fact the crests of a mountain chain protruding from the sea; it has numerous active volcanoes. The Japanese complex is comprised of 3,000 small islands.

Land is at a premium in Japan; every square metre available and suitable is terraced and cultivated; it is so cleverly and extensively irrigated that it produces one of the World's highest agricultural yields. Forestry and fishing add to the Country's resources. Crops include rice, potatoes, cabbage an fruit.

Honshu, the main island which contains three quarters of the population and the most populated city in the world, Tokyo with 11,6 M, has swift rivers providing hydro-electric power.

Although the island group has a fairly moderate climate. Honshu is hot and humid in summer. Japan is influenced by monsoons. The mountains of the western range receive a great deal of snow which results in torrents and floods in the spring. Earthquakes, tidal waves and typhoons are commonplace and very destructive.

There are few minorities in Japan and only two main religions, Buddhism and Shintoism. The japanese have their own distinctive art, colorful and gay. Sumo wrestling, Judo and Karate are Japan's contribution to the realm of sports. At one time, the Japanese excelled at swimming; currently, gymnastics, table-tennis and baseball are to the fore.

Japan's governmental system is the result of its WWII defeat; the Emperor, a descendant of the Yamato Dynasty of 200 AD, has had his governmental powers taken away and had to renounce his sovereign right to make war. The Diet runs the Country.

The USA were instrumental in "emancipating" Japan in the 1800s. The latter's imperialist ambitions began to show in the war with China in 1894-1895, whereby it gained Formosa and an influence in Korean affairs. The 1904-1905 war with Russia gave Japan the southern part of Sakhaline and freedom of action in both, Manchuria and Korea, the latter being annexed in 1910. On the side of the Allies in 1914-1918, Japan gained German possessions in Asia. It seized Manchuria in 1931 and in 1932, helped create the new independent state of Manchukuo. In 1937, Japan invaded China, occupying Nanking, Canton, Hangchow and Hai-Nan. Siding with the Nazi-Fascist Block in 1941, it launched the war in the Pacific by attacking Pearl Harbour in December of that year. After initial successes, Japan was decisively beaten in 1945, when atomic bombs dropped on Nagasaki and Hiroshima sealed its fate. Paradoxically, the Country has recovered to become one of the World's leading industrial powers.

Kashmir

Area: 222,237 km²
Pop.: 5,700,000
Cap.: Srinagar & Jammu

NE of Pakistan and North of India, Kashmir is a beautiful, mountainous region of the Himalayas, the subject of disputes between India & Pakistan. The Country has the rare distinction of having two capitals, Srinagar in the highlands during the summer and Jammu in the lowlands in the winter when the higher regions become too cold.

Agriculture and textiles are the main activities.

Korea

Korea is a peninsula jutting out of Manchuria into the Sea of Japan, West of the Japanese island of Kyushu. Under Chinese influence until 1910 when taken over by Japan, Korea was liberated in 1945 only to find itself divided into North and South Korea, both republics. The decision to partition the Country along the 38th parallel did not, it seems, take into account the fact that most of the industrial capability was in the North, while the bulk of the agricultural activity lay South, a situation which was bound to create problems. Here was an opportunity for Communist intervention. In 1950, North Korea, supported by Russia and China, attacked the South and rolled back the ill-prepared South Korean forces. Led by the USA, the UN came to the rescue, pushed back the invading forces until a stalemate of sorts was achieved, followed by a truce in July 1953. A demilitarized zone was created, and since then, opposing sides have been "observing" each other, with minor clashes occuring from time to time.

The Strait of Korea which contains 3000 islands, separates it from Japan. The Korean mainland is made up of rugged mountains, higher in the East where they are in fact a projection of a Chinese range. The climate is monsoon-affected; hot and humid in summer and quite cold in the winter, especially in the North where the wind blows from the Manchurian plains.

North Korea

Area: 120,538 km²
Pop.: 16,246,000
Cap.: Pyongyang.

North Korea is considerably industrialized and has the richest mineral beds in Asia (graphite, magnesium, tungsten, iron ore, lead, zinc, coal and petroleum); it also has ample hydro-electric power. The Communist Regime has instituted collective farming to step up grain output. Trade is mostly with China & the USSR.

South Korea

Area: 98,431 km²
Pop.: 35,860,000
Cap.: Seoul

South Korea has a president-led republican government. Essentially an agricultural region, the South has nevertheless developed a powerful manufacturing industry built around chemicals, textiles, plywood, electronics and plastics. The hub of this activity is Seoul. South Korea has well over 2M small farms which produce rice, wheat, barley and beans. Fishing is of importance. Trade is mostly with the USA, Japan and Canada.

Laos

Area: 236,800 km²
Pop. Grth.: 2.5
Pop.: 3,630,000
Cap.: Vientiane

For 50 years, Laos was under French rule, but in 1954, it became a constitutional monarchy. Political division set in and in and in 1960, US-backed Royal Laotian troops were pitted against North Vietnam-supported Pathet Lao. Inevitably, Laos was drawn into the Vietnamese conflict. In 1973, the Pathet Lao emerged victorious and monarchy had ceased to exist.

Laos is a hot, humid, heavily forested, mountainous country, squeezed between Vietnam to its NE and Thailand to its SW. The main geographical features of the Country are the Annamese Mountains and the Mekong River Delta. The economy of Laos is based on forestry (teak), rice, tobacco, cotton, opium and some corn. There are also potentially rich reserves of gold, silver, zinc, tin, lead and gypsum under ground.

The climate of Laos is monsoonal. Railways are non-existant and roads, as with most monsoon-affected countries, are passable only during the dry season. Transportation is consequently limited to the Mekong system.

Malaysia

Area: 329,749 km^2
Pop. Grth.: 2.9
Pop.: 13,300,000
Cap.: Kuala Lumpur

Malaysia is a parliamentary democracy and is made up of 13 states, two of which are Sarawak and Sabah, former British protectorates. The Country has retained strong links with Britain which administered and developed it for over 100 years. Malaysia is a complex of territories which spread into the China Sea. Most of the land is jungle-covered, hot and damp except in the highlands where it is cooler. Sabah and Sarawak which lie on the NW Coast of Borneo, are mountainous. Heavy rains drench the land from November to March.

Most Malaysians live in coastal areas and through resources such as rubber, tin and timber, enjoy a high standard of living. Oil is also being produced and refined in the region.

The Malay population is mainly urban. There is tension between the Malays and the Chinese who are the main operators of businesses and the source of skilled labor. Japan is the main trading partner.

Mongolia

Area: 1,565,000 km^2
Pop. Grth.: 3.0
Pop.: 1,620,000
Cap.: Ulan Bator

The very name of Mongolia and that of its capital evoke the memory of great Barbarian invaders such as Genghis Khan, Kublai Khan and Tamerlane. Under Chinese and Russian domination until 1924, Mongolia emerged as an independent communist state with a government patterned after that of the USSR. Located between the USSR and China, Mongolia with its Altai mountain range and its Gobi Desert, is a country of extremes where summer temperatures often reach 38°C (100°F) and where the thermometer dips to (-) 40°C, (-) 54°F in the winter. It is a land of nomads who raise cattle including camels. Its soil contains petroleum, tungsten, uranium, coal, copper and gold. Mongolia is the most thinly populated nation of the Globe.

Nepal

Area: 140,797 km^2
Pop. Grth.: 2.2
Pop.: 13,710,000
Cap.: Katmandu

An independent constitutional monarchy, Nepal is the land of the Sherpas, the most famous one being Tensing who assisted Sir Edmund Hilary in his conquest of 8,848 m Mount Everest in 1953. The lofty Himalayan peaks offer extremely cold temperatures, but as one proceeds down to the valleys, the climate gets warmer, in fact quite hot in the swampy plains bordering India. Half the land is covered with forests which provide good timber. Sugar, jute, rice, millet and maize are grown. Terrace-farming is commonplace, but the crops are used to support native life. Tourism is increasing.

Pakistan

Area: 803,943 km^2
Pop. Grth.: 3.0
Pop.: 76,770,000
Cap.: Islamabad

When the British left India in 1947, the Country was confronted with a politico-ethnic problem caused by its large Moslem element which was eventually accomodated by the creation of two territories separated by northern India itself. One such territory, West Pakistan, is on the NW shoulder of India, the other, East Pakistan, much smaller, on its NE tip. It is this latter territory, which after the civil war of 1971, became Bangladesh.

Pakistan's resources are few and ill-developed. The population which grows at a fast rate, is poor, largely illiterate (80%) and lives mostly off farming and herding in the Indus-irrigated Punjab region.

To the North rises the Hindu Kush Range, scene of heavy fighing during the civil war. The climate is monsoonal with extremely hot temperatures in the summer (40°C or 110°F), followed by the usual Asian rainy season. Winter is fairly cold, especially at the higher altitudes.

Crops include sugar-cane, rice, cotton and tobacco. Fishing is of importance for local subsistence. The soil yields iron ore, natural gas, oil, salt, limestone and low-grade coal. Hydro-electric energy is plentiful.

The Islamic faith unites the disparate elements of this nation of many cultures and languages. The socialist government has nationalized transportation, banking and some industries. Main cities are: Islamabad, Karachi, Quetta and Hyderabad.

The Peoples' Republic of China

Area: 9,596,961 km^2
Pop. Grth.: 1.4
Pop.: 1014,000,000
Cap.: Peking

History and Culture:
China which accounts for a good fifth of the World's population, is the third largest country of the World. Its history goes back 2000 years and at one time, under the Han Dynasty, (206-220 BC), its size rivaled that of the Roman Empire. Briefly under Mongolian domination during the 13th Century, China emerged independent once more with the Manchus (1644-1911). In 1900, the Boxer War saw the Chinese trying to shed occidental influence, which necessitated the intervention of European powers; this ended in the occupation of Tientsin and Peking. Sun Yat Sen appeared and launched the revolution from which emerged modern Chinese nationalism in 1921. A war with Japan from 1937 to 1945, was immediately followed by a civil war, pitting communists against nationalists and which resulted in Chiang Kai-Shek seeking asylum on Taiwan. Mao Tse Tung took over the reins of government in 1949 and China was on its way. From a humble agricultural nation, it rapidly became a vast and powerful industrialized power with a nuclear energy capability and generally, the wherewithall to "deal" with the Western World as well as the USSR with whom it broke over communist ideology. Chinese troops fought the UN in Korea during the latter part of the war (1950-1953). It was admitted to the UN Council in 1971 over a great deal of opposition on the part of many nations of the Western Block. (see Vietnam)

Ancient China's advanced civilization produced such commodities as paper, silk, porcelaine, the compass and printing. The World also "owes" the discovery of gun-powder to China. Lack of space forces us to omit one aspect of China's culture, as volumes would be required to adequately deal with its art and literature. Finally, Buddhism, Taoism and Confucianism all flourished in Cathay.

90% of China's population, which grows by 14 million per year, lives in the eastern region, concentrated around the estuaries of the Yangtse-Kiang and Yellow rivers; of this number, 33% live in urban areas. Over 30 Chinese cities have populations over 1 M. Rice and wheat are the staple diet of the Chinese, supplemented by fish and vegetables. Religion has been played down by the government, like the cult of the family. Communes have replaced private holdings. A sort of uniform is worn by millions to emphasize social uniformity. Accomodation varies from mud and straw huts to modern dwellings. Consumer goods are scarce and the cult of the erudite and scholarly has been replaced by that of the hard-working, energetic peasant-labourer. Life is austere and ascetic, but the People seem genuinely behind the Government. There are para-military youth organizations spear-heading social reforms. The Country is a one-party state, ruled by the Communist Party, with a premier and a chairman at its head.

Geography, Land and Climate:
China is bounded by the Pacific in the East, the USSR in the North, India, Pakistan, Bangladesh, Nepal and Kasmir in the S.W. It is comprised of 21 provinces, five autonomous regions, one of which is Tibet, and three special cities: Peking, Tientsin and Shanghai.

The bulk of China is made up of plateaus, and is divided into: Western China which includes Tibet, the Altai, Kunlun, Tien Shan and Pamir mountains, and the Gobi and Taklamakan deserts; Northern China which includes the Manchurian Plain, the North China Plain and the Yellow river; Southern China with the YangtzeKiang which traces its source to Tibet.

The climate varies from hot and humid in the South, to harsh in the North and West.

Economy:
The huge population of China eats all it can produce and the Country must import wheat to subsist.

Floods and erosion vie with droughts to make agriculture a difficult undertaking. Technology, though improving, is still inadequate and sheer muscle-power has to supplement. Farm machinery and fertilizers are lacking. Despite these handicaps, China leads the World in rice production. Tea, vegetables, maize and cotton are also produced in vast quantities. The Country's sub-soil is rich in minerals but largely unexploited. Iron, steel, coal, zinc, copper, manganese and lead are mined, and oil production is increasing. Industry is concentrated in the East, around Peking, Tiensin and Nanking. A growing manufacturing industry turns out chemicals, textiles, and aircraft. Handicrafts account for 25% of all manufactured goods. China's trading with the USSR has diminished and it has turned to Japan and West Germany. Hong-Kong is the transfer point for much of the exporting.

The Philippines

Area: 300,000 km^2
Pop. Grth.: 2.9
Pop.: 47,720,000
Cap: Quezon City

The Spaniards ruled the Philippines for over 350 years and the USA added another 50 until in 1946, independence was achieved. It is not therefore, surprising to find a strong Ispano-American influence permeating the Philippine's way of life; for instance, Christian missionary work is evidenced by the fact that 80% of the population is Roman Catholic.

Geographically, the Philippines consist of over 700 inhabited islands and thousands of coral atolls concentrated in an area between Borneo and Taiwan. It is a land of thick rain forests, active volcanoes and rushing waters; many of its villages are built on bamboo stilts at the edge of lakes or swampy areas.

It seems that the Philippines have everything going for them as well as against them; for instance, typhoons are commonplace and monsoons add to an already hot and wet climate (Manila gets 80 inches of rain a year). On the credit side, the combination of heat and moisture results in rich crops of rice, sweet potato, maize, bananas and coconuts, as well as hundreds of colorful flowers.

Volcanic ranges occupy most of the land forcing the inhabitants to hug the coastal areas. Fishing also plays an important role in the economy. Most of the World's copra comes from the Philippines; àbaca is also grown. The dense rain forests of the region yield hardwoods such as mahogany. With American technological know-how and capital, hydro-electric power is being developed as well as various industries.

Sikkim

Area: 7,298 km^2
Pop.: 205,000
Cap.: Gangtok

Sikkim is a small mountain kingdom associated with India since 1974. Its population is mostly Nepalese. Grain farming is the main occupation.

Singapore

Area: 581 km^2
Pop. Grth.: 1.2
Pop.: 2,360,000

This island republic at the crossroads of Asian ocean traffic, is one of the largest ports in the World. A causeway accross the Johore Strait links Singapore to the southern tip of the Malaysian Peninsula.

The population of the Island is mainly Chinese; dynamic, skillful and resourceful. Singapore enjoys a high standard of living. The bulk of the population lives in Singapore itelf, working in industries related to port activities such as ship-building, or located in its vicinity (oil refineries, petro-chemical plants, mills and factories).

The climate is hot, but cooled by sea breezes; rainfall is adequate. Despite intense cultivation of its low-lying, wet terrain, Singapore has to import most of its food.

Singapore was a British Colony until 1959, when it secured its independence.

Sri Lanka

Area: 65,610 km^2
Pop. Grth.: 2.0
Pop.: 14,740,000
Cap.: Colombo

Formerly Ceylon, Sri-Lanka is situated at the southern tip of the Indian sub-continent. It is a luxuriant, tropical island with a hot climate tempered by sea breezes. The land consists of a coastal plain encircling a mountainous interior. Rainfall is heavy in the SW where tea and rubber crops are grown. The soil yields graphite and gemstones whilst forests produce hardwoods. Ceylon was a British Colony from 1802 to 1948. It is now a socialist republic.

Taiwan

Area: 35,961 km^2
Pop.: 14,990,000
Cap.: Taipei

Taiwan or Formosa, is the home of the Nationalist Chinese. 140 km off the East Coast of China, it includes the Pescadores Islands which lie in the Taiwan Strait, as well as Quemoy and Matsu, islands turned into fortresses by Chiang Kai-shek who sought safety there with over one million troops and followers after his defeat by Communist forces, on the mainland, in 1949. There, backed by American aid and moral support, Nationalist China prospered and achieved one of the best balanced economies in the East. Communist China made some half-hearted attempts at intimidating Chiang Kai-shek, bombing Taiwan from its mainland battery positions, but gave up in 1958.

The Chingyang Range covers the Island, presenting steep cliffs on the Eastern side, while in the West, the hills, terraced for cultivation, run down to the Coastal plain where most people live and grow rich crops of soybeans, rice, vegetables and sugar-cane. A hot and humid sub-tropical climate and several rivers which flood during the monsoon, stimulate growth. Typhoons are frequent. While important mineral reserves exist on the Island, the main resource is timber which is exported, as are textiles, clothing and many other items resulting from an imposing manufacturing and processing complex of industries. Tourism is growing.

Thailand

Area: 514,000 km^2
Pop. Grth.: 2.5
Pop.: 46,140,000
Cap.: Bangkok

Geographically, Thailand has the shape of a flower with its stem taking root in the Malay Peninsula; it is surrounded by Burma, Laos, Cambodia and the Gulf of Siam. the NW part of the "flower" presents steep mountains while in the East, the land becomes a high, dry plateau. The land is densely forested, and its fertile valleys produce cotton, rice, and rubber in large quantities.

The "stem" part is mountainous. Tin, silk, lacquer and handicrafts add to the Country's resources. Manufacturing and hydro-electric power are being developed with the help of foreign investment. The climate is tropical with perennial monsoons. Means of transportation are more or less confined to the river systems. Thailand's forests yield teak and other types of wood, and teem with abundant bird and animal life.

Siam, as it was formerly called, was founded in 1350; it has a constitutional monarchy. Its capital is the headquarters of SEATO.

The USSR

Area: 22,402,200 km²
Pop. Grth.: 0.9
Pop.: 264,110,000
Cap.: Moscow

Geography:
To describe the boundaries of the USSR is tantamount to listing a sizable number of countries, due to its enormous size and extent. Suffice it to say that it stretches from the Baltic to the Pacific and from the Arctic to China in the South. A study of the map will reveal the existence of a large upland plateau in the East and two immense plains, the West Siberian Lowland and the East European Plain. The former lies East of the Urals which separate Asia from Europe, while the latter which contains 75% of the Russian population, extends from Poland to the Urals. The Central Siberian Uplands lie East of the Yenisei River and continue into Eastern Siberia.

Some of the main geographical features of the USSR are: the Pamir Knot, on the Afghan border, a region of high peaks, glaciers and extremely cold climate; the Caucasus which runs between the Caspian and the Black Seas; the Altai, Sayan and Tien Shan ranges on the border of China and Mongolia; the Kara Kum and Kyzyl Kum, two huge deserts; the Karagiye Depression, near the Caspian Sea, 132 m below sea level; the Caspian Sea, largest body of inland water in the World; Lake Ladoga, near Leningrad, largest lake in Europe; Lake Baykal, deepest fresh water lake in the World; the Volga River, longest in Europe; the Don, near Stalingrad (Volgograd), the Dnieper and, in Siberia, the Yenisei, the Lena of Napoleonic fame and the Ob; and finally, the Kamchatka Peninsula.

Climate:
By virtue of the immensity of the land, the climate has to be one of extremes. Eastern Siberia has recorded an unbelievable -68°C (-90°F), whilst some of the central Asian regions have very warm temperatures. Basically, summers are warm and winters are long and cold; snow covers the ground for many months with the European Plain getting most of it. The Pacific Coast gets both, rain and snow in abundance.

Economy:
Russia's Donets and Kusnets Basins contain huge reserves of coal, whilst the Volga, Urals and Caucasus regions hold billions of tons of oil. Natural gas is also available. Iron ore, gold, beryllium, mercury, diamonds, manganese, lead, zinc and titanium are mined in the Ukraine, the Urals, the Altai Range, Kazakstan, Uzbekistan and Armenia. The forests are an almost inexhaustible source of timber.

The economy of Russia is state-controlled with cooperatives being the method used to harness production. Self-employment is limited to the fields of literature, the theatre and craftmanship. Free enterprise is virtually non-existent, being found only in the fruit and vegetable sectors where farmers sell produce from private plots cultivated under the collective farm system. The land is worked under two systems; the Collective or Cooperative system where the workers, after meeting a certain quota for the Government, share the rest (planning and decision-making is their own); or the State Farm system, where planning and control is effected by the Government & wages are paid. Produce includes wheat, maize, rye, barley, oats, beets, potatoes, cotton & tea. Despite the magnitude of Russia's agricultural capability, frequent droughts have forced it to buy grain abroad.

New strains of weather-resistant grains are being tested for Siberian cultivation and growing in some of the semi-arid regions of the Country. Livestock-raising is increasing to meet a growing demand for meat. More and more people are leaving the country to work in urban areas.

Heavy industry is still of paramount importance, consumer goods generally coming second; in this latter area, however, output of utilities such as refrigerators, radios, TVs, etc, has been increased. Main products of the heavy industry are machinery, transport equipment, machine tools, chemicals, fertilizers, synthetic fibers and plastics.

The USSR trades mostly within the Eastern Block, western trade being limited to certain raw materials.

Transportation is a problem in the USSR due to distances and extremes of temperature. This applies in particular to motor transportation. The many rivers of the Country provide all the hydro-electric power needed.

Culture:
As with climatic conditions, diversity of cultures in Russia is a corollary of its huge size. Many ethnic groups live within the boundaries of the USSR, each with its own language and culture; Russians, Ukrainians, Byelorussians, Turkics, Ugro-Finnish, Tatars, Japhetics, Iranians, Armenians etc; these numerous and diverse races are all united under the one central government which rules from Moscow. As with China, Religion is played down and is separated from the State. Roman Catholicism, Lutheranism, Buddhism and the Islamic Faith are the main religions of the Country.

Athletics are popular and government-backed. Soccer, ice-hockey, swimming, track and field and weight-lifting are the favorite sports of the people. Russia's rivalry with the USA in the pursuit of Olympic and other gold medals, is well known. Chess, opera and ballet are popular and Government-encouraged.

Russia has produced superb authors such as Pushkin, Dostoyevsky, Tolstoy, Chekov, Pasternak and Solzhenitsin - great musicians like Rimsky-Korsakov & Prokofiev. Permissive philosophy and criticism of the Government and Communist idealogy is not permitted and can result in imprisonment or banishment.

History:
The Scythians are believed to have been the first inhabitants of Russia; they lived on the northern shores of the Black Sea. Then came the Slavs of the northern forests who settled along the Volga and Don rivers; they founded Novgorod and Kiev, the latter seeing the introduction of Christianity to the land as a result of contacts with the Byzantine Empire. In the 1200s, Mongolian herds led by Batu Khan, over-ran Kiev and established the seat of their empire in Stalingrad. Two hundred years later, Moscow emerged as the main power and reached its zenith under Yvan the Terrible. After the latter's death, Russia was invaded by Poles, had to face frequent uprisings of the Don region Cossacks as well as a great deal of domestic intrigue. The arrival of Peter the Great in 1696 resulted in the taming of the intrigue-prone Boyars, modernization of the Country, the building of a navy and the acquisition of a port on the Baltic after 21 years of war with Sweden. Catherine the Great (1762-1796), continued the modernization begun by Peter and followed his example by securing an opening on the Black Sea, by defeating the Turks.

With modernization came sophistication, power and eventually, involvement. Thus it was that Russia joined England, Austria and Prussia against Napoleon who was defeated in 1812, more by the size of the Country and its climate, than by the military action of his enemies. Later in the 19th Century, came territorial expansion which added Turkestan, Kazakstan and Uzbekistan to the Empire. There followed in 1905, a disastrous war with Japan, Russia losing its entire fleet. The stage was set for the revolution! When Russia's unfortunate participation in WWI resulted in increased misery and discontent, not only in the ranks of the army, but at home, Lenin and his Bolsheviks overthrew the Tsar, executed him as well as his family and took over the reins of power.

Stalin followed Lenin in 1924 and proceeded to purge the Country of his rivals. Stalin played an important role in WWII as one of the Big Three. During this major conflict and after initial set-backs reminiscent of Napoleon's drive toward Moscow in 1812, the Russians rallied, turned back the German invaders and compressed their armies against those of the Allies advancing from the West. WWII was followed by a "cold war" resulting from differences of opinions and idealogies between Russia and the Western Powers. After Stalin's death, matters improved while relations between Russia and China reached the breaking point over the two countries' approach to Communish; this resulted in border incidents between the two nations.

In 1961, Russia which was engaged in a space race with the USA, scored the first success; Gherman Titov and Yuri Gagarin were the first humans to orbit in space. In the following decade, an effort was attempted at cooperation with the USA, in space ventures.

Government:
The USSR is a federation of 15 republics under Moscow (Russia, Armenia, Azerbaijan, Byelorussia, Georgia, Kazabkstan, Kirgizia, Moldavia, Tadzikistan, Turkmenistan, Uzbekistan, the Ukraine, Estonia, Latvia, Lithuania) 20 autonomous republics, 8 autonomous regions, 10 districts and six territories. The President of the Supreme Soviet is Head of State.

Vietnam

Area: 329,556 km²
Pop. Grth.: 2.3
Pop.: 51,080,000
Cap.: Hanoi

From 1945-1975, Vietnam was torn by a civil war resulting in a divided country with an industrial North and an agricultural South. Georgraphically, Vietnam is a fringe of land on the East Coast of the peninsula formed by Cambodia, Loas and Thailand; it extends from the Gulf of Tongking to the South China Sea. The main features of Vietnam are the Mekong and Red River deltas, and the Annamite mountain range which runs from China to Hanoi in the SW, home of the rugged Montagnard tribe.

Collective farming produces mainly rice and other subsistence crops. Rubber is grown and is the Country's main export. Iron, lead, gold, tin, tungsten, zinc, and phosphates are mined. Cement, paper and textiles are produced in the South, despite a dearth of energy sources.

Having ruled the Country from the turn of the century, France had to abandon it to Communist forces in 1954, following a stinging defeat at Dien-Bien-Phy.
in 1954, following a stinging defeat at Dien-Bien-Phu.
This left Vietnam split between a communist North and an anticommunist South until 1975, when despite massive financial, material and military aid from the USA, the Communists emerged victorious. In January 1979, Ho-Chi-Minh's dream of a Vietnamese Indo-China was nearly realised by a large-scale invasion of Cambodia. A month later, China came to the latter's rescue by launching a punitive "raid" on Vietnam, ostensibly intending to pull out when the slap had been administered. In support of its Vietnamese allies, the USSR massed troops on the Chinese border and enjoined China to withdraw.

CENTRAL AMERICA

The region crowded between Mexico and the NW edge of South America and known as Central America, is in fact a narrow land bridge connecting the two Americas. Seven small countries make up Central America; Guatamala, Belize (formerly British Honduras), El Salvador, Honduras, Nicaragua, Costa-Rica and Panama.

The climate of Central America is generally hot and humid; the land is heavily forested, with volcano-dotted highlands subject to volcanic eruptions, earthquakes and hurricanes. The climate and geographical features of Central America's component countries being more or less similar, we will simply list their pertinent statistics and show in what respect each country stands out. Boundaries will be readily identified by perusing the map.

Belize

Area: 22,963 km²
Pop.: 126,000
Cap.: Belmopan

Self-governing colony of Britain until Sept. 21, 1981 when it gained its independence.

Costa Rica

Area: 50,700 km²
Pop.: 2,190,000
Cap.: San José

Famous for its coffee.

El Salvador

Area: 21,041 km²
Pop. Grth.: 3.0
Pop.: 4,660,000
Cap.: San Salvador

Smallest of all Central American countries, its main exports are coffee and cotton.

Guatamala

Area: 108,889 km²
Pop. Grth.: 2.9
Pop.: 6,620,000
Cap.: Guatamala City

Has more pure-blooded descendants of Maya Indians than any other country of Central America.

Honduras

Area: 112,088 km²
Pop. Grth.: 3.6
Pop.: 3,560,000
Cap.: Tegucigalpa

World's third largest banana producer.

Nicaragua

Area: 130,000 km²
Pop. Grth.: 3.6
Pop.: 2,640,000
Cap.: Managua

Lake Nicaragua is the largest in all the Americas and the only one in the World to contain sharks.

Panama

Area: 75,650 km²
Pop. Grth.: 3.1
Pop.: 1,880,000
Cap.: Panama City

Its canal is a short-cut between the Caribbean and the NE Pacific; leased to the USA at a cost of $2,100,000 a year. Source of contention.

MEXICO

Area: 1,972,547 km²
Pop. Grth.: 3.6
Pop.: 69,380,000
Cap.: Mexico City

The Republic of Mexico, though much smaller than the USA, has similar boundaries. To, its North, Mexico shares its border with the USA, much as the latter does with Canada, whilst to the West and East, oceans wash its shores, just like those of the USA; the Pacific bathes its western shores while the Gulf of Mexico rolls its breakers on its eastern beaches. To the SE, two Central American countries, Belize and Guatamala mark Mexico's southern limits.

Geographical Features — Climate:

Mexico can be divided into readily identifiable regions from West to East: Baja California (Lower California), an elongated peninsula running parallel to Mexico, West of the Gulf of California; sparsely populated, it has little to offer beyond lonely, arid mountains and deserts — The mainland, overrun by two mountain ranges running in a north-westerly direction and meeting roughly in the middle of the Country; the western range is called the Sierra Madre Occidental, while its eastern neighbor is known as the Sierra Madre Oriental — numerous volcanoes are found among the rugged peaks of this region —Between these two sierras, fertile land and a mild climate have resulted in two thirds of the population electing to live there. — Finally, the far-flung Yucatan Peninsula, a projection of land in the Gulf of Mexico, towards Cuba; a rather inhospitable land of steaming swamps and heavily forested mountains. Mexico's climate is generally hot and dry, with the central region benefitting from a more temperate climate. Again, some areas such as the Yucatan, have very hot and humid temperatures which make life unpleasant.

Economy:

Farming, fishing, breeding livestock and forestry are the Country's main industries. Another important source of income is its minerals (copper, zinc, iron, lead, mercury, sulphur, and silver, where it leads the World in production), and a tourist industry which flourishes as more and more affluent people come to Mexico to view its pre-Colombian archeological treasures and the ruins of a once thriving and most advanced civilization.

History and Constitution

Spanish conquistadores were the first Europeans to land in what is now Mexico; from 1517 to the mid 1500's, is all it took Cortez and other Spanish conquerors to first plunder and then eradicate the leading civilization of the region, the Aztecs. It was not until 1821 that Mexico was able to shake Spanish rule.

Mexico is a one-party federal democracy with an all-powerful president heading its destiny.

THE WEST INDIES

Colombus landed in the West Indies in 1492, thinking he had set foot on Indian soil. This pleasant region of the New World, is made up of a group of tropical islands known as the Greater Antilles which bask in the bright sunshine of the Caribbean Sea. Its islands are Cuba, Puerto Rico, the Bahamas, Jamaica, Haiti, the Dominican Republic, the Leeward Islands, the Windward Islands, Barbados, Trinidad and Tobago.

What these islands have in common is a delightful climate which attracts millions of vacationers trying to escape the rigors of less favourable climes. Many, also have a common lot; political instability with its inevitable consequences on their economy, including of course, their main industry, tourism. Again, we will list hereunder, the main, relevant statistics:

The Bahamas
Tourism is the main industry.

Area: 300 islands representing 13,935 km²
Pop. Grth.: 3.4
Pop.: 220,000
Cap.: Nassau

Barbados
Most easterly island of the Caribbean group, this coral island obtained its independence from Britain in 1966. Tourism is its main industry.

Area: 431 km²
Pop. Grth.: 0.7
Pop.: 250,000
Cap.: Bridgetown

Cuba
Largest island in the Caribbean. Big exporter of sugar and tobacco. Socialist state modelled after that of the USSR. Fertile plains; gently rolling hills.

Area: 114,524 km²
Pop. Grth.: 1.4
Pop.: 9,850,000
Cap.: Havana

Dominican Republic
Occupies eastern half of Hispaniola; population is of Spanish-Negro descendants. Sugar and coffee are main exports. History is marked with violence and revolutions.

Area: 48,734 km²
Pop. Grth.: 3.0
Pop.: 5,280,000
Cap.: Santo Domingo

Haiti
Haiti shares the western part of Hispaniola with the Dominican Republic. It was the first Negro republic established in the 1800s, when its population, descendants of African slaves, rebelled against France. The Economy is in bad shape. Due to over-crowding, the land produces barely enough for its inhabitants to subsist.

Area: 27,750 km²
Pop. Grth.: 1.8
Pop.: 4,920,000
Cap.: Port au Prince

Jamaica
Tourism, bauxite, sugar, bananas, cocoa and coffee, all contribute to fill the coffers of this Caribbean island. The land is mountainous.

Area: 10,991 km²
Pop. Grth.: 1.5
Pop.: 2,160,000
Cap.: Kingston

Leeward Islands
A group of islands made up of: The Virgin Islands, St Kitts, Nevis, Anguilla, Barbula, British Domenica, Montserrat, Redonda, Guadeloupe, St. Barthelemy, Marie-Galente, Desirade, Les Saintes, Netherland Antilles and North St. Martin.

Puerto Rico
A Commonwealth of the USA with self-government, tourism and sugar processing are its two most important activities. A mountainous, and fertile land.

Area: 8,891 km²
Pop.: 2,770,000
Cap.: San Juan

Trinidad and Tobago
These two islands lie only 11 km off the coast of Venezuela. Negro-East-Indian population with English as official language. Because of its pitch lake, it is the largest producer of natural asphalt in the World. Petroleum is its most important source of revenue.

Area: (two islands) 5128 km²
Pop. Grth.: 1.6
Pop.: 1,130,000
Cap.: Port of Spain

Windward Islands
Comprised of Grenada, Santa Lucia, St. Vincent, La Martinique and the Grenadines, some 600 islets of rocks, located between Grenada and St. Vincent.

EUROPE

Generalities — Component Nations

The fall of the Roman Empire in 476, was offset by the "birth" of Europe as the World's focal point for the development of a new civilization.

The power of Europe peaked in the 19th Century, but was severely eroded by two world wars and a wave of nationalism which swept the World and steadily brought about the fall of colonial empires. Like-wise, the rising importance of China and the USA and their marked influence in world affairs, were to gradually relegate Europe to a position of lesser importance in this domain. Nonetheless many of the World's most prosperous nations are still found in Europe and when it comes to art, science and history, the eyes of the entire Universe turn towards the Cradle of Western Civilization.

A look at the Eurasian Continent shows Europe as a vast peninsula issueing from it and projecting itself to the S.W. as far as the Atlantic Ocean. To the North, the Arctic marks its limits, while the warm Mediterranean provides many of the inhabitants of its 34 component nations, some of the World's most beautiful resorts. The area with which we are concerned and which starts at the Ural Mountains in the East, represents 10,523,000 km², and a population of 680 millions.

The climate of Europe is temperate, but while it is warm in the South, and S.W., influenced by the Mediterranean and the proximity of North Africa, it gets considerably colder as one progresses towards the East and North (Poland, European Russia, Norway, Sweden, Finland etc.).

Agriculture accounts for about 25% of the land usage and employs one quarter of the work-force.

Geographically, Europe is sub-divided into four main regions: the N.W. mountain system; the Great European Plain; the Central Uplands and perhaps the best known of all, the Alpine Mountain System which begins with the Pyrenees in Spain, becomes the Alps in Northern Italy and in Switzerland, to end in the Balkans.

Iron Curtain Nations as well as France, England, Germany and Belgium, boast a well developed heavy industry to which has been added in latter years, a highly developed manufacturing capability.

Socio-politically, and despite great efforts of diplomacy, despite bold economic ventures designed to unite its nations through trade, Europe remains divided, mainly because of the different ideologies which pervade East and West. On one side, the European Economic Community, a Western World initiative, on the other, the socialist countries-supported Council for Mutual Economic Aid, the Iron Curtain countries answer to the former association.

Austria
History - Culture:
Prior to 1200, Austria was part of the Holy Roman Empire, the House of Hapsburg ruling until 1806 when the empire ceased to exist. The ruling Monarch then became the emperor of both Austria and Hungary. Vienna, the capital, became the cultural centre of Europe, excelling in the artistic field and in particular, in music. After WWI the empire was broken up, Czechoslovakia and Yugoslavia becoming independent. The declining strength of the empire which began in the 1800's, further accentuated by this dismemberment, resulted in a weak Austria, unable to bar the way to ambitious Germany in 1938. "Allied"

Area: 83,849 km²
Pop. Grth.: —0.12
Pop.: 7,510,000
Cap.: Vienna

with Germany in W II, it shared its fate in defeat and only regained its independence in 1955, when it became a federal republic.

Geography — Climate:

70% of Austria is covered by mountains - the Alps. Loftier in the West, these beautiful mountains attract tourists and the World's top skiers. The Wien Basin to the East, is heavily populated and, as in Switzerland, is of great importance.

Manufacturing is increasing, and tourism is likely there to stay. The Brenner Pass cannot ensure adequate communications between North and South; the Alps thus have a negative impact on the Country.

Agriculture and cattle-raising are not able to ensure Austria's self-sufficiency in the food area; it must import much of its needs. Wood, pulp, paper and furniture industries are supported by forests of good quality timber. Main cities are: Vienna, Linz, Graz and Salzburg. The climate is similar to that of Switzerland.

Azores
A group of islands under Portugese rule, it contains nine islands, the main ones being: Fayal, Tirceira, San Miguel. Of strategic value to the Western Block nations. Located in the Atlantic, NW of the Moroccan Coast.

Area: 2313 km²
Pop.: 330,000
Main city: Angra

Balearic Islands

Area: 5014 km^2
Pop.: 558,000
Cap.: Palma de Mallorca

The fifteen islands of this group, off the Gulf of Valencia, form a province of Spain. Majorca, Minorca and Ibiza, the largest, are essentially tourist resorts.

The Balkans

The Balkan Peninsula is an essentially mountainous region of Southern Europe, favoured with a continental climate which becomes very mild on the Mediterranean Coast. The Balkans are comprised of Romania Bulgaria, Greece and Turkey. Under Turkish rule from the end of the 14th Century, the region was reconquered mainly by Austrian and Russian monarchs in the 18th Century. Taking advantage of splits between Central European powers and the decline of the Ottoman Empire, the nations of the Balkans gradually secured their independence during the 19th and 20th centuries.

Albania

Area: 28,748 km^2
Pop. Grth.: 2.5
Pop.: 2,670,000
Cap.: Tirana

History:
Like many other European nations, Albania's history really begins as a Roman province. Subsequently, it passed under Byzantine rule to later become part of the Bulgarian and the Serb Empire. During the 15th Century, Turkey had Albania under its domination; in 1912, after many uprisings, Turkey had to recognize Albania's independence. From then until 1946, Albania has been a principality, a republic (1925), a Kingdom (1928), a subject of Italy (1939) and finally, after WW II, a people's republic. The "honey-moon" with the USSR dit not, however, last long for in 1961, the two governments having had serious differences of opinion, Albania sided with China. Albania has the dubious honor of being the first country of the World to officially declare itself an atheist state.

Geography - Climate:
A very mountainous country interspersed with fertile regions on the coasts and in the Korca Basin. The climate is temperate.

Economy:
The sub-soil of Albania contains a good mineral potential but its development is slow; such was the case for chemical and engineering industries as well as steel works now being developed under government direction. Oil fields are beginning to produce and have justified the establishment of a pipe-line to the port of Vlora in the SW. Cement, petroleum products, textiles and food industries are government-controlled. All land in Albania is state-owned and farming is collective. Sugar-beet, potatoes, fruit, grain, cotton, tobacco are grown. Half of Albania being covered by forests with a great variety of trees, forestry is of major importance to its economy. Most of Albania's trade is with Red China (70%); the remainder going to Eastern Block countries.

Bulgaria

Area: 110,912 km^2
Pop. Grth.: 0.4
Pop.: 8,810,000
Cap.: Sofia

History:
Contempory Bulgaria corresponds to Ancient Thrace. The original Asians of Turco-Mongolese origin who conquered the Slavs inhabiting the land, were eventually absorbed by them. Bulgaria became an important state of the Balkans in the 12th and 13th centuries, but a Turkish invasion kept the Country under Ottoman rule until the 19th Century. In 1908 Bulgaria became an independent realm; it chose to side with Germany in WW I and through the latter's defeat in 1918, lost territory. Again in 1941, it picked the losing side, to find itself occupied by Russia in 1944; the stage was set for yet another people's republic (1946).

Geography — Climate:
In the North and descending towards the Danube, the Balkan mountains cover the land. To the SW, the lowlands, Bulgaria's most productive region. The climate is of the mild Mediterranean type in the South, but continental in the North. Main cities are Plovdiv, Varna, Ruse & Burgas.

Economy:
Copper, zinc, iron, manganese and coal are mined. Oil is now drilled off-shore, in the Black Sea. Manufacturing which employs 33% of the work-force, is nationalized (fertilizers steel, textiles, coke). A large merchant fleet uses the ports of Varna and Burgas on the Black Sea. What farm land there is, is collectivized. Wheat, sundry cereal grains, tobacco, cotton, vegetables & fruit are grown, while live-stock is raised. Fishing is only for local consumption. Forestry is increasingly important. Most of Bulgaria's trade is with COMECON countries, the USSR coming first.

Greece

Area: 131,944 km^2
Pop. Grth.: 1.1
Pop.: 9,440,000
Cap.: Athens

History:
It is ironical that the cradle of the World's most brilliant culture should now be one of modern Europe's less prosperous countries! After reaching peaks of excellence in the areas of art, literature, philosophy and science, Greece now has to be content with the appreciation of a tourist world, ogling the ruins of its once magnificent past!

Greece's geographical position placed it in the orbit of developing civilizations. To the North, Greece is bounded by Balkanic countries such as Albania, Bulgaria and Yugoslavia, whilst in the South, the Mediterranean gives it beauty and a delightful climate. The Aegean and Ionian Seas, East and West of Greece respectively, do nothing to mar the Country's reputation of ancient beauty and serene climate. It can easily be seen that Hellas, as it is called in Greek, jutted across the sea lanes of a developing and increasingly inquisitive mercantile world.

The near barbarian tribes, which occupied early Greece (Pelasgi), got from the Aegeans and Cretans (Minoans) the first seeds of civilization. In the 15th Century BC, Acheans merged with the Minoans to form the Mycenian civilization, an era which produced the Trojan War. Dorians, Eolians and Ionians followed to leave the imprint of their architectural genius for the modern world to admire. The Country was then broken into military and commercial cities, vieing with one-another for supremacy. Out of this was born the rivalry between Sparta and Athens, and later, Thebes. Threatened by the Medes and Persians in their Aegean sea possessions, the Greeks, under the leadership of Athens, defeated the Persians in the 5th Century BC. These wars, coupled with internal troubles such as the Peloponnisos war which saw Sparta beaten by Athens, weakened Greece and made it an easy prey for Philip II of Macedonia who subjected the Country in 338. His son, Alexander the Great, was to consumate the total defeat of the Persians, push the boundaries of the empire as far as the Euphrates in Asia, only to die young, seemingly catching pneumonia following a swim. His generals split and weakened the empire sufficiently to allow the Romans to step in and reduce it to the state of a Roman province (146BC). Despite these stinging defeats, Greece retained its moral, literary and intellectual influence over the civilized World. A vassal of the Eastern Empire in the Middle Ages, subjected to barbarian forays and in the power of the Crusaders in the 13th Century AD, Greece then came under the rule of Turkey and suffered a great deal; it regained its freedom with the help of France and England who defeated the Sultan's forces at the Battle of Navarin (1827). Its freedom was officially proclaimed in 1830. In 1912-13, Greece profited by wars in the Balkans, gaining territories. Again Greece profited by siding with the Allies in 1917, obtaining Thrace and Epirus, but ground was lost to the Turks in 1922. Greece became a republic in 1927. Invaded by Mussolini's forces in 1940, it held its own but succumbed when the former's German allies came to the rescue. (Greece had returned to the monarchy in 1935).

Geography and Climate:
Greece is a vast peninsula whose southern tip, the Peloponnisos, has "broken off" as it were, to allow the Gulf of Corinth to flow between it and the mainland, aided by the Corinth Canal. The Peloponnissos is mountainous whilst the mainland offers coastal plains which gradually build up to mountain ranges as they near the Balkans. The climate varies from Mediterranean in the South, to wet in the Northern highlands. Summers are hot, winters are moist and mild. The ruins of Sparta and many classical cities are found in the Peloponnisos and in Eastern Greece.

Economy:
Half of Greece's population works in agriculture to produce sustenance crops, but only one third of the land is arable. Manufacturing centres such as Athens, the port of Piraeus and Salonica contribute to industries which currently export nearly 50% of their production (wine, raisins, olive oil, figs); chemicals, metals, shipping and tourist industries add to Greece's revenues.

Main Possessions and Cities:
Crete, the Ionian Islands (Corfu), the Thracian Islands, the Sporades, the Cyclades, Santorin, Thira, the Aegean Islands (Lesbos, Chios, and Samos) the Dodecanese (Rhodes). Main cities: Athens, Salonica and Patrai.

Romania

Area: 237,500 km^2
Pop. Grth.: 1.0
Pop.: 21,850,000
Cap.: Bucharest

In the 2nd Century AD, Roman Emperor Trajanus conquered and developed Dacia (Romania today) as a Roman province. Thus we find at its origins, Dacians and Romans, as well as elements of Slavs and Turks. For six centuries from the 4th Century AD, Romania watched the tide of barbarian invasions flow westward, as Goths, Huns, Slavs, Avars and Tatars rampaged and plundered the land. After being compressed toward the plains of the Danube by the Hungarian conquest of Transylvania, Romania came under Turkish rule in the 15th Century. Between the 15th and 20th centuries, there followed a series of events such as the governing of the Romanian principalities of Moldavia and Valachia by Greeks, their passing under Russian protectorate in 1774 and the eventual taking over of Moldavia by Russia in 1812. An internal revolt took place in 1821 and in 1877-78, Romania went to war against the Turks, at Russia's side. The war of 1914-18 saw Romania in the Allied Camp and benefitting from it by the acquisition of territory. A twist of fate forced King Michael to place pro-Axis Premier Antonescu at the helm in 1940, a move which could only result in Romania siding with Germany against Russia. Defeated in 1944, Romania decided to throw its lot in with the Allies. Communists took over in 1946 and forced the abdication of Michael. Another People's Communist Republic was born.

Geography — Climate:
Romania is made up of a central plateau overlooked by the Carpathians in the East, the Transylvanian Alps in the South and the Bihor Massif in the West. Hot summers and cold winters are characteristic of its continental climate.

Economy:
The oilfields of the Ploesti region have become supportive of a large manufacturing industry; steel, and its by-products, chemicals, textiles and processed farm products, add to the yield of a fertile land which includes cattle-raising. Farms are, for the greater part, state run.

Main cities are: Bucharest, Ploesti, Craiova, Galati, Brasov, Timisoara, Cluj and Oradea.

Turkey

Area: 780,576 km^2
Pop. Grth.: 2.3
Pop.: 44,310,000
Cap.: Ankara

History — Culture:
Although Ankara is its capital, the city of Turkey with the most impact on the World, certainly has been, and still is, Istambul. This strategically located city on the Bosphorus, between the Sea of Marmara and the Black Sea, had its importance in the days of the Persian invasions of Greece, as it had during WW I, for it controls access to the land bridge which is Turkey, between Europe and Asia. Formerly called Constantinople, it was the capital of the Byzantine (1025) Empire. As Istanbul, the city was the centre of the Moslem World.

Ancient Troy stood on the western shores of Turkey while on its eastern borders, Noah is said to have "beached" the Ark after the deluge.

Taking Asia Minor away from the Byzantine Empire, Seljuk Turcs founded a sultanate in Turkey in 1025 (Anatolia), while Turcomans settled near the Bosphorus. In 1300, Ottoman Turks took over from the Seljuks and later, seized Constantinople (1453); they then proceeded to increase their empire by adding to it parts of North Africa, the Balkans and Persia; they even made inroads into Central Europe, going as far as Hungary. At one time, the Ottoman Empire included Bulgaria, Macedonia, Thessaly, Salonika, Serbia, Bosnia and Bagdad.

The zenith of the Ottoman Empire was therefore reached in the 16th Century. From then on, and except for a slight resurgence in the 19th Century, it declined. Algiers and Tunisia were lost, Hungary went the same way in 1687, followed by Albania, Belgrade, Dalmatia and Herzegovina. The control of the Black Sea went to Russia in 1774; Greece regained its independence in 1829, Rumania in 1856, Serbia and Bulgaria in 1878. Tripolitania was lost after a war with Italy in 1911, and the rest of Turkey's European possessions went following the Balkan War of 1912-13. Turkey was also on the losing side in 1914-18. A victory over Greece in 1922 saved its national existence.

Made into a republic in 1923, Turkey saw its leader, Mustafa Kemal, abolish the sultanate, reform education and the legal system and contribute to the emancipation of women. In more recent times, Turkey became an ally of the USA and a NATO member, securing from rich America, considerable aid aimed at offsetting Russian influence.

Geography — Climate:
European Turkey is made up of rolling plains which run into low mountains in the NW, on the Bulgarian border. Anatolia, which is located in the centre of the land bridge which is Asia Minor, is a rugged plateau ringed in the North by the Pontic Range and in the South by the Taurus Mountains. A narrow coastal strip runs along the Mediterranean. The climate of the Central Plateau is dry and hot in summer, but cold in winter; the coastal areas are mild and maritime. Turkey is known for its frequent and violent earthquakes, especially in the region of Anatolia.

Economy:

Agriculture, cattle-raising, fruit, cotton, tobacco are of importance. Fishing thrives in the Black Sea. Industries such as glass, cement, textiles, chemicals, and paper are developing. Steel and iron plants, petroleum production and other endeavors are receiving governement backing. Main cities are: Ankara, Istanbul, Izmir, Adana.

Yugoslavia

Area: 255,804 km²
Pop. Grth.: 0.9
Pop.: 22,160,000
Cap.: Belgrade

History:

After WW I, delegates of all Slavic nations of the Balkans wanted annexation with victorious Serbia. Thus was born a Kingdom of the Croats, Serbs and Slovenes in 1918. In 1931, after going through internal troubles, the new state took the name of Yugoslavia. However, political events turned the Country toward Moscow; the German invasion of Russian territory in 1941, more or less inter-locked Russia and Yugoslavia. When the Germans invaded their Country, Yugoslavs fought fiercely, some under nationalist Mikailov, others under Broz (Tito) the communist. The latter eliminated his rival after the war and took over the Country, ipso facto eliminating the monarchy. Again, it was a case of the Communist "honey-moon" not lasting, as Tito, although favoring communist philosophy, had no intention of submitting to the dictates of Moscow. Present day government of Yugoslavia is an independent form of communism and federal republic which rules over six small communist republics: Serbia, Croatia, Montenegro, Slovenia, Macedonia and Bosnia-Herzegovina.

Geography — Climate:

It seems that through the ages, the coast of Yugoslavia has sunk beneath the sea. The result: narrow islands are all that is left of coastal mountain ranges; harbors allow vessels to sail over what used to be valleys. The climate is Mediterranean, the kind which attracts millions of tourists. Inland, the Dynaric Alps run parallel to the coast and rolling plains are ideally suited to agriculture. The climate is continental.

Economy:

Yugoslavia possesses considerable mineral wealth (coal, iron, copper, gold, lead, manganese etc.); its industry is found mainly in the NW, (steel, iron, textiles, cement, fertilizers). Agriculture is its mainstay and includes livestock-raising. Forestry is also of major importance. Main cities: Belgrade, Zagreb, Sarajevo, scene of the assassination of Archduke Franz- Ferdinand of Austria in 1914, which triggered WW I.

Belgium

Area: 30,513 km²
Pop. Grth.: 0.1
Pop.: 9,890,000
Cap.: Brussels

One of the most densely populated countries of Europe, its Flemish population speaks a Dutch dialect, while its Walloons, in the South, speak French.

Freed from Dutch rule in 1830, Belgium was occupied by Germany during both world wars. Like other colonial empires, Belgium felt the impact of world nationalism and lost many of its overseas territories, the most important of which was the Belgian Congo, now known as Zaire.

A short coastal strip delineates a flat fertile marshland reclaimed from the sea. These coastal plains include the estuary of the River Scheldt. Near the border of Holland lies the Compine Region, Belgiums's main coal-mining district. To the South, highlands reach as far as the Ardennes Plateau.

Coal, iron, engineering and metal working are the main industries. Wool, linen and lace add to the Country's resources, as does its rich agriculture.

The climate of Belgium is mild and coastal; its main cities are: Brussels, Liege, Antwerp, Bruges and Ghent.

The British Isles

Across the English Channel which separates it from Continental Europe, lies the United Kingdom of Great Britain and Northern Ireland, which includes Wales, Scotland and the N.E. portion of Eire, known as Northern Ireland. The second island of the British group is Eire itself. A number of small islands off the N.W. coast of England, complete the British Isles complex.

The United Kingdom

Area: 244,046 km²
Pop. Grth.: 0.0
Pop.: 55,820,000
Cap.: London

History and Culture

Thousands of years ago, a race of "barbarians" called the Iberians settled in the British Isles. Later, invading bands of mainland barbarians such as Celts, Saxons, Jutes and Norsemen joined Roman legions in leaving their imprint on British soil. In time the true Britons were to survive and give English civilization the basic ingredients of greatness - the coolness of the English, the fiery nature of the Scot and the fine human qualities of the Welsh were to produce a race where individualism would not stifle the spirit of human cooperation; a race from which some of the world's greatest explorers and most famous writers, soldiers and statemen were to come.

The many incursions mentioned, culminated in the Norman invasion of 1066. After a period of adjustment, Normans and Anglo-Saxons merged into one language and one culture. In the 11th and 13th centuries, Norway ceded the Hebrides, the Shetlands and the Orkneys to Scotland. Later James VI of Scotland succeeded Elizabeth I and took the title of James I of England (1603). England, Wales and Scotland were one, but official ratification had to await the Act of Union of 1707. To this Union known as Great Britain, Ireland was added in 1801.

Britain is perhaps best known for the empire it built in the 17th and 18th centuries and over which the sun was said to "never set". The Empire included India, Canada, many countries of Africa and for a while, colonies in America; Australia and New Zealand were also part of the Empire as they are now part of the Commonwealth.

After reaching a peak of power in the second part of the 19th century, Britain impoverished itself by taking part in the first World War from which it emerged crippled and exhausted. Not fully recovered, it plunged headlong into a second world conflict, this time against Nazi ideology (1939-45). Although with its Allies, Britain emerged victorious, it was downhill from then on - its economy was shattered; its role as a world power declined steadily, its constitution, industry and labor force were in need of up-dating. The stage was set for Socialism. To boot, the empire started to disintegrate following a wave of nationalism sweeping a world which had become sated with the idea of domination. Although most of England's colonies chose to remain within the British Commonwealth of Nations, Britain would no longer hold sway over them.

A thorn in Britain's side was, and still is, the violence-marked struggle between Northern Ireland's Protestants and Catholics, the latter demanding a share of the power in running the Country with the Protestant majority. This demand, backed by a terrorist campaign led by the outlawed Irish Republican Army, has resulted in frequent bombings and shootings which forced Britain to intervene by sending troops to keep law and order. Despite these measures, no settlement is in sight.

The modern era brought to the UK, a different kind of invasion, a wave of immigration from Britain's former empire, in particular, Africa, Pakistan, India and the West Indies; this created serious problems in the housing, employment and educational areas. Racial discrimination was brought to the fore.

Britain's population is mostly urban. The Country has free, compulsory education to the age of 16 and a social welfare system which looks after its citizens "from the cradle to the grave", with all its resulting short-falls and blessings.

Mainly a Protestant nation, Britain has two main official churches, the Church of England and Scotland's Presbyterian Church.

Spectator sports such as soccer, cricket and rugby are popular, as is horse and dog racing. Wimbledon has long been the undisputed mecca of amateur tennis.

Economically, Britain seemed to have only one chance of salvation in the 1970's, and that was admission to the European Common Market; this was achieved in 1973. Nevertheless, the ill was too deeply ingrained and severe inflation coupled with devalued currency, further eroded its economic health. North Sea oil, however, seems to offer hope for a healthy come-back.

Constitution — Main Cities:

Britain is a constitutional monarchy. It has a House of Commons (635 elected members) which is the governing body, and a 1000 member House of Lords. Although the monarch is the Head of State, it is the Prime Minister and his Cabinet who govern the nation.

The main cities of the UK are London, Capital of England and the UK. Cardiff (Wales), Edinburgh (Scotland) and Belfast (Northern Ireland). Leeds, Manchester, Hull, Liverpool, Birmingham, Bristol, and Portsmouth are among the main cities of England, while Oxford and Cambridge are synonymous with high education, the World over.

Land and Climate:

From rolling, green country-side in the South, the land becomes rugged, desolate and windswept as one moves North. The Scottish Highlands are craggy and cut by deep fjords known locally as lochs. The main regions of the UK are: the Scottish Highlands, with the Grampian Mountains; the Central Lowlands (valleys of the Tay, Clyde and Forth Rivers) with their coalfields, farmlands and industries; the Southern Uplands (sheep- grazing country - Valley of the Tweed, and the Cheviot Hills on the border between England and Scotland); the Penine Mountain range which runs into Derbyshire from Scotland and whose flanks are rich in coal; the Lake District, a resort area; North and South Wales with their coal mining; the Southwest Peninsula (Somerset, Devon and Cornwall), a beautifully rugged area favored by tourists and holiday makers, and finally, the English lowlands with their gently undulating greenland, their downs in the S. and S.E. and the industrial Midlands.

Northern Ireland, for its part, is known for its saucer-like shape, its low hills and fertile lowlands.

The Severn (Wales) the Thames (S.E.) the Clyde, the Mersey and the Humber are the main rivers.

Despite its northern latitude, Britain's climate is temperate although the proximity of the surrounding oceans makes the atmosphere highly impregnated with humidity, giving one the impression of greater cold than is actually the case in winter. Precipitation, the lot of many islands of similar latitudes, is a way of life; it rains during every month of the year, especially in the NW.

Economy:

Coal is Britain's main natural resource and the basis of its electrical power. Iron ore is of a lesser quality. Clay, chalk and limestone are plentiful in the SE. Natural gas and oil deposits have been located in the North Sea, the latter auguring well for Britain's self-sufficiency in this latter domain, in the next decade.

There is not enough farming land in Britain, to enable its farmers to feed the Nation; what is available lies in the East and is fully exploited to produce barley, oats, potatoes and wheat. The western part of the Country is given to livestock rearing (beef and sheep). The milder South produces fruit, vegetables and flowers. Fishing is very important to Britain which has many fishing fleets covering the seas, often causing diplomatic impasses as was the case with Iceland in 1976. A large part of Britain's gross national product is accounted for through manufacturing; engineering products lead the field and steel production ranks fifth in the World. Birmingham, Wolverhampton and Coventry are main centres of industry (cars, cycles and tractors) - Manchester and Liverpool play a large role in Britain's economy. Apart from the products mentioned, textiles, ship-building, chemicals, printing, glass-making, chinaware and cutlery add to Britain's sources of income.

Northern Ireland boasts the largest ship-building yards in the British Isles and is known for its dairy and farm products. Irish linen is world-famous. The industrial complex around Glasgow produces steel, iron & ships. Scottish wool is also world-famous.

Trade and International Affairs:

Its insular position makes Britain one of the World's leaders in trading; of necessity, it imports more than it exports. Banking, shipping insurance and an actively pushed tourist industry add money to Britain's coffers.

Its long experience in the realms of statesmanship and diplomacy result in the World looking to England for support in many delicate situations.

Eire

Area: 70,283 km²
Pop. Grth.: 1.2
Pop.: 3,370,000
Cap.: Dublin

(The Republic of Ireland)

The Irish are descendants of the Gaels, a branch of Celts. Patriotic, religious to the point of fanaticism, artistic, musical and a complete extrovert, the Irishman is both loved and made light of by those who have not bothered to get to know him.

The people of Eire are 90% Roman Catholic; their history is one of hardships, of struggle for survival and the pursuit of freedom. The struggle against British rule, marked by sporadic violence between 1916-1921, was not to cease when Eire was granted dominion status - Civil war raged for a couple of years and ultimately, Ireland severed its ties with the British Commonwealth in 1949. To this day, the activities of the IRA, which is against the partition of Ireland, are a serious problem, not only for Britain and the official government of Eire, but for a good segment of its population, which has had its fill of bloodshed.

Political strife is not the only ill to have beset the Irish Nation. In 1845, a blight which lasted four years, wiped out the entire potato crop on the Island, condemning one million people to death by disease and starvation. Thousands who decided to leave Ireland to escape misery, still found it in the form of inhumanly crowded emigrant ships and more disease, such as small-pox epidemics. And then there were the innumerable problems associated with re-settling in a new country; adjustment, periods of unemployment, family separations, loneliness and even rejection!

The mild climate of the Emerald Isle, as it is called, is much like that of Northern Ireland: plenty of rainfall makes the land green and fertile, ideal for agriculture, including the rearing of live-stock. An injection of foreign capital has helped Ireland develop industries such as oil refining, fertilizers, textiles, and ship-building. Tourism also adds to the Island's revenue.

The River Shannon is the longest in the British Isles while Killarney Lake is a favourite with tourists.

The Republic of Ireland is divided into four provinces - Connaught, Leinster, Munster and Ulster. Provinces are sub-divided into counties.

Canary Islands

Area: 7,273 km²
Pop.: 1,139,000
Main cities: Palma and Tenerife

There are thirteen islands in this group which form two Spanish provinces - half of these islands are uninhabited. Located in the Atlantic, West of the Sahara.

Cape Verde Islands

Area: 4,033 km²
Pop. Grth.: 1.9
Pop.: 320,000
Cap.: Praia

These fifteen islands of different sizes secured their independence from Portugal in 1975. The group lies West of Senegal.

Crete

This island, South-East of the Peloponnisos, in the Mediterranean, has a population of only 400,000, but its claim to fame is that it was, in ancient times, the seat of a civilization which dominated the Eastern Mediterranean. Important geological finds at Knossos have thrown a great deal of light on the Pre-Mycenian civilization.

Czechoslovakia

Area: 127,869 km²
Pop. Grth.: 0.8
Pop.: 15,250,000
Cap.: Prague

History — Culture:

Czechs and Slovaks were part of the Austro-Hungarian Empire until 1918, when they were united by the creation of Czechoslovakia. Hitler's ambition led him to take over the Country's Sudetenland in 1938. At the same time, Poland and Hungary helped themselves to pieces of the cake by grabbing Teschen and Southern Slovakia respectively. The dismemberment of Czechoslavakia was consummated when Germany took what was left in 1939. Six years would pass before Russian troops would liberate Czechoslovakia from German rule. Communism, not surprisingly, gained ground rapidly until in 1946, it emerged as the largest party. Pressure continued and in 1968, the Country was occupied by Soviet and other Eastern Block troops - the stage was set for a 20 year Czech-Soviet pact. The present form of government is a federal-socialist republic. Czech and Slovak socialist republics have equal authority, but both are subordinated to the dicta of the Communist Party.

Czechs, who account for over 60% of the population, occupy the western provinces of Moravia and Bohemia, while the Slovaks, (30%) live in the East. The remainder of the population is a mixture of Hungarians, Poles, Russians and gypsies. 75% of the population is Roman Catholic.

Geography and Climate:

The western part of the Country is a rich agricultural basin surrounded by mountains. To the East, except for the valley of the Danube, are forested mountains. The climate is cold in winter and warm in summer - rainfall is adequate.

Economy:

The work force is split between mining (coal, some metals), manufacturing, agriculture and forestry. All industries are nationalized. Like other Eastern Block nations, Czechoslovakia's trade is within the Block, although it does business with Austria, West Germany and the U.K. Main cities are: Prague, Brno, Ostrava, Plzen and Bratislava.

Finland

Area: 337,009 km²
Pop. Grth.: 0.3
Pop.: 4,750,000
Cap.: Helsinki

History - Geography:

One third of Finland lies inside the Ărctic Cirle, East of Sweden. Incredibly enough, it boasts 80,000 islands on its rugged coastline, and 60,000 lakes of various sizes inland; over 70% of its land is covered with coniferous forests. Most of Finland is a low-lying plateau where winter conditions prevail from December to May. Finnish territory extends into Lapland.

A Swedish possession until the early 19th Century, Finland was ceded to Russia in 1809. The history of this Country is replete with narratives of wars with Russia, all ending in defeat, but never in discouragement on the part of this proud people. In 1917, Finland secured its independence, following the confusion of the Russian Revolution. The Russians attacked and subdued Finland in 1940, despite a heroic resistance. However, the lull was not to last long, as in 1941, the Finns attacked the USSR but had to conclude an armistice in 1944. Finland then turned against its former Allies, the Germans. Through the Treaty of Paris, in 1947, Finland lost land to the USSR.

Economy:

The economy of the Country is based on its large forests. Despite a short season, agriculture has its importance. Metal processing, textiles and the design of glass, ceramics and furniture add to Finland's economy. The hardy Finns are known for their rugged sports endeavors, such as cross-country skiing, which is usually followed by immersion in their beloved saunas.

France

Area: 547,026 km²
Pop. Grth.: 0.7
Pop.: 53,480,000
Cap.: Paris

History

Bounded in the NW by the English channel and surrounded by Belgium, Luxembourg, Germany Switzerland and Italy, France traces its origins to the Roman conquest of Gaul by Caesar, in the first century B.C. Under the Romans, Gaul became a prosperous province, though it was constantly threatened by border clashes with barbarian tribes such as Germans, Wisigoths, Vandals, Huns and Franks. In 486 A.D., Clovis the Frank, taking advantage of the declining power of Rome, overthrew the Roman governor of Gaul and established the Frankish realm. Merovingians were to be followed by Carolingian monarchs, the most famous being Charlemagne (742-814) who was crowned emperor by Pope Leo III in 800 A.D. Under this powerful soldier-emperor, the kingdom grew in wealth, size and power; its very size brought about its collapse, after the death of the monarch. In 987 the stage was set for Hughes Capet to install the Capetians on the throne of France; with help from the all-powerful Church, the new kings strove to restore order among the many feuding dukes of the realm, bring one and all to recognize and submit to the Crown and, generally speaking, increase the kingdom. In 1328, their King, Philip VI, fought the English whose Edward III, challenged his right to the throne; there followed wars which saw the English defeated by Joan of Arc at Orleans in 1429, and the eventual demise of English influence in France, in 1453.

The Bourbon dynasty, with its first ruler Henry IV, followed and became most powerful. Among famous monarchs of this line, were Louis XIII, who, with Cardinal Richelieu, was the real creator of absolutism, and Louis XIV, the Sun King. The reprehensible personal conduct of Louis XV, and the reverses suffered by his administration which saw France lose India and Canada, were to set the stage for the French Revolution of 1789 and the subsequent overthrow of Louis XVI and the monarchy in France.

During his reign, Napoleon I consolidated the gains and conquests of the Revolution and brought land and glory to republican France. Defeated by Wellington in 1815, Napoleon made way for 33 years of restored monarchy. In 1848, however, the French chose to go back to the Republic but not for long. In 1851, Louis Napoleon restored the empire, only to be defeated in the Franco-Prussian War of 1870-71, which resulted in France losing Alsace and most of Lorraine to Germany.

France sought alliances with England and Russia (Entente Cordiale and Triple Entente). The two World Wars of 1914-18 and 1939-45 brought it enormous suffering, though with its Allies, it emerged triumphant.

Like other nations, France felt the impact of the financial crisis of the 30's. The end of WWII was to bring about the Fourth Republic whose progress was hindered by expensive and unpopular colonial wars in Indochina and Algeria. Dien-Bien-Phu was to spell the end of French colonial power in Asia, whilst Algeria obtained its independence. The Fifth Republic resulted in considerably increased presidential powers. Under its war-time leader, Charles de Gaulle, France disengaged itself from colonialism and achieved a degree of political stability.

Land - Climate:

France offers a variety of geographical features: mountain ranges, hills, plateaus and plains. Its mountain ranges are the Massif Central, the Massif Armoricain, the Vosges, the Ardennes, the Maures, the Esterel and the Corsican Massif (old ranges), and, more recent ones geologically, the Pyrenees and the Jura. Half of France is covered by plateaus and plains, rolling hills and fertile valleys; these are divided into the Paris Basin, the Aquitaine Basin, the Rhone and Saône Plains, the Alsatian Plain as well as the Northern and Southern Plains. The coasts vary from flat, dune-covered shores to steep cliffs that rise out of the sea. Rocky beaches and marshy shores alternate. Many rivers born in the mountainous areas supply the Country with ample fresh water, (the Seine, Loire, Garonne, the Rhone and part of the Rhine). France has a temperate climate influenced by the surrounding seas; it varies with the regions - cold winters with snow, hot and dry summers, in the East; balmy winters with hot and dry summers on the Mediterranean.

Economy:

Agriculture is France's main resource. Livestock-raising is also of importance, as is the mining of coal, iron ore and bauxite. Hydro-electric power is not sufficient for its modern needs. Other industries are: textiles, chemicals, foodstuffs, wood, leather and tourism. Wine-growing is a most lucrative industry for France, and characteristic of its culture.

Departments and Possessions:

The tiny principality of Monaco with its fewer than 500 acres, and its world-famous casino in Monte Carlo, is on French soil. The rugged Island of Corsica, with a population fo 219,000, covers 8,723 km². Both are departments/territories of France. The French Alps and their highest peak, Mont-Blanc are worthy of mention. Martinique, Guadeloupe and Guiana are other French departments.

Among the main cities are: Paris, Reims, Rouen, Le Havre, Brest, Rennes, Orleans, Nantes, Bordeaux, Toulouse, Marseilles, Nice, Lyons, Dijon and Nancy.

Germany

Initially occupied by Finns, the country known today as Germany was host to Celts and then to a group of tribes known as Germans; these founded the first Frankish realm which later, Charlemagne was to enlarge into the Empire of the Occident, and later still, the Holy German Empire. The Emperor's sway was rather academic; in fact, the many states which the German Empire contained were quite independent. Throughout the years, Germany was caught in wars, political and religious strife.

The division of the Country into North and South eventually led to the emergence of a strong Prussian Empire (1701). Napoleon I, having eliminated the Holy German Empire and replaced it by a Confederation of the Rhine, was to see it dissolved in 1815 and reorganized into a German Confederation.

The King of Prussia, with the help of Bismarck, threw Austria out of the Confederation (Sadowa, 1866) and had himself crowned emperor at Versailles at the end of the war of 1870-71, which saw the defeat of France. There followed a period of rapid economic, military and industrial expansion - Germany became a colonial power.

The main-stay of a strong central European Alliance in 1914, Germany fought against the Allies and lost in 1918; this defeat was to mean the loss of most of its colonial empire and a considerable amount of land in Europe. Of importance was the creation of the Polish Corridor which separated Germany proper from Prussia.

Like most nations, Germany had its serious recession; this came in the wake of WWI, when inflation and unemployment were at their peak. The stage was set for Adolph Hitler and the National Socialist Party, to "pick-up" the Nation and drive it irrevocably towards dictatorship (1933) and a second world war (1939-45) which was to prove as devastating as the first and result in the defeat of Germany and its ruling Nazi Party. During this conflict, countries were annexed or overrun (the Rhineland, 1936, Austria, Memel, the Baltic port, and Czechoslovakia in 1938-39 and Poland in Sept. 1939), but these territories were lost after the Fall of Germany and a huge displacement of population took place. Then came the partition of Germany into four zones of occupation. Berlin itself had four similar zones. In time, three of the zones occupied by the Allies became West Germany; the Russians held the fourth which became East Germany. Thus, after a victory for which all four "partners" had paid dearly, there came a period of "peace" which opposing ideologies turned into a "Cold War" (1949). From then on, East and West Germany developed and recovered, each in its own way, each with its life-style, its own philosophy and sense of values.

As facts must be recognized, we will deal briefly with East and West Germany, as political entities.

West Germany

Area: 248,577 km²
Pop. Grth.: —0.3
Pop.: 61,340,000
Cap.: Bonn

History:

Although the blood of many races runs through German veins, its people are basically from two main stocks: the tall, fair-complexioned Nordic types, and the shorter, darker, southern or Alpine types.

West Germany is a federal republic which, came into being in 1949; it is made up of states, each with its own government. Its population, bolstered by refugees, grew fast after WWII. Thanks to a swiftly recovering industry, this sudden growth did not cause unemployment and other assorted problems because refugees, for the most part, were absorbed in the work force.

Geography, Land and Climate:

Geographically, West Germany is made up of a northern plain, central and southern uplands and a S.W. region. The land varies from glacial deposits & sandy infertile soil, to drained valleys, low hills, plateaus and mountain ranges, such as the Swabian and

Franconian Jura and the better known Alps on the Austrian border. Thickly forested uplands in the South contain the Black Forest; the climate in the high regions of the South is severe, but elsewhere, it is mild and pleasant.

Economy:

West Germany is a major industrialized nation. The revival of its crippled industry, after WWII, was facilitated by a large influx of refugees which offset the shortage of man-power caused by the war, by American capital, by a world demand for coal and, mainly, by the courage and industriousness of its people who wanted to rebuild their nation.

The Ruhr, the Saar and the Aachen areas are synonymous with coal-mining, but iron ore, potash and other metals are found in this region. Manufacturing, machinery, engineering of all types, textiles, forestry and chemicals, all contribute to West Germany's industrial complex. Farming and fishing are also important economic assets.

Strategically, West Germany is the hyphen between East and West. A member of NATO, it has a key role in Western defense strategy. Main cities are: West Berlin, Hamburg and Munich.

East Germany

Area: 108,178 km²
Pop. Grth.: —0.2
Pop.: 16,790,000
Cap.: East Berlin

History

Politics and the resulting way of life are about the only things that distinguish East Germany from her western sister. The German Democratic Republic set up in 1949 is merely the Eastern Block's answer to the western political entity set up in the West, the Federal Republic of Germany.

Geography — Land and Climate:

East Germany is on the Baltic coast; much of its land is sandy. To the South, one finds hills, lakes, marshes and rivers born from melting glaciers, eons ago. In the South where the bulk of the population lives, agriculture and industry thrive and there are magnificent forests (the Hartz mountains, the Thuringer Wald and the Erzgeberg, near Czechoslovakia). A continental climate prevails in East Germany, but winters are colder here than in West Germany.

Economy:

Lignite is produced in greater quantity in East Germany than anywhere else in the World. The Country also has potash and many metals, including uranium. Petroleum must be imported. Having evolved from a mainly agricultural nation, East Germany now has a manufacturing industry which yields two thirds of its income. Chemicals, fertilizers, synthetic rubber, textiles, leather, cement, steel and iron, all form part of its industrial spectrum. Most industries are government-regulated. Agriculture takes second place to manufacturing although farming is more important here than in West Germany. Forestry and fishing are also of importance. East Germany's main trading partners are those countries which form the Communist Block.

Hungary

Area: 93,030 km²
Pop. Grth.: 0.5
Pop.: 10,700,000
Cap.: Budapest

History:

Many tribes of "barbarians" such as the Celts, Lombards and Avars preceded the Magyars in Hungary. In Budapest, visitors can see the vestiges of Roman occupation. After Charlemagne destroyed the Avar empire founded in 568, there followed the christianization of the Country, Turkish domination, the advent of the Hapsburg Dynasty and a gradual germanization which led to the formation of the Austro-Hungarian empire of 1867, dismembered in 1918. After a brief flirtation with Communism (1919), Hungary went back to monarchy without a king, with Admiral Horthy acting as Regent from 1920-45. ProNazi influence grew and resulted in the acquisition of Slovakia, part of the Ukraine and two thirds of Transylvania. Having elected to side with Germany in 1941, Hungary saw its territory ravaged by the Russians. War was followed by serious economic and financial troubles. Since 1946, Communism has made considerable progress in the Country, though in 1956, a large segment of the population revolted against the regime. Despite frantic appeals to the West which was obviously not disposed to confront the USSR on the issue, the rebellion was crushed by Russian armor called in by Janos Kadar. Hungary therefore continues as a Communist-controlled country, although there are signs of liberalization.

Geography - Climate:

With the exception of the Bakony Forest ridge and Transdanubia, (the mountainous land between Lake Balaton and the Danube), Hungary is a vast plain, ideal for agriculture. The Hungarian Plain is famous as the invasion route used by barbarian hordes from the East. The climate is hot and dry in the summer, and windy and very cold in the winter.

Economy:

Except for a small amount of coal, iron ore and bauxite, Hungary has no mineral wealth to speak of and has to rely on imports for its rapidly expanding manufacturing industry. Natural gas deposits and petroleum offer potential. The back-bone of the economy is agriculture which employs 25% of the entire work-force. Cattle-raising, forestry and fishing are also of importance. The Country produces good wines and excellent gypsy music. Close to 70% of Hungary's trade is with the USSR and Communist Block countries.

Main cities are: Budapest, Dubrecen, Szeged, Miskolez and Pecs.

Iberian Peninsula

Andorra

Area: 453 km²
Pop.: 28,400
Cap.: Andorra la Vella

Nestled in the rugged Pyrenees, on the border between France and Spain, NW of Barcelona, Andorra is a principality administered jointly by French and Spanish politico-legislative bodies. Tourism is the Country's main industry.

Gibraltar

Area: 6 km²
Pop.: 30,000

Formerly known as the Pillars of Hercules, Gibraltar guards the entrance to the Mediterranean. Though part of the Iberian Peninsula, it has been a governor-administered British possession since 1804. Spain's claims to this strategic part of its peninsula have gone unheeded, even by the population of the Rock which declared itself overwhelmingly in favor of retaining British rule in the late 1960's. An airfield and a shipyard were of great value to England and its Allies in WWII.

Portugal

Area: 92,082 km²
Pop. Grth.: 1.3
Pop.: 9,870,000
Cap.: Lisbon

History - Culture:

The first inhabitants of this part of the Iberian Peninsula eventually had to share their land, language and customs with North Africans, Coastal European tribes, (Alts and Germans) as well as Greeks and Romans.

Recognized as independent in 1385, Portugal was able to initiate explorations such as the reconnaissance of the African coastline and de Gama's search for the elusive sea route to the East (the noted explorer finally landed in India, in 1498). Following these feats, Portugal's colonial empire grew; Brazil was won and lost and many lands were acquired in the Far East. In 1910, the Republic was proclaimed, but political unrest and the ambitions of Oliveira Salazar turned the Country into a dictatorship. Salazar died in 1970. Political turmoil and even bush wars in Portugal's overseas territories, hurt the Country; it was civilian authority versus military juntas. Currently, Portugal is governed by a Supreme Revolutionary Council and a Constituent Assembly. Socialism holds the lead among the political parties.

Geography and Economy:

Portugal is a wide strip of coastline on the western edge of the Iberian Peninsula. Generally adequate mineral resources are under-exploited. Portugese wines are growing in popularity, while its textiles and canned sardines industries hold their own on the market. Manufacturing offers potential and tourism is popular. Main cities are: Lisbon on the Tages River, Oporto on the NW Coast, Situbal, Braga, and Coimbra.

Spain

Area: 504,782 km²
Pop. Grth.: 1.4
Pop.: 37,180,000
Cap.: Madrid

History - Culture:

Behind the Spaniards of today, one finds the very same tribes and peoples who invaded Portugal in the early days of its existence. The Basques who are neither French nor Spanish live in the Bay of Biscay area and on the French border; they have been and continue to be a source of concern to the government, because of their attachment to their own culture and their separatist philosophy.

From the 5th Century BC, Spain had to cope with invasions by Carthaginians and Moors from North Africa, as well as German raids from the East. The Moorish invasion of 711 AD was to leave its imprint on many aspects of Spanish culture (physical characteristics, music and architecture). Under the leadership of a Castilian king, the many independent kingdoms of Spain were gradually united and thus acquired the necessary strength to overthrow Moorish domination in 1492 (Fall of Granada to the forces of Ferdinand and Isabella.)

Colombus' voyages of 1492, opened up a large part of the New World to colonization and gave Spain a niche in world affairs. Spanish sea power came to an end with the battle of Trafalgar in 1805. Before this, Spain had suffered a stinging defeat when its Armada, sent against Elizabeth I of England to avenge the death of Mary Stuart, foundered in a storm off the English coast (1588). Stripped of her power, Spain had to reconcile herself to the lot of a poor agricultural land. But the fiery Spaniards were not about to accept their fate without a fight. Monarchy shared a dictatorship with the Military in 1923. By mutual consent of King and military leaders, a republic was proclaimed in 1931. Sharp political divisions led to the Civil War of 1936-39, which brought Franco to power as a dictator. Franco went through the motions of restoring monarchy but did not act decisively on this issue until the eve of his death, when his successor designate, Don Juan Carlos, Prince of Bourbon, took over the reins of government. But political troubles were not to be settled that easily!

Geography:

The Spanish meseta or plateau, is surrounded by rugged mountains and is broken by sierras. Whilst the Andalusian plain of the SW is rich and fertile, that of Aragon in the NE, is the direct opposite. Coastal plains are fertile. The climate and rugged beauty of coastal Spain have made its Mediterranean Coast a tourist mecca.

Economy:

The mineral potential of Spain is much in the same position as that of Portugal, largely undeveloped, though the Country leads the World in the production of mercury. Manufacturing and engineering are increasing; textiles, woollen goods and cotton, coupled with low labor costs, are having a serious impact on world markets. Ship-building, paper, cork, iron and steel, are of importance to Spain's economy. Spanish wines are gaining in favor the World over; its fruit crops are unequalled. Agriculture, still the mainstay of the Nation's economy, adds to Spain's resources, as do forestry, fishing and cattle-raising. Although Spain trades with Western, Block nations, its generally un-democratic government has kept it out of the E.E.C.

Iceland

Area: 103,000 km²
Pop. Grth.: 0.8
Pop.: 230,000
Cap.: Reykjavic

Located in the North Atlantic, just South of the Arctic Circle, Iceland is a rugged island of volcanoes, fjords and glaciers. Its coastal low-lands support livestock-raising and dairy-farming. Despite its northern position, the island is warmed by the North Atlantic Drift which tempers its climate; it is subject to earthquakes and volcanic eruptions.

Settled in 850 by Norwegians, Iceland was for a while, under Danish rule; it has been an independent republic since 1944. Although Iceland has a few industries such as mining and aluminium smelting, its main source of revenue is its fishing industry which is at the same time, a source of friction with other fishing nations, Britain in particular, because of the catch limitations imposed. The West has its cold war; Iceland has its "cod war".

Italy

Area: 301, 225 km²
Pop. Grth.: 0.5
Pop.: 56,910,000
Cap.: Rome

Geography and Climate:

The historical back-ground of Italy, its contribution to the world of arts and music, the beauty of its land and its very shape give it a special place in any atlas or history book. The well-known boot-shaped peninsula juts out into the Mediterranean, and is flanked in the East by the Adriatic and in the West by the Ligurian Sea. Whereas the toe and heel of the "boot" bathe in the Mediterranean, its top rests on the Alps which separate it from the rest of Continental Europe. The coast-line which in places is of incredible beauty, varies from gold, sandy beaches to steep, rocky escarpments; inland, rolling plains give way to rounded hills which in turn become rugged mountains, snow-capped in the North. South of the Alps, lie the regions of Piedmont, Lombardy, Veneto and Giulia. The broad Lombardy Plain is densely populated and contains important industrial centres such as Milan and Turin. As one moves South, one runs into the Apennines, which form the back-bone of peninsular Italy and extend into Sicily. The main rivers of Italy are the Po which flows through Lombardy into the Adriatic, the Arno which has its source in the Apennines and finds its way to the Mediterranean through Pisa and Florence, and finally, the Tiber, the river of Rome, which has witnessed so much history. There are many other rivers , such as the Volturno and the Metauro on the banks of which, Hasdrubal, brother of Hannibal was defeated and killed by the Romans in 207 BC; many of these minor water systems are dry in the summer. The lakes of Italy such as Como, Garda and Maggiore, were excavated by nature during the Ice Age, while others, like Trasimeno, scene of Hannibal's victory over

Flaminius in 217 BC, are water-filled volcano craters. (Lakes Bolsena and Bracciano are in the same category). Italy has three active volcanoes, the most famous of which is Vesuvius which buried Pompeii, Herculaneum and their civilizations in 79 AD; the other two are Etna in N.E. Sicily, and Stromboli on the island of the same name which lies in the Tyrrhenian Sea, North of Sicily. The Vatican, home of the Pope and seat of roman Catholicism, is a small, independent state of the Italian mainland, as is the Republic of San Marino, S.W. of Rimini, on the Adriatic Coast.

The Italian Islands (Sicily and Sardinia)

The two main islands of Italy are Sicily which lies at its toe, like a football about to be kicked, and Sardinia, to the West. Sardinia is situated South of Corsica, in the Tyrrhenian Sea; it is a mountainous and infertile island, where sheep are grazed and where cheeses and cereal grains are produced. Fishing is also active off its coasts. Coal, lead, zinc and iron are mined.

Sicily is a hot and rugged plateau encircled by a narrow coastal plain where cities such as Messina, Palermo and Catania have developed since early history. Pantelleria, between Tunisia and Sicily, was used briefly by the Military during WWII.

Climate:

The climate of Italy is very warm in the South to progressively if slightly cooler as one moves North. Summers are long and hot and the seemingly forever-lasting blue skies, occasionally make one yearn for a cloud or two; The "temporali" or squalls which hit at times, are the answer to the foregoing prayer. Winter s range from mild in the South to cold and miserable in mid-Italy, to quite cold with plenty of snow in the higher regions and in the North.

Economy:

Characteristically, the agricultural South is poor and the industrial North is rich. Italy is poor in minerals, but makes up for it through heavy industrialization. Genoa, Turin, and Milan are important centres of industry, the latter two, in the motor industry. Large oil-refining plants favor the development of petro-chemical manufacturing. Textiles, leather and synthetic fibres are among other industrial products. Important sources of energy, such as oil and coal, must be brought in.

Glass-work and lace-making are among Italy's famous products. Ship-building is also important. Last but not least, the tourist industry adds to Italy's coffers.

Although two thirds of the land is under cultivation, including every square metre available, even on steep mountainsides, the terrain is not overly propitious to much else than growing grapes and other fruit such as olives, tangerines and cherries. Italy is thus not self-sufficient in food, any more than it is in timber. Italian wines however, are growing in popularity. It is of interest to note that Italy is the *largest* wine producer in the World!

Main cities are: Rome, Milan, Turin, Genoa, Venice, Florence, Pisa, and Bologna.

History:

Italy's history starts in legend, with Romulus and Remus in 753 BC. There followed seven kings until 510 BC when the Republic of Rome was proclaimed. The Age of the Caesars started in reality with the dictature of Julius Caesar, but was historically marked by the ascension of Octavius to the throne, as the first emperor of Rome, in 31BC. There followed many emperors of various dynasties (Julio - Claudians, Flavians & Antonines); some were magnanimous, some were monstruous. There were soldier-emperors and there were effete nullities. The reign of these emperors was followed by the era of African-Syrian emperors (192-235), the era of military anarchy (235-268) and the age of Illyrian emperors (268-284). Under Constantine (322-337) Christianity became the official religion of the Empire and the firm hand of the Emperor stayed the oncoming decline of the Empire. Those who followed him, however, were not able to stem the barbarian invasions. At the death of Theodosius in 395, the gates of Rome were open; it had written its history, it was time for the so-called barbarians to begin laying down the foundations of modern civilization in Europe.

The Middle Ages saw Rome fought over by Barbarians and Byzantines; gradually it began to feel the influence of Central European geo-politics and the impact of the French Revolution. Italy was declared an independent Kingdom in 1861.

WWI saw Italians fighting on the side of the Allies against Austria and Germany; whilst on the victorious side, Italy suffered considerably. The period of discontent that followed facilitated Mussolini's Facist takeover in 1922. In 1936, Mussolini added to Italy's colonial empire, by conquering Ethiopia. Casting its lot with Germany in WWII, Italy was administered stinging defeats in Greece and in North Africa. Mussolini abdicated, was "rescued" by Germany, reinstated only to be captured and executed by Italian partisans in 1945. The dictature was over, but so was monarchy. Italy is now a president-led republic.

Culture:

To properly do justice to Italy in the sphere of culture, one would have to devote book upon book to the subject. Suffice it to say hat no country has done more for the arts and sciences. (Dante, Michelangelo, Da Vinci, Raphael, Galileo, Colombus, Verdi, Stradivarius and Fermi). Italy saw the beginning of one of the greatest faiths in the World, Christianity; it was also the seat of the Renaissance in the 14th Century. Italy is synonymous with classical beauty; it is a country where ancient, historical monuments co-exist in all their splendor with the greatest feats of modern architecture, neither losing in the exchange. In Italy, one can re-live the past, enjoy the present and contemplate the future, yet feel part of all three.

Liechtenstein

Area:	**160 km²**
Pop.:	**21,350**
Cap.:	**Vaduz**

A small, neutral principality which goes back to the 14th Century; through a customs union with Switzerland, it has a Swiss currency.

Luxembourg

Area:	**2,586 km²**
Pop. Grth.:	**—0.2**
Pop.:	**360,000**
Cap.:	**Luxembourg**

Luxembourgeois are a fine mixture of French, Belgian, German and Dutch people. The Grand Duchy, as it is known, was occupied by Hitler's forces from 1940-44.

Iron ore, steel and farming are its main sources of revenue, while tourism adds to its income.

Malta

Area:	**316 km²**
Pop. Grth.:	**1.2**
Pop.:	**350,000**
Cap.:	**Valetta**

This hilly land with its cultivated terraces, is located 92 km South of Sicily. Malta is the largest of a group of five islands There is evidence of stone and bronze age human occupation of Malta, but in later years, the land was visited and settled by a mixture of Greeks and Phoenicians, Spaniards, Italians, North Africans, and even Normans who ruled it as did the Romans, the Spaniards and the French. In 1814, Malta went under British rule, and thus the stage was set for the Island to become Britain's WWII fortress and outpost in the Mediterranean. Bombed and strafed by Axis planes, the courage of its civilians was to earn it the George Cross. in 1964, Malta became independent. Its main value is trategic. (naval and airforces refuelling point, fine harbour) Dearth of space and an arid soil hamper cultivation, despite a most favorable climate. The Island's only mineral resource is limestone.

Most foods must be imported. Tourism is of importance as a revenue. Some industries were launched in the late 1960s, but it is too early to assess their impact on Malta's dwindling population and on its economy.

The Netherlands

Area:	**40,844 km²**
Pop. Grth.:	**0.7**
Pop.:	**14,030,000**
Cap.:	**Amsterdam**

Located in the delta region formed by the estuaries of the Maas, Rhine and Scheldt rivers, the Netherlands is a land of canals, dykes, polders, and tulips. Much of its land has been recovered from the sea; in fact, 2/5th of it lies below sea level, relying on dykes for protection. This kind of soil is most propitious for agriculture, though the eastern part of the Country is covered with glacial deposits and peat bogs. The climate is typical of a coastal area.

Like Belgium, the Netherlands has felt the impact of the wave of nationalism that has been sweeping the World over the last 30 years.

Holland (which in the early history of the Netherlands, was the main province), belonged to Spain until the death of Philip II in 1598, but had to wait 50 years for the ratification of its independence (Treaty of Aix-la-Chapelle). Gradually, the Netherlands became an important maritime and commercial nation, competing with France and England, even laying down in India, the basis of a colonial empire. Futile wars against France followed; eventually the Country fell to its revolutionary neighbor and became a convenient realm for Louis Bonaparte. Freed from French rule in 1813, it acquired Belgium in 1815, and kept it until 1830.

Strictly neutral in W WI, the Netherlands was not able to maintain the same position in 1939-40, when Nazi Germany over-ran and occupied it. The post WWII period was to bring it problems in its Indonesian possessions.

Natural gas, petroleum, the processing of food products, chocolate and cocoa are at the basis of the Country's economy. Dairy products, fishing and cattle-raising are of major importance. The strength of the Dutch industry lies in its highly skilled work-force.

Main cities are: Amsterdam, Rotterdam, Utrecht and The Hague, as well as Arnhem and Nijmagen of WWII fame. The Netherlands is a monarchy.

Orkney Islands

The islands number 67 and form an archipelago North of Scotland. Their 21,500 inhabitants are engaged in cattle-raising and in fishing. The largest island of the group is Mainland, South of which lies Scapa Flow of WWII fame.

Poland

Area:	**312,677 km²**
Pop. Grth.:	**1.0**
Pop.:	**35,440,000**
Cap.:	**Warsaw**

History

Surrounded by Germany, Czechoslovakia and the USSR, Poland has perhaps been the most tortured country of History. The Slavonic tribes which founded the Polish nation in the 5th Century A.D., had to cope with Teutonic knights bent on usurping their lands for colonization purposes. There followed a period of internal disorders resulting in the land being split into numerous dukedoms. Then came war with the Russians, brushes with the Turks, the Swedes and Ukrainian Cossacks. In more recent times (1772), major powers such as Austria, Prussia and Russia split the Country into three parts; a second partition of the "Polish Cake" came in 1793 and yet a third in 1795. This latter partition was to lead to the total elimination of Poland from the map of Europe. Poland had been the pawn between big, greedy nations. In 1807, Napoleon I gave Poland back her independence by creating the Duchy of Warsaw, but in 1815 the previous partition was back in effect. In 1830-31 and in 1863, revolts took place but were crushed. In 1914-18 and again in 1939-45, Poland was invaded and over-run by Germany. The latter conflict was to be particularly hard, Poland suffering heavily in all areas of life. When, in the latter part of the war, the Russians steam-rollered their way West through Poland, pursueing retreating German Troops, they "liberated" Poland, but only for a while. Following this liberation in 1945, Poland was soon to discover that what she was to gain in the West at the expense of Germany, was to be offset by her losses to Russia on the Eastern border. It was also a foregone conclusion that the Country would become Communist-controlled, although fervently Catholic Poland was never to be crushed religiously by an atheistic philosophy. Collectivization also met with failure, 80% of the land remaining privately owned. In summary, Poland is a Communist People's Republic, Communist Party-controlled.

Land - Climate - Economy:

The soil of Northern Poland is mainly glacial deposits and is infertile. The Central Lowlands and the Southern Plateau are better able to support agriculture and cattle-raising. One quarter of the land being covered with coniferous trees, forestry and its resulting industries are an important asset to Poland's economy. Deep sea fishing is also worth mentioning. Whilst prior to WWII, Poland was essentially an agricultural nation, it has developed into an industrialized power. Its manufacturing industry covers steel, locomotives, fertilizers, ships and textiles. Like other Communist Block nations, Poland trades mostly within that sphere.

Main cities are: Warsaw, Lodz, Lublin Poznan, Gdansk, Gdynia and Bialystok.

Scandinavia

That peninsula of Northern Europe which contains Sweden, Norway and Denmark is designated as Scandinavia. Its western country, Norway, is bounded by the Norwegian Sea, a northern extension of the North Sea, whilst its right boundary rests on the Gulf of Bothnia which separates Sweden from Finland. To the South, Denmark is itself another peninsula pointing North towards Norway and Sweden; it separates the Baltic from the North Sea.

Denmark

Area:	**43,069 km²**
Pop. Grth.:	**0.3**
Pop.:	**5,120,000**
Cap.:	**Copenhagen**

Geography - Climate:

Islands, lakes and small farm holdings are characteristic of this land, one of Europe's most prosperous countries. A leader in agriculture and in engineering, Denmark boasts numerous Nobel prize winners. the Country is made up of 500 islands, the largest one being Zealand which houses its capital, Copenhagen. Denmark separates the Baltic Sea from the North Sea and looks upon Sweden to its near North. South is Jutland, on the borders of West Germany. Among its overseas possessions are Greenland and the Faröes in the North Atlantic. Whilst beaches adorn the Jutland West coast, Denmark is known for its undulating plains and the fjords of its other coasts which form natural harbors. Denmark has a mild, damp, insular climate.

Economy:

Poor in natural resources, Denmark is essentially an agricultural nation, also engaged in fishing and forestry. Cattle-raising and dairy-farming are of importance to this population of Scandinavians, 30,000 of whom are of German ancestry. Danes are permissive in their life-style.

Denmark's manufacturing industry is mainly based on its imports, turned into skillfully manufactured goods which find their way to the export markets.

History:

From the 8th to the 10th centuries, Danish Vikings raided Western Europe. From a loose grouping of bands, they united in the 10th Century and eventually accepted Christianity. Powerful Danish Kings ruled England from 1013-1042 - Denmark united with Sweden and Norway in 1397 and held the power among the three countries. After Sweden seceded from the union in 1528, wars with Denmark reduced the latter's power and caused it to lose territory to its former ally. Napoleonic wars further eroded Denmark's importance by taking away Norway. Despite an active underground, Denmark had to cope with German occupation during WWII.

Like Sweden, Denmark recovered with the help of the US. Gradually, manufacturing gained in importance over agriculture. A member of the EEC, Denmark is a constitutional monarchy with a prime minister and a house of parliament. Its main cities are Copenhagen, Aarhus, Aalborg, Odinse, Randers and Horsens.

Norway

Area: 324,219 km²
Pop. Grth.: 0.4
Pop.: 4,070,000
Cap.: Oslo

Geography - Climate:
A land of rugged mountains, beautiful valleys and deep fjords, Norway is a mountainous plateau of bare, glacial-age rock. 150,000 islands lie off its coast. Despite its northern latitude, Norway's climate is mild, being warmed by the North Atlantic Drift which keeps the ports free of ice. Inland and in the mountains, it is colder and snow is plentiful.

Economy:

As nearly three-quarters of the land is rocky mountains and moors, there is little left for agriculture. Norwegians have therefore concentrated on developing industries such as processed foods, pulp and paper, chemicals, forestry, fishing and, more recently, tourism. Norwegians have also harnessed many of their swift streams to produce hydro-electric power and have built a large commercial fleet for import and export purposes. North Sea oil is also a promising resource.

History: Prior to the 9th Century, Norway's history is mixed with that of other Scandinavian races. The Vikings and their sagas started Norway on a history all its own. During the reign of King Olaf V, (1330-1387), it was united with Denmark; this union lasted until 1814, Sweden having been added from 1397 to 1528. During the 300 year period prior to 1814, Norway gradually became a Danish province but was eventually ceded to Sweden by the Treaty of Kiel. One century of legal and political hassle followed and ended in the Treaty of Karlstad (1905), which recognized the separation of the two countries. Haakon VII became Norway's monarch. Norway's neutrality during WWI was not to prevent it from suffering heavy losses to its merchant navy. 1940

saw the country occupied by Germans until freed by the Allies in 1945. The proud character of its people and American aid helped Norway recover from the suffering and losses of WWII. Today, Norway has a constitutional monarchy, a prime minister and a house of parliament. Its main cities are: Oslo, Bergen and Trondheim.

The people of Norway are a seafaring race, proud and patriotic. They are also outdoor types known the World over for their skiing ability. Among the many Norwegians who have achieved world fame, are Lief Eriksson, who is said to have landed on NA soil around 1000 AD; Amundsen, the first to reach the South Pole; Thor Heyerdahl (Kon-Tiki); Ibsen the playwright, and Grieg the composer.

Sweden

Area: 449,964 km²
Pop. Grth.: 0.3
Pop.: 8,290,000
Cap.: Stockholm

Geography - Climate:
As already stated, Sweden shares the Scandinavian Peninsula with Norway to its West. Sweden has a terrain somewhat similar to that of Norway, except that, not benefitting from the North Atlantic Drift, most of its northern coast freezes from November through April. When winters are severe, the whole Country is locked in snow and ice. unlike Norway, Sweden does not have many fjords, but boasts thousands of small coastal islands which shelter it, to some extent, from the weather. Moving from South to North, one runs into most of the Country's 96,000 lakes dotting a gently rolling, green country-side, and then, into huge forests, to finally end up in glacial ridges in the North. Summers are pleasant and an asset to tourism.

Economy:

Sweden's natural resources are timber, hydro power and iron ore, (world's largest producer). Engineering accounts for one third of the Country's industrial production, (ships, automobiles and farm machinery). Pulp, plywood and the well-known Scandinavian furniture add to Sweden's resources, as does her tourist industry.

History:

The Roman historian Tacitus was the first to talk about Sweden or at least, about the Svears, a tribe of Barbarians who inhabited the land in the First Century AD. A study of the histories of Norway and Denmark reveal that Sweden's origin is closely linked with theirs: Swedish Vikings struck East and South towards Russia and Byzantium; Sweden joined Norway and Denmark in a Scandinavian Union in 1397, but dissatisfied, broke away in 1520; acquisition of Norway in 1814, and the separation of the two countries in 1905. Following the wars of the French Revolution, Gustav IV was deposed, and Sweden "adopted Bernadotte, a marshal of Napoleon Bonaparte, as its new king (1810). Following these years of intense geo-politics, Sweden pursued a course of domestic evolution; it remained neutral in both World Wars of the 20th Century which helped it progress as one of the most socialized and industrialized countries in the World.

Main cities are: Stockholm, Malmo, Goteborg, and Uppsala.

Scilly Isles

British islands of slightly more than two thousand inhabitants at the S.W. tip of England.

Shetland Islands

The main city of this archipelago of 19,000 people, North of Scotland, is Lerwick. Other towns are Bressay, Whalsay, Yell, Unst and Fetlar.

Switzerland

Area: 41,288 km²
Pop. Grth.: —0.4
Pop.: 6,350,000
Cap.: Bern

Switzerland usually evokes thoughts of breath-taking Alpine scenery, skiing championships, tourism, international banking and philanthropy (the Red Cross), not to omit a most successfully achieved atmosphere of internationalism. All are characteristic of this country of geologically young mountain ranges and of deep, picturesque and unforgettable valleys.

History - Culture and Constitution:

The desire of nations and individuals for independence is not an aspiration peculiar to the 20th Century. Indeed, late in the 11th Century, people of this region got together, formed a loose league of cantons and declared themselves independent from Austria; this independence was finally recognized in 1648. One hundred and fifty years later, France conquered the area, but had to relinquish its hold in 1805.

Switzerland is a federal republic made up of 22 cantons, each with its own government for regional matters. Communications, defence and currency are the responsibility of the Federal Government.

The Swiss are largely of Latin and German origin; German, French and Italian are the three official languages.

Geography and Climate:

The well known alps which cover more than half of the Country, are not Switzerland's only mountains; the Jura in the West and the Central Plateau also contribute their peaks, cols and valleys to the natural beauty of the land. Switzerland's climate varies with the elevation; warm summers are commonplace in the central plateau, but the mountainous regions are of course, colder and have more precipitation.

Economy:

Highly industrialized but lacking in minerals, Switzerland had to concentrate on industries such as watch-making and the production of precision instruments where skilled labor offsets the lack of raw materials. Food processing and textiles are important to the Country's economy as is its hydro-electric capability. The very nature of the land restricts farming while it favors livestock-raising and dairy-farming.

Switzerland's geographical position, its long-standing neutrality and basically sound government, have made it a respected if small nation.

The Madeira Islands

Area: 789 km²
Pop.: 270,000
Cap.: Funchal

A Portugese possession West of Morocco and known in particular for its wines. Also grows tropical fruit and sugar-cane.

THE MIDDLE EAST AND NEAR EAST

Afghanistan

Area: 647,497 km²
Pop. Grth.: 2.5
Pop.: 15,490,000
Cap.: Kabul

Between India and Iran, Afghanistan lies across an almost traditional invasion route - The British in India, looked upon it as a buffer state against attacks from the East. The Country is traversed by the towering Hindu Kush range and watered by such rivers as the Kabul, the Helmand and the Amu Darya. Despite these water-systems and the Hari Rud River in the NW., Afghanistan is a generally infertile land with a rugged climate.

The raising of cattle, sheep and goats for meat and wool is one of the nomad, mostly illiterate population's main source of sustenance and revenue. (Karakul lambskins are exported in large quantities.) Some sugar-beet and fruit are cultivated. Textiles are another source of revenue. Some natural gas is exported to Russia whilst Lapis Lazuli, mined here in great abundance, finds its way to world markets. Most of the Country's mineral wealth remains untapped.

Afghanistan broke away from Persian rule in the 8th Century, lost its independence in 1880, but regained it in 1919.

The main cities are: Kabul, Kandahar and Herab.

Bahrain

Area: 622 km²
Pop. Grth.: 10.5
Pop.: 350,000
Cap.: Manama

Located west of Qatar in the Persian Gulf, Bahrain is known for its petrol and its pearls.

Cyprus

Area: 9,251 km²
Pop. Grth.: —0.1
Pop.: 620,000
Cap.: Nicosia

Cyprus is located in the Eastern Mediterranean, in the angle formed by Turkey to the North and Syria to the East.

The 20th Century in particular, has been one of political turmoil for this former colony of Britain, granted its independence on the basis of equal rights for both the Greek majority and the Turkish minority. The Greeks who wanted enosis or union with Greece, were seemingly preparing to invade the Island when the Turks moved faster. The UN stepped in with a peace-keeping force which managed to keep the opposing sides apart. The Turks have been relegated to the NE part of the Island. Diplomatic efforts are being deployed to bring about a formal partition of Cyprus, acceptable to both Greeks and Turks.

Cyprus is an island of rugged beauty with awe-inspiring peaks and beautiful, sandy beaches, popular with

tourists. The Country has two mountain ranges, the Kyrenia in the North, and the Loftier Troodos in the SW. Between these, lies the Mesaoria Plain. Winters are mild, summers hot and dry. Citrus fruits and grapes are cultivated. Some mining is done in the Troodos.

Iran

Area: 1,648,000 km²
Pop. Grth.: 2.2
Pop.: 35,510,000
Cap.: Teheran

Formerly known as Persia, Iran's history evokes memories of huge hordes launched against ancient Greece, its conquest by Alexander the Great in 331 BC and Arab rule during the 7th Century. Lying across the invasion route from the East, it endured attacks by Turks, Mongols & Russians. In 1925, the Pahlavi family took over; the Shah was overthrown, reinstated and again overthrown in 1979, by the religious leader Ayatullah Khomeinei, determined to bring Iran back to Islamic ways.

Geographically, Iran consists of high plateaus encircled by mountains. Strips of lowlands occur along the Gulf and the Caspian Sea - it also has two large desert areas in the East. Farming, cotton, tobacco growing and livestock-raising are the main activities.

Iraq

Area: 434,924 km²	
Pop. Grth.: 3.5	
Pop.: 12,770,000	
Cap.: Bahgdad	

The Tigris-Euphrates region of modern Iraq (Mesopotamia) was the cradle of civilization. Alexander the Great reached the Euphrates in his conquest of the East in 331 BC, and like Iran, the Country fell to the Arabs in 652 AD, to later come under Turkish domination. The British took over in 1918, but Iraq became an independent monarchy in 1932. During the 1950's, a military coup made it a republic. There followed two more army revolts and considerable trouble with the fiercely independent Kurds of the northern regions; this lasted from the 1960's through 1974. Iraq has engaged in Arab-Israeli wars, with no positive benefits accruing from it.

Iraq is partly a dry, sandy plain, swampy in the south, with deserts in the West and mountain ranges in the NE. The absence of adequate rainfall is compensated by the fine irrigation provided by the Tigris and Euphrates. Farming and cattle-raising are the population's main endeavour. Iraq produces cotton, tobacco, and a great deal of petroleum.

Israel

Area: 20,770 km²	
Pop. Grth.: 2.2	
Pop.: 3,780,000	
Cap.: Jerusalem	

Israel is a young nation, the realization of a dream shared by Jews the World over, from time immemorial. The new nation was born in 1948 as a result of a UN decision to partition Palestine into a Jewish and an Arab state. It goes without saying that the decision was not acceptable to the Arab World; there followed four wars - 1948, 1956, 1967 and 1973 - out of which Israel's fierce determination to protect its home-land, its spirit of unity and its military leadership, were to prevail. Much land was acquired - some was given back under diplomatic and political pressure; largely, however, Israel emerged better off than it had been prior to the engagements. A settlement reached in 1975, restored some of the Sinai desert to Egypt and part of the strategically important Golan Heights to Syria. The suez Canal was open to Israeli ships.

The Palestinian Liberation Organization still agitates; violence and terrorism are evidenced through murderous ambushes, dramatic highjackings etc. the signing of a peace treaty with Egypt on March 26, 1979 in Washington, after months of negociations ended a state of war that had persisted over the last 30 years.

Israel is divided into four geographic regions; The Coast, Galilee, the Emek and the Negev. The coastal area and Galilee are propitious to farming, whilst the Emek yields a number of crops. The Negev contains natural gas & petroleum deposits. The Israelis have done a superb job of reclaiming land from the sea and the desert; they live and work in Kibbutzims, where the collectivity owns all property. Israel's ancient, biblical history draws a large tourist flow from all over the World and adds to the Country's revenue. The Dead Sea area has mineral resources. Main cities are: Jerusalem and Tel Aviv.

Jordan

Area: 97,740 km²	
Pop. Grth.: 3.2	
Prop.: 3,090,000	
Cap.: Amman	

Most of Jordan is sandy, inhospitable desert and rock-strewn plains. The only fertile area lies West of the River Jordan. One can well appreciate the telling blow administered to the Country, when the Israelis occupied the West bank in the 1967 war, depriving Jordanians of most of their native food sources. Like Israel, Jordan derives some benefits from Dead Sea area mining.

Created by Britain after WWI, TrasnsJordan became an independent monarchy in 1946. In the 1948 war, it fought Israel, conquered Palestine and part of Jerusalem. Half a million Palestinian refugees then found their way within its borders; thus it is that Jordan's population is largely made up of Palestinians. For a while, in the 1960's, there was political trouble with Palestinian Nationalists, but the problem was solved when the guerillas pulled out of the Country. Nonetheless, Jordan is still in the explosive orbit of the Arab-Israeli feud.

Kuwait

Area: 17,818 km²	
Pop. Grth.: 6.1	
Pop.: 1,270,000	
Cap.: Kuwait	

Kuwait is an Arabian emirate at the very top of the Persian Gulf. Its main resources are petroleum and pearls.

Lebanon

Area: 10,400 km²	
Pop. Grth.: 2.5	
Pop.: 3,090,000	
Cap.: Beirut	

Lebanon is a narrow coastal plain, squeezed between Syria and the Eastern Mediterranean. It was ruled by France from 1918 to 1943, when it became a republic. Politico-religious fighting frequently occurs between Christians and Moslems, each accusing the other of dominating the Country; such conflicts took place in 1975, 76 and 78. Under such conditions, peace in Lebanon is always precarious.

Oman

Area: 212,457 km²	
Pop. Grth.: 3.1	
Pop.: 860,000	
Cap.: Muscat	

Oman is located at the very tip of the Arabian Peninsula, on the Arabian Sea; it has a narrow coastline which is fertile in the NW, and boasts lush date gardens. Elsewhere, it is arid. Behind the coastline, mountains and a plateau, look down onto the Saudi-Arabian desert. Camels are bred in large numbers and oil production is growing as new wells are bored. The strategic position of Oman at the entrance to the Gulf bearing its name, was bound to cause trouble. The Persians were particularly bothersome during the 11th Century and the Portugese controlled its ports for a hundred years prior to 1700. Although the present Government controls the interior of the Country, it has to cope constantly with dissident factions.

Qatar

Area: 11,000 km²	
Pop. Grth.: 5.7	
Pop.: 210,000	
Cap.: Doha	

Qatar is a small peninsula jutting out into the Persian Gulf, off the Eastern coast of Saudi-Arabia. Oil, dates & camels are the sources of revenue.

Saudi-Arabia

Area: 2,149,690 km²	
Pop. Grth.: 3.1	
Pop: 8,110,000	
Cap.: Ar Riyad	

Saudi-Arabia, a land of coastal mountains and rock-strewn, sandy hinterland, occupies most of the Arab Peninsula, between the Persian Gulf and the Red Sea.

Economically, Saudi-Arabia has a modicum of agriculture and its natives, including the nomadic Bedouins, raise cattle, camels, horses & sheep. Petroleum is its greatest source of wealth.

Saudi-Arabia's claim to fame lies in the fact that two of its cities, Mecca, (birthplace of the prophet Muhammad) and Medina, are the holiest places of the Muslim World. Mecca is the site of Moslem pilgrimages every year.

Prior to WWI, Saudi-Arabia was under Turkish rule; however, Saudi leaders fought for the Allies and secured the independence of their Country. SaudiArabia has also joined other Arab nations against Israel.

South Yemen

Area: 332,468 km²	
Pop. Grth.: 2.7	
Pop.: 1,850,000	
Cap.: Aden	

South Yemen is located on the shores of the Gulf of Aden, between Oman to the NE and Yemen to the SW. Its northern boundary, like Oman's, looks upon the Saudi-Arabian desert of Ar Rab'al Khali. It is a country of burning desert sands and of rocky, arid mountains. Agriculture fishing are the main activities. The People's Democratic Republic of Yemen has been free since 1967 when nationalists abolished the sultanates which had been ruling the Country.

Syria

Area: 185,180 km²	
Pop. Grth.: 3.3	
Pop.: 8,090,000	
Cap.: Damascus	

A land of rolling plains and deserts between Iraq to the East, and Turkey to the North, Syria looks down upon the Eastern Mediterranean. Behind the high mountains of the coastal range, lies and arid land with little rainfall. Irrigation is essential to farming. Fruit are the main product.

Like many other Middle-East nations, Syria was once a Roman province; then it came under Arab domination, but was wrested from Muslim rule by European Crusaders who came and stayed for 200 years.

Prior to WWI, Turkey claimed authority over Syria, but France took over and ruled until 1946, when Syria achieved its independence. Syria was a member of the UAR from 1958 to 1961. It is a socialist republic.

United Arab Emirates

Area: 83,600 km²	
Pop. Grth.: 8.4	
Pop.: 750,000	
Cap.: Abu Dhabi	

An inverted "L" shaped country on the Persian Gulf, between Oman to the East, and Kuwait to the West. Petroleum, fish, dates & camel-raising are the basis of the economy.

Yemen

Area: 195,000 km²	
Pop. Grth.: 2.3	
Pop.: 5,790,000	
Cap.: Sana	

Due South from Kuwait, right across Saudi-Arabia and in its SW corner, lies the Yemen Arab Republic with its fertile coastal belt and its mountainous hinterland. Yemenis are a nation of farmers and fishermen. Agriculturally, Yemen is most productive - millet, mokha coffee, cotton and qat are products of the land.

Between the days when Yemen was ruled by Imans, and today, the Country has witnessed many struggles for power. Currently, Yemen is ruled by a nine-member Command Council.

NORTH AMERICA

North America is comprised of Canada, the USA, Alaska and Greenland; it has an area of 23.4M km², and a population of close to 300 million. Its northern limits are the Arctic whilst in the South, it touches Central America; its eastern and western shores are washed by the Atlantic and Pacific oceans.

It would seem redundant to describe the geography of North America, deal with its terrain, vegetation and climate when this is covered in detail under the headings "Canada" and "USA".

The original inhabitants of North America are believed to have come from Asia in the wake of migrating herds of animals, crossing the land-bridge which linked Asia and Alaska 30,000 years ago; they were the ancestors of North American Indians.

Discovered by Colombus in 1492, the new Continent was to be visited, claimed, fought-over and colonized by France, England and Spain over the centuries. Eventually, Anglo-Saxons emerged as the main "landlords", both in Canada and in the USA.

Canada

Area: 9,976,139 km²	
Pop. Grth.: 1.1	
Pop.: 24,088,900	
Cap.: Ottawa	

Generalities:

This vast country, stretches from the Atlantic Ocean in the East, to the Pacific in the West; its southern edge is delineated by a common border with the U.S.A., whilst to the North, it loses itself in the Arctic ice mass.

Its relatively small population is generally concentrated along the St. Lawrence River which takes its source in the Great Lakes complex, roughly mid-way across the land and runs into the gulf bearing its name. This population concentration within 300 or so kilometres of the U.S. border, is of economic significances. West of Ontario, the largest of ten provinces, population concentrations are patchy until one reaches Vancouver, the Country's West Coast metropolis of 1,100,000 inhabitants. As for the huge stretches of land in the North and North West, they are completely uninhabited except for small communities in places such as Yellowknife in the Northwest Territories and Whitehorse in the Yukon.

History - Ethnic Components:

Although John Cabot was the first to land on Canadian soil in 1497, Jacques Cartier gets the real credit for discovering Canada in 1534. Following this later discovery, Canada was to be the object of a half-hearted colonial development effort by France, involved in many wars in Europe in the 16th and 17th centuries. When the British started eyeing the New World at the turn of French Canada's first century of existence, it was to be only a matter of time before the "flirtation" would end in the British conquest of 1759. Thus it is that Canada has two main founding races. Inevitably, and with the maintaining of a "hands-off" policy for France, the English element was to grow faster than the francophone component, until it now constitutes approximately 50% of the Country's population, those of French origin claiming about 30%, whilst the remainder is shared by people of European extraction such as Germans, Poles, Ukrainians, White Russians, a smattering of American Indians (300,000) and Eskimos (17,000). English and French are the main languages.

Geographical Features - Climate:

With the exception of Greenland and Alaska, Canada covers the whole northern half of North America. The land is divided into seven major regions.

In the East, the Appalachian range covers Newfoundland, the 10th province, extends S.W. across Prince Edward Island, "hops" onto the mainland to New Brunswick and Nova Scotia, to finally "backtrack" over the Gaspé Peninsula, Quebec's easternmost, mainland extension. - Lying not quite mid-way between the Atlantic and Pacific Oceans, the Great Lakes complex is comprised of Lake Superior, Lake Michigan, Lake Huron, Lake Erie and Lake Ontario. All of Lake Michigan lies in the U.S.A., while roughly half of all the others span the border of Canada and the U.S.A..-The St. Lawrence Valley low-lands, the most heavily industrialized and populated region. -The Canadian Shield with untold mineral wealth, lying beneath pre-historic rock. -The Hudson Bay Low-land with its swamps. -The Region of the Plains which encompasses Manitoba, Saskatchewan and parts of Alberta and British Columbia.- The Western Mountain Ranges, comprised of the Rockies, the Coastal Range, and the islands within the Arctic Circle.

Canada's hydrography accounts for one third of all the fresh water supply in the World. The main rivers are the MacKenzie, the Frazer and the St. Lawrence. The Colombia and the Yukon both flow on through the U.S.A..

Although Canada's climate is deemed to be continental, its winters, more so in the N and NW, produce arctic temperatures, whilst summers yield near tropical heat, especially in the Niagara Peninsula and in the Prairie Provinces.

Provinces, Territories and Main Cities:

Canada has ten provinces: British Columbia, Alberta, Saskatchewan, Manitoba, Ontario, Quebec, New Brunswick, Nova Scotia, Prince Edward Island, Newfoundland, and two territories: The Yukon and the North-West Territories. Main cities, from West to East are: Vancouver, Victoria, Edmonton, Calgary, Regina, Winnipeg, Toronto, London, Hamilton, Ottawa, Montreal, Quebec, Fredericton, Halifax, Charlottetown and St. John's, and in the N and NW, Yellowknife and Whitehorse.

Constitution:

Canada is an independent, self-governing, constitutional monarchy and a member of the British Commonwealth of Nations. As nominal head of state, the reigning British monarch is represented in Canada by a governor-general in Ottawa, whilst lieutenant-governors perform relatively the same functions at the provincial level. Each province is self-governing and the territories have a measure of self-government. The Federal Parliament sits in the capital, Ottawa, and has a Senate made up of 102 members as well as a 264 member House of Commons. A Prime Minister heads the Country. The main political parties are the Liberals, the Conservatives, and the National Democratic Party. Social Credit has a measure of popularity.

Economy:

Most of Canada's farm products, with grain crops leading, come from the Great Lakes area, the St. Lawrence Valley and the Prairie Provinces. Livestock-raising as well as ranching and dairy-farming constitute an important source of income. The Niagara Peninsula and British Colombia yield important fruit crops and Prince Edward Island, in the East, is well known for its high grade potatoes.

The Grand Banks off Newfoundland are world-renowned commercial fishing grounds as are the offshore waters of B.C. where salmon, and shellfish abound. From a sports standpoint, the lakes and rivers of Canada teem with trout, bass, sturgeon, pike, muskelonge and other species of fresh water fish.

Canada's forests supply the wood pulp needed to cater to half the word's newsprint requirements, and its numerous rivers and water-falls, all the hydro power needed domestically.

Asbestos and nickel are two of Canada's most important minerals. Iron ore is mined in Ontario, Quebec and British Colombia; uranium and potash in Ontario and Saskatchewan; copper, zinc, lead and platinum in the Canadian Shield. Gold, which was the cause of the Klondike Gold Rush of the 1860's is still mined in the North, as is silver. Finally, the all important commodity, oil, is pumped from the rich oilfields of Alberta.

It is of interest to note that over 60% of Canada's products are manufactured, with processed foods leading, followed by transport equipment and machinery. Canada's main trading partners are the U.S.A. (60%) Japan, England and West Germany.

The United States of America

Generalities - Ethnic Component - Culture:

Area:	**9,363,123 km²**
Pop. Grth.:	**0.7**
Pop.:	**220,099,000**
Cap.:	**Washington**

Perhaps the most varied land in the World, the USA is one of the richest and most powerful. With its large population of many racial origins, the USA ranks fourth in the World in area, after the USSR, Canada and China, and again, fourth in population, this time, after China, India and the USSR.

Most European countries are represented in America's ethnic profile, whilst people of all other regions of the World will be found in this melting pot of races. It is of interest to note that ethnic groups emigrating to the US, expressed a marked preference for some regions over others. Thus it is that we find a preponderance of Spanish blood in California and Florida, while in New York State, one finds unmistakable evidence of Dutch and Irish predilection for that state. German emigrants seem to have favored Pennsylvania, whilst Frenchmen preferred Louisiana. Italian immigrants, for their part, seem to have shown a great deal of flexibility, for there is an important proportion of them in just about every corner of the US.

A fact worthy of note, descendants of African blacks "imported" to the US in the early days of the Colony to work on southern plantations, have now reached over the 22 million mark!

For all this diversity of races and cultures, and although each ethnic group seems to have retained its life-style, there has been an amazing integration into American life.

Americans are education-conscious. Education is compulsory to the age of 16 - public schools offer free education and there are numerous private establishments run by the Church and by Private Enterprise. The educational spectrum contains all levels of learning from elementary schools to universities. The USA has possibly the highest degree of literacy in the World, with over 98% of its adult population able to write and read.

The trait which is most likely to identify an American is his belief in the equality of men of all walks of life, irrespective of birth. While he is fascinated by many aspects of the monarchy or other autocratic forms of government into which his ancestors were born and have lived, he decidedly prefers his democratic way of life where his passion for independence ranks supreme.

Spectator as well as individual sports play a large role in American life.

History - Culture:

Although Spaniards were the first to set foot on the shores of what is now Florida, the first Anglo-Saxon settlers came to found Jamestown in Virginia, in 1607. Years later, (1620), the Mayflower delivered its pilgrims on Plymouth Rock, in Massachussetts and American colonization was on its way.

In the years which followed, the number of colonies grew with the population. The existence of these British colonies was constantly threatened by England's traditional enemy; in fact, the French managed to push steadily down from Canada and reached as far as Louisiana where they founded New Orleans. It was only the end of the Seven Year War in 1763, which eliminated the French threat for good.

However, the American, even in those early days, was developing a fierce passion for independence and the basic rights of the individual. In the early 1770's, the colonists rebelled against British rule, refusing outright, taxation without representation. The War of American Independence was launched, near Boston, in 1775. By the following year, thirteen colonies had declared themselves free and independent states. Their skillful general, George Washington, eventually defeated the British and the Treaty of Paris (1783), recognized their independence.

After a constitution was drawn up in 1787, the first American President, George Washington, (1789) proceeded to put his house in order; Louisiana was purchased from France, and Florida from Spain, (1819). There was a brief renewal of hostilities with England from 1812-1814.

There followed in 1846, a two year war with Mexico which yielded Texas, New Mexico, Utah and California.

In 1861, one of the bloodiest conflicts on record broke out between the North, under Lincoln, and the South under Jefferson Davis. The right to own slaves was not so much the determining factor in launching this war, but rather the right of the Southern states to secede from the Union in order to retain this right. After bitter fighting and unspeakable slaughter, the war ended in 1865 when Lee surrendered to Grant at Appomatox.

In the ensueing years, important date-lines were 1917, when the US joined the Allies of world War I to swing the tide of battle in their favor; the Great Depression of the early 30's, and then the period of economic resurgence under Franklin D. Roosevelt who was elected President in 1933.

The Japanese attack on Pearl Harbor in December 1941, resulted not only in a tremendous war effort by the US in the Pacific, but determined its entry into World War II againt Axis powers in Europe. The collapse of the Axis occured in May 1945, while the atomic bombs dropped on Hiroshima and Nagasaki sealed the issue, a few months later.

A police action in Korea from 1951 to 1954, as part of a UN force, and a bloody involvement in Vietnam, from 1964 to 1973, contributed to disillusion Americans about the sense and value of armed conflicts.

The fighting over, America turned its eyes towards outer space. Thus it was that the US engaged in a technological race with the USSR, with space achievement being the objective. Although the USSR's Yuri Gagarin and Gherman Titov were the first humans to orbit the earth in August 1961, America's Collins, Aldrin and Armstrong were the first to set foot on the moon in July 1969.

Paradoxically, the glory of America's technological achievements was to be tarnished by the unsavoury overtones of the Watergate affair of 1974, which eventually brought about the resignation of then President, Richard Nixon and the political demise of several of his associates.

Influence in World Affairs:

The wealth of the USA, its philanthropic disposition towards its fellow men and the realization that it behove it to play a major role in world affairs, for economic as well as for strategic reasons, prompted America to literally pour money and material aid into the economy of strategically located, developing countries.

Relations with Russia and China, once most delicate, gradually became less tense; this thaw in the ice of the Cold War was accentuated by a split between Russia and China over ideological differences in the two countries' approach to World Communism. The willingness of the USSR to co-operate with America in space ventures was always threatened by differences of opinion of the two countries at UN level or by the taking of diametrically opposed positions in the geo-political field. Such was the case for the civil war in Angola in 1976, and for the long-standing, explosive situation between Israel and the Arab States in the Middle East.

The wealth and philosophy of the USA are not the only reasons why this Country is called upon to play a major role in international affairs. Indeed, its involvement in the world scene is a direct corollary of its size as well as its acknowledged presence throughout the world Over & above its land mass which takes in Alaska to the North, the US is present in Panama where it leases the canal and its zone. Puerto-Rico which lies off its SE coast, is a commonwealth of the USA. Hawaii, the last state to join the Union (1959), is an important outpost in the Pacific. Finally, the influence of the USA in Japan and in the Far East cannot be overemphasized.

Geographical Features:

Reduced to its most simple expression, the USA is made up of three main regions: the western mountain ranges, the eastern highlands and, in between, the plains.

Within the western range, lies the Pacific Mountain System, the westernmost part of which is designated as the Coastal Region (low mountains and deep valleys). East of this range, runs the Cascade Range which originates in Canada, runs through Northern California and ends as the Sierra Nevada. Skipping the Intermountain Plateaus as relatively uninteresting with their dry uplands and deep gorges, one comes upon the magnificent Rocky Mountains, now bare rock and snow-capped, now heavily forested, rich in game and minerals and always wealthy in incomparable alpine scenery.

Moving East, one finds the Coastal Plain which seems to act as a "cordon sanitaire" between the Atlantic and the Appalachian Highlands whose shadows stretch from the Gulf of St. Lawrence to Alabama. The latter are made up of low, worn-out ridges.

Westward from the Appalachians, one gets literally lost in the enormous regions which account for half the USA; the Central Lowlands and the Great Plains regions with their fertile land.

With such water systems as the Mississipi-Missouri,

the Ohio, the Colombia and the Colorado, the USA does not go wanting for fresh water ! Some-one living in or around Death Valley or in certain arid regions of the West and South West, such as the Mojave Desert or the Arizona Badlands, may not readily agree. Worthy of mention is the master-piece of natural art produced by the great Colorado, the Grand Canyon. This most photographed crevice which the river has carved in the earth's crust (1.6 km deep and of unspeakably savage beauty) needs no expert to interpret its message of grandeur.

Many of the lesser rivers such as the Hudson, the Potomac and the Delaware have been harnessed for hydro power, navigation and irrigation. As for the Great Lakes, they have already been dealt with under the heading "Canada". May we just mention the border falls, Niagara, one American, one Canadian, both sharing equal popularity with hydro-electric authorities, amateur photographers and honey-mooners.

Climate and Vegetation:

The immensity of the USA calls for considerable variations in climatic conditions. Whilst Hawaiians will generally be comfortable wearing sarongs the year round, Alaskans will need parkas to survive during the winter. Whereas coastal regions will have ample rainfall, some of the SW plateaus may not see rain for years on end; whilst Montana faces bone-chilling temperatures with wind-chill factors reminiscent of the arctic tundra, the swamp-lands of Florida yield extremely humid heat. On the whole, however, and going by the average center, the climate of the USA can be deemed to be continental.

The States

The USA breaks down into 10 economic regions, each readily identifiable by its life-style or its contribution to the over-all American economy. The New-England states where industrial development is most evident; — New-York State which is synonymous with population density, banking, trading and transportation; — the Mid-Atlantic states where the stock in trade is agriculture and mining; — the Mid-West states which have an important mining capability and add farming to their spectrum of activity (Chicago, the World's cross-roads, and Detroit, the Motor City, stand out as the two main cities in this group); — the Prairie states, the World's granary; — the Mountain states where pasturelands vie for recognition with the mining industry; — the Pacific states, known the World over for their film industry (Hollywood), their oil producing and refining and their aircraft production (in 1849, California was the site of a gold rush, while latterly, Alaska yielded some of the precious metal); — the Southern states, which apart from evoking painful memories of bloodshed, call to mind tobacco and cotton growing; — the South-West states, land of cowboys and oil wells; — the Border states which are more or less on the dividing line of the Civil War of the last century and which divide the main economic areas of the Country - these are known for their "blue grass", horse-breeding and American country music.

Economy:

With well over three million farms in operation, the US has enough left over to supply a good portion of the World. The Mid-West produces wheat, 50% of the World's corn and most of its cotton. Its dairy-farming and fruit industry are also of major importance (the former on the coasts and the latter in the SW), while cattle, sheep and horses graze huge tracts of land, also in the SW.

The lumber industry, largely because of over-felling, has gradually moved West until it reached the coastal forests. Surrounded by oceans, except in the North where untold numbers of lakes and rivers substitute for the larger bodies of water, the USA does not suffer from a lack of sea-food.

One fifth of the World's crude petroleum comes from the USA where new oil fields are constantly being discovered and put into operation to meet the Nation's enormous demand for this source of energy. After the USSR, the US is the biggest coal producer, with most of this commodity coming from the regions East of the Mississipi. Iron, steel, textiles and paper constitute major industries. Bauxite, nickel, tin and manganese are mined and much of it is exported. As for industry, trade and commerce, the USA is the World's leading manufacturing nation; food processing, chemicals, machinery of all types, packaging, all appear on its manufacturing "menu".

SOUTH AMERICA

Generalities - Geography:

The fourth largest continent of the World, South America contains 13 countries, and because of its enormous size, has perhaps the greatest variety of races, cultures and climates. In some of its remotest areas, Stone Age life is still evident whilst in the Argentine and in Brazil, the main cities offer a life-style which is among the most sophisticated.

South America has some out-standing geographical features such as the size of Brazil, its largest country, compared to that of French Guiana to the North. Next comes the abundance of hydrographic resources which contribute to the fertility of the land; the Orinoco, the Amazon River, with its many tributaries in the northern part of the Continent, and the Rio de la Plata system, its most important inland waterway. The Andes Cordillera runs along the West Coast; its forbidding peaks have long hindered the development of the Country. A number of parallel ranges stretch from Colombia to the N.W., right down to Tierra del Fuego on the southern tip of the Continent - many of the peaks are active volcanoes. In the East, roll the gentle and well-watered hills of the Brazilian Highlands; their fertile soil favors the growth of coffee and cotton.

From the North-East, the Guiana Highlands run through Venezuela, skirt the Guianas and end in Brazil; many of the tropical forests of this area remain largely unexplored.

Between these two mountain systems, lie the interior lowlands which stretch from the Amazon Basin in the North, to the Rio de la Plata in the South. There, one finds tropical rain forests where huge anacondas occasionally engage in mortal combat with fearless jaguars, and schools of ferocious piranhas strip the flesh off the carcass of a water-buffalo in a matter of minutes! The llanos, or grassy plains of the lower Orinoco, with the exception of those of the inhospitable Gran Chaco, are quite fertile, more so further South where they run into the Argentine Pampas.

Despite its size, the South American Continent does not display the same drastic differences in climatic conditions as found in North America. The highest temperatures are experienced in the Argentine, while the coldest are registered at the southern tip (Tierra del Fuego) which is, of course, subjected to cold air currents from Antarctica. The western coastal area is cooled by the Humboldt Current, while the East is warmed by the Brazil Current.

Component Nations:

Argentina

Area: 2,776,889 km²
Pop. Grth.: 1.3
Pop.: 26,730,000
Cap.: Buenos Aires

After Brazil, the Argentine is the largest country of the Continent, both in area and in population. Like Chile, it stretches from the "Tropic of Capricorn to its North" to the Island of Tierra del Fuego which it shares with that Country in the extreme South. Paraguay's Gran Chaco reaches across the northern regions of the Argentine. The main rivers are the Uruguay & Parana, whilst Aconcagua is the tallest mountain of the Hemisphere. Buenos Aires is the largest city of the Southern Hemisphere. The pampa and its fertile grasslands, devoid of trees, is known for its large cattle herds and romantic gauchos. And then, there are orchards, vineyards and important mines nestled in the foot-hills of the Andes. Finally, Patagonia to the South, is synonymous with bleakness and sheep-raising; it is also the scene of oil-drilling.

As with Chile and for the same reasons, the climate of the Argentine varies with its regions.

Beef exports are the most important, with sheep and wool running close. Main industries are meat processing and canning. Agriculture and mining add to the Country's economy.

The true natives have long been replaced by Spaniards and Italians who settled the land years ago.

Here again, political turmoil has been a trade mark of life with civilian and military governments following one-another with monotonous regularity. Inflation, and political crimes are rampant.

Bolivia

Area: 1,098,581 km²
Pop. Grth.: 2.7
Pop.: 5,430,000
Cap.: La Paz

Bolivia is land-locked and straddles the Andes where its peaks are the loftiest. The Altiplano in the West, a high and cold plateau where most of the population lives is one of the main features of the Country. Elsewhere, forest-covered lowlands run into scrub and desert areas. On its border with Peru, Bolivia boasts Titicaca, the highest, navigable lake in the World (3810 m).

The population is made up of Amerindians, Mestizo and Europeans.

The climate ranges from hot in the West to cool in the East with the usual tropical rain season from December to February.

Despite its mineral wealth, Bolivia is the poorest of all South American countries: this is mainly because of stiff competition in the tin industry and labor troubles in the mines.

The mid 70's were marked by peasant revolts brought about by food shortages and the increased cost of living. Nonetheless there is potential in the area of petroleum development.

Brazil

Area: 8,511,965 km²
Pop. Grth.: 2.8
Pop.: 118,650,000
Cap.: Brasilia

Brazil could be called the "Texas" of South America. The largest in South America, both in area and population, it has the World's largest water system (the Amazon) and some of the World's largest forests; it also produces more coffee than any other country. The capital, Brasilia is an extravaganza.

Brazil is divided into three regions: the humid and heavily forested basin of the Amazon which the Government is trying to open up, the scrubland of the N.E. where unbelievably long droughts are commonplace and the fertile plateaus of the central and southern parts of the Country where most of the life-sustaining activites take place.

Brazil's immense, mineral wealth is almost untapped though the Country supplies a good portion of the World's requirements in quartz, crystal, mica, manganese, iron ore, colombium and beryl. Oil deposits add to its potential as do, gold and diamonds.

Gaining its independence from Portugal in 1800, Brazil became a republic in 1889. Like other Latin-American countries, its history has been marked by civil strife and political turmoil.

Chile

Area: 756,945 km²
Pop. Grth.: 2.3
Pop.: 10,920,000
Cap.: Santiago

Chile is the "outer-lining" of South America, as it were; it is a narrow strip of land compressed between the Andes and the Pacific, from Peru down to the Antartic, a distance of 4500 km. This very length accounts for the different climatic conditions of the many zones covered by Chile. The South, influenced by the proximity of Antarctica, presents bleak forests, islands, fjords and glaciers by contrast, the North is much warmer and its lakes, forests and mountains, most favored by holidaymakers. Between these two zones lies a pleasant, temperate region where most people live and where industry thrives. The North, however, contains desert land which nevertheless has a great value since its soil yields copper and nitrate.

The 150,000 Arauca Indians of Chile, were never conquered either by the Incas or the Spaniards.

Inflation in the 1960's and political turmoil in the 1970's have been the cause of serious concern.

Colombia

Area: 1,138,914 km²
Pop. Grth.: 2.7
Pop.: 25,640,000
Cap.: Bogota

Main features: Four mountain ranges run parallel, across the Country, creating narrow fertile valleys where the bulk of the population lives. These rugged mountains have impeded development until modern aviation solved the problem. Whereas the South and East of Colombia are covered by fertile, grassy plains, these are replaced by steaming rain forests in the Amazon Basin.

Apart from its population of mixed ancestry (Negro - Indian - European), Colombia has several hundred tribes of Amerindians within its borders. Its economy is sustained by the production of coffee, bananas, crude petroleum, gold and close to the entire supply of emeralds in the World.

The current form of government is a presidential democracy. Anarchy and civil war marred the existence of this nation from 1948-58.

The 1970's brought inflation and a population explosion of staggering proportions.

Ecuador

	Area: 283,561 km²
	Pop. Grth.: 3.4
	Pop.: 8,080,000
	Cap.: Quito

On the N.W. Coast, North of Peru, Ecuador is one of the smallest countries of South America. Two ranges of the Andes Cordillera cross it, forming a fertile valley which accomodates most of its population. Numerous volcanoes dot these regions, including Cotopaxi, the tallest active volcano known. Off-shore, but part of Ecuador, lie the Galapagos Islands of turtle fame.

Main products are bananas, coffee, cocoa, sugar, oil and straw. Ecuador also has potential mineral wealth which only awaits development.

French Guiana

	Area: 91,000 km²
	Pop.: 60,000
	Cap.: Cayenne

Smallest and least populated of the three Guianas; its mainly Negro population hugs the coastal strip where the land is hardly developed. Timber, sugar and rum are the only exports. An overseas department of France, it had the dubious honor of housing, until 1944, a French penal colony, Devil's Island, notorious for its history of brutality and inhuman treatment of its inmates.

Guyana

	Area: 214,969 km²
	Pop. Grth.: 1.7
	Pop.: 860,000
	Cap.: Georgetown

Its Negro-East-Indian population lives in a flat, swampy strip of land on the Atlantic Coast - the remainder of the Country is covered with equatorial rain forests. Independent from England in 1966, it became a republic in 1970.

Sugar and rice are at the basis of the economy. Bauxite is mined in large quantities.

Paraguay

	Area: 406,752 km²
	Pop. Grth.: 3.0
	Pop.: 2,970,000
	Cap.: Asuncion

Like Bolivia, Paraguay is land-locked at the centre of South Ameria, but has two important water-ways, the Paraguay and Parana rivers which give it access to the South Atlantic via the Rio de la Plata. The inhospitable Gran Chaco, found West of the Paraguay River, is little more than scrub and swampy jungles. To the East, rolling grasslands rise to a forested plateau where the bulk of the population enjoys a pleasant, sub-tropical climate.

Despite the poverty of its inhabitants, the military dictators of the Country did not hesitate to launch Paraguay into two disastrous conflicts with its neighbors over the last century.

Agricultural products are the main exports, as is the extract of quebracho, used in tanning leather.

Peru

	Area: 1,285,216 km²
	Pop. Grth.: 2.8
	Pop.: 17,290,000
	Cap.: Lima

Land of the Incas, Peru is controlled by criollos or descendants of the original Spaniards, who settled there.

Ranges of the Andes form, with the Pacific, a narrow fertile strip of land with desert stretches here and there. A generous water-system covers the Country - jungles, largely unexplored, occupy the plains which lie at the foot of the eastern slopes. The tropical climate is cooled by the Humboldt Current.

Peru has one of the World's largest fishing industries. Apart from a good supply of metal wealth (gold, copper, zinc, silver and lead), Peru's chemical industry is of growing importance. Its agricultural industry is also worthy of note.

Suriname

	Area: 163,265 km²
	Pop. Grth.: 0.8
	Pop.: 440,000
	Cap.: Paramaribo

Formerly a Dutch possession, Surinam, with Guyana to the West and French Guiana to the East, is known as one of the Guianas. It has a low, swampy plain on the coast, mountains and tropical forests in the hinterland. Its main source of wealth is high grade bauxite. Its

population is made up of Negroes, East Indians and Chinese.

Uruguay

	Area: 176,215 km²
	Pop. Grth.: 0.6
	Pop.: 2,910,000
	Cap.: Montevideo

One of the smallest republics of the Continent, Uruguay has a pleasant climate, adequate rainfall, superb hydrography as well as rich, rolling, grassy plains. Separated from the Argentine by the Rio de la Plata estuary, and the Uruguay River, its eastern flank rests on the South Atlantic; its climate and soil cannot but result in agriculture being its main industry.

The people of Uruguay are mostly all descendants of Spaniards and Italians.

It is of interest to note that Uruguay introduced social democracy in the Southern Hemisphere. Although known for its political stability prior to the mid 1940's. Uruguay unfortunately followed the example of other Latin-American countries in the realm of civil strife and urban terrorism; this was to lead, inevitably, to strict censorship and curtailment of freedom.

Venezuela

	Area: 912,050 km²
	Pop. Grth.: 3.0
	Pop.: 13,520,000
	Cap.: Caracas

Rich petroleum deposits on the shores and in the bed of Lake Maracaibo have made Venezuela the wealthiest country of South America. It owes its name, which means "little Venice", to Spanish explorers who in 1499, upon seeing an Indian village built on stilts in Lake Maracaibo, saw there a resemblance to the Venice of Europe.

Independent from Spain since 1811, Venezuela suffered a century of violence and dictatorships; it is now becoming more democratic.

Four regions divide the Country. The Venezuelan Highlands, a spur of the Andes, in the N.W.; - the lowlands around Lake Maracaibo, hot and humid; - the Guyana Highlands, South of the Orinoco river, a wild and sparsely populated land, and the fertile llanos of the Central Plain.

Venezuela has the highest water-fall in the World, Angel Falls (979 ms).

Its main source of wealth nearing exhaustion, Venezuela is looking to other minerals like iron, bauxite, asphalt and nickel to fill the void.

OCEANIA

Oceania is that part of SE Asia which is comprised of MELANASIA (Australia, Tasmania, Fiji, the Solomons, the New Hebrides, New-Zealand, New-Caledonia, Papua-New-Guinea, etc), also known as MICRONESIA, which is made up of the numerous Pacific islands located North of the Equator (Hawaii, Samoa, Timor, the Mariannas, Carolinas, Marshalls, Marquesas, etc). Of the above divisions, more ethnic than geographic, the first is sometimes called AUSTRALASIA.

The islands of Oceania are either volcanic or the result of eons of coral deposits; their climate is hot and humid, frequently wet, which makes for a luxuriant vegetation. The coasts are surrounded by coral reefs which make their approaches dangerous. Most major powers have either extended their colonial empires to Oceania, or have at least exercised an influence over its development (England, France, Germany, Portugal and the USA). The population is either Melanasian or Polynesian.

Note:

Many small islands, not elaborated on in this text, will be found on the maps of Oceania; among these are Guam, an important U.S. military base, the Gilbert and Ellice Islands, Easter Island, Wallis, Futuna, Tonga and Nauru. The same applies to other areas of the World.

Australia

	Area: 7,686,848 km²
Geography:	Pop. Grth.: 1.1
	Pop.: 14,420,000
	Cap.: Canberra

Geography:
A very large island of Oceania, Australia's shores are washed by the waters of the Tasman Sea to the East and those of the Indian Ocean to the South and West. Geographically, it seems that Australia is what is left of an ancient continent - a vast plateau with several large desert areas (Great Sandy, Gibson, Great Victoria, Simpson & Sturt deserts), a few mountain ranges (Gawler, Mt Lofty, Grey, Selwyn, Gregory and Great Dividing ranges, the latter on the East Coast), and a very hot and dry climate. Some areas of the hinterland (outback), are literally parched by an implaccable sun and a total absence of rain. Australia also has the Great Barrier Reef off its NE Coast, in the Coral Sea of WWII fame.

The population is established mostly along the coastal areas, the hinterland being reserved for cattle and sheep-grazing.

Economy:
Australia is essentially an agricultural country where sheep are grazed by the million for their meat and their wool, and cattle are raised for meat and dairy products, most of the former going to export. Wheat, sugar-cane, fruit and latterly, wine, are part of Australia's agricultural spectrum. Gold, silver, iron ore, tin, bauxite, copper, zinc, lead and coal are mined. The most important industries are textiles and metallurgy. Main cities are Canberra, Darwin, Melbourne, Sydney, Brisbane, Adelaide and Perth.

History and Culture:
Australia was colonized by England, but a good sprinkling of European races has been added over the years; the sub-continent has some 140,000 aborigines.

Australians are out-going, informal and straight-forward - they "knock" royalty but will die for it as they did so valiantly in both World Wars. True out-door types, they excel at sports and once dominated swimming, tennis, cricket and sailing.

Australia is comprised of six states (see map) and two territories. In 1770 Capt. James Cook "took the land over" from the aborigines - there followed the establishing of a penal colony, near present-day Sydney; later true colonization started, and was soon accelerated by the discovery of gold in the 1850's.

Australia administers many islands in the Indian Ocean: the Christmas, Coco, Heard and McDonald Islands, Norfolk and Lord Howe islands, as well as Pitcairn, the small tropical rock which sheltered the Bounty mutineers.

Carolinas

An archipelago of Oceania of over 30,000 inhabitants, East of the Philippines and North of New-Guinea, the Carolinas went from Spanish then to German and finally to American control, in 1947.

Fiji

	Area: 18,272 kr
	Pop. Grth.: 2.1
	Pop.: 610,000
	Cap.: Suva

A British colony of Melanasia until 1970, its main islands are Fiji, Levu and Vanua Levu. Main products are rice, sugar-

cane, tropical fruit, copra and wood.

Hawaii

	Area: 16,705 km²
	Pop.: 770,000
	Cap.: Honolulu

An archipelago of the Pacific, Hawaii was a US overseas territory from 1898 until 1959 when it became the last state to join the Union. Its economy is based on tourism, fruit, coffee, sugar-cane, rice and fishing. It is a major US military base and boasts shipbuilding and oil processing industries. Hawaii is the largest of the 132 islands of the archipelago and has an active volcano, Mauna Laua.

Mariannas

Except for one of its islands, Guam, this archipelago of the Pacific, East of the Philippines, had been under Japanese rule when it passed to American administration in 1947. Population 50,000.

Marquesas

A French archipelago of Eastern Polynesia, also known as Mendena or Nouka-Hiva. It has a population of 45,000.

Marshall Islands

An archipelago of Oceania, SW of Hawaii, this group of islands was first under German domination from 1906 to 1914, under Japanese mandate from 1914 to 1944, and under American administration since 1947. Site of a naval battle between the American and Japanese fleets in 1944. Approximately 12,000 inhabitants.

New-Caledonia

Area: 18,997 km²
Pop.: 135,000
Cap.: Noumea

An island of Melanesia discovered by Capt Cook in 1774, it went to France in 1853 and was used to "accomodate" convicts and political prisoners until 1898. The Island is mountainous and surrounded by treacherous reefs; its forests and sub-soil are its main assets. Gold, iron, nickel, copper and coal are mined. Tobacco, coffee, copra, rice and sugar-cane are cultivated.

New-Hebrides

A 37 island group of Melanesia with a population of 84,000, it is situated between Fiji and New-Guinea and covers an area of 14,760 km².

New-Zealand

Area: 268,676 km²
Pop. Grth.: 0.4
Pop.: 3,140,000
Cap.: Wellington

East of Australia, New-Zealand is made up of a North and a South island, separated by Cook's Strait, thus named in honor of British Capt. James Cook who explored Oceania in the 1700s. The northern island is of irregular shape and has many volcanoes, whilst its southern sister is more massive and considerably larger. Both islands are traversed by a volcanic mountain range which in the South, is known as the Southern Alps.

Although Capt. Cook charted the coasts of the Island in 1769, Abel Tasman discovered it in 1642. The natives were Maoris who now account for 10% of the population; they are deemed to be descendants of Polynesians who reached the Island in the 900s.

This land of rolling hills, orchards, forests, thermal springs and volcanoes, has a mild climate tempered by sea breezes. The Country offers beautiful sights and a great deal to tourists and sportsmen alike.

Mainly self-governing since the 1860s, New-Zealand became an independent dominion in 1907; it administers Niue, the Tokelau and Cook islands.

New-Zealand is basically an agricultural nation where cattle are bred for meat, dairy products and wool; as a result, the Country's main industries are textiles and food products. Like Australia, New-Zealand is part of the British Commonwealth of Nations and has helped England in both World Wars of the 20th Century.

Papua - New-Guinea

Area: 461,691 km²
Pop. Grth.: 2.9
Pop.: 3,080,000
Cap.: Port Moresby

New-Guinea (Papua) is a large island directly North of Australia, under whose mandate it formerly operated. New Britain (Bismark Archipelago) and Bougainville Island of WWII fame, are part of the complex. It has been independent since 1975.

The climate is hot and wet and the land covered with mangrove swamps and tropical forests where one finds tribes still very close to the Stone Age. Two main ranges cross the Island, the Owen Stanley and the Bismark. The cultivation of sago and yams and the raising of pigs and poultry are at the basis of the limited economy.

Samoa

A former German possession of 34,000, this Polynesian archipelago is partly under American and New-Zealand administration. The main product is copra.

Solomon Islands

Area: 29,785 km²
Pop.: 210,000
Cap.: Homara

Another Melanasian archipelago East of New-Guinea, the Solomons were shared by England and Germany prior to WWI, the German area of Bougainville going to Australian administration subsequent to 1914. The Solomons were the site of a major naval battle between the Japanese and American fleets in 1942; the islands secured their independence in 1978. Wood and copra are the main products.

Tahiti

Part of the Society Archipelago, Tahiti has a population of over 80,000. Its capital is Papeete. Sugar-cane and tobacco are produced.

Tasmania

An island immediately South of Australia from which it is separated by the Bass Strait, Tasmania has a certain mineral wealth and is popular with tourists.

Timor

Area: 14,925 km²
Pop.: 500,000
Cap.: Dili

This island of Indonesia, North of Australia and East of Flores, was shared by Indonesia and Portugal until 1975 when it gained its independence. Its main products are copra and sandal wood.

POPULATION BY STATE AND PROVINCE
UNITED STATES AND CANADA

UNITED STATES:

Alabama	3,769,000	Ohio	10,731,000
Alaska	406,000	Oklahoma	2,892,000
Arizona	2,450,000	Oregon	2,527,000
Arkansas	2,180,000	Pennsylvania	11,731,000
California	22,694,000	Rhode Island	929,000
Colorado	2,772,000	South Carolina	2,932,000
Connecticut	3,115,000	South Dakota	689,000
D.C.	656,000	Tennessee	4,380,000
Deleware	582,000	Texas	13,380,000
Florida	8,860,000	Utah	1,367,000
Georgia	5,117,000	Vermont	493,000
Hawaii	915,000	Washington	3,926,000
Idaho	905,000	West Virginia	1,878,000
Illinois	11,229,000	Wisconsin	4,720,000
Indiana	5,400,000	Wyoming	450,000
Iowa	2,902,000		
Kansas	2,369,000	CANADA:	
Kentucky	3,527,000		
Louisiana	4,018,000	Alberta	2,135,900
Maine	1,097,000	British Columbia	2,687,000
Maryland	4,148,000	Manitoba	1,027,000
Massachusetts	5,769,000	New Brunswick	709,100
Michigan	9,207,000	Newfoundland	583,600
Minnesota	4,060,000	Northwest Territories	42,800
Mississipi	2,429,000	Yukon	21,500
Missouri	4,867,000	Nova Scotia	856,100
Montana	786,000	Ontario	8,600,500
Nebraska	1,574,000	Prince Edward Island	124,000
Nevada	702,000	Quebec	6,325,200
New Hampshire	887,000	Saskatchewan	975,700
New Jersey	7,332,000		
New Mexico	1,241,000		
New York	17,648,000		
North Carolina	5,606,000		
North Dakota	657,000		

MAPS OF THE WORLD

LEGEND

MAP PAGES

SIZE OF NAMES ON MAP PAGES IS RELATED TO THE POPULATION

LOS ANGELES	More than 1,000,000
Toulouse	250,000 to 1,000,000
Darlington	100,000 to 250,000
Acomb	50,000 to 100,000
Levens	10,000 to 50,000
Lorton	1,000 to 10,000
Earby	less than 1,000

INDEX

Ubangi; river, 78 A 4
Ubangui; river, 76 L 3
Ubaque, 112 G 5
Ubate, 112 F 5
Ubauro, 68 H 3
Ubekenot; isl., 87 C 12
Ubinas, 114 K 8

University City, 102 E 5
University; river, 99 B 13
Unnao, 69 I 9
Unst; isl., 46 B 9
Unzen, 63 M 1
Upano; river, 112 L 3
Upata, 113 E 13

The index reference system refers to the page number and section of the map defined by corresponding codes (side and top or bottom).

The World (Political)

1:75,000,000 1" = 1200 mi 1 cm = 750 km

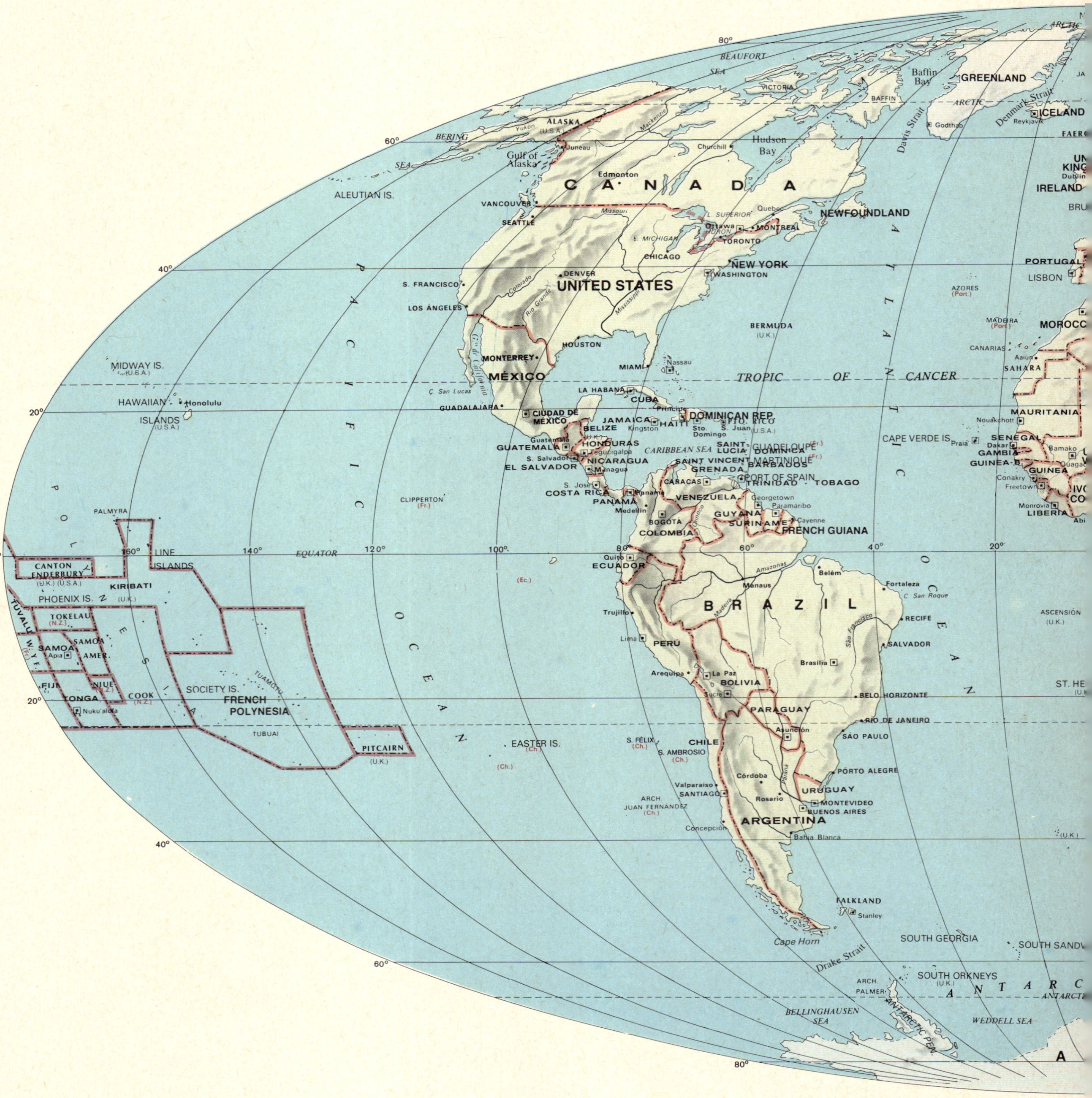

ARCTIC

BEAUFORT SEA

VICTORIA

Baffin Bay

GREENLAND

BAFFIN

Godthab

Denmark Strait

ICELAND

Reykjavik

FAER

80°

60°

BERING SEA

ALASKA (U.S.A.)

Yukon

Juneau

Gulf of Alaska

Mackenzie

Churchill

Hudson Bay

Edmonton

C A N A D A

Davis Strait

Arctic

UN KING

IRELAND

Dublin

BRU

ALEUTIAN IS.

VANCOUVER

SEATTLE

L. SUPERIOR

Quebec

NEWFOUNDLAND

Ottawa

Montreal

ATLANTIC

40°

S. FRANCISCO

DENVER

Colorado

UNITED STATES

Missouri

L. MICHIGAN

L. HURON

TORONTO

CHICAGO

NEW YORK

Washington

PORTUGAL

LISBON

AZORES (Port.)

LOS ÁNGELES

Rio Grande

Mississippi

HOUSTON

BERMUDA (U.K.)

MADEIRA (Port.)

MOROCC

MONTERREY

MEXICO

C. San Lucas

MIAMI

Nassau

LA HABANA

CUBA

TROPIC OF CANCER

CANARIAS

SAHARA

Aaiun

MAURITANIA

Nouakchott

20°

MIDWAY IS. (U.S.A.)

HAWAIIAN ISLANDS (U.S.A.)

Honolulu

GUADALAJARA

CIUDAD DE MÉXICO

Guatemala

JAMAICA

Kingston

BELIZE

HONDURAS

Tegucigalpa

CAPE VERDE IS.

Praia

SENEGAL

Dakar

GAMBIA

GUINEA-B

GUINEA

PUERTO RICO

HAITI

Sto. Domingo

DOMINICAN REP

S. Juan (U.S.A.)

GUADELOUPE

SAINT LUCIA

DOMINICA

MARTINI

BARBADOS

GUATEMALA

S. Salvador

EL SALVADOR

NICARAGUA

Managua

COSTA RICA

S. José

PANAMA

Panamá

CARIBBEAN SEA

SAINT VINCENT

GRENADA

PORT OF SPAIN

TRINIDAD

TOBAGO

CARACAS

VENEZUELA

Georgetown

GUYANA

Paramaribo

SURINAME

Cayenne

FRENCH GUIANA

Medellín

BOGOTÁ

COLOMBIA

IVORY CO

Conakry

Freetown

Monrovia

LIBERIA

Ouaga

PALMYRA

LINE ISLANDS

EQUATOR

CLIPPERTON (Fr.)

Quito

ECUADOR

(Ec.)

0°

CANTON ENDERBURY (U.K.) (U.S.A.)

KIRIBATI

PHOENIX IS. (U.K.)

Manaus

Belém

C. San Roque

Fortaleza

ASCENSIÓN (U.K.)

Trujillo

BRAZIL

RECIFE

TOKELAU (N.Z.)

TUVALU

SAMOA

SAMOA AMER.

Apia

Lima

PERÚ

São Francisco

SALVADOR

Brasília

ST. HE

FIJI

NIUE (N.Z.)

COOK (N.Z.)

SOCIETY IS.

TUAMOTU

Arequipa

La Paz

BOLIVIA

Sucre

BELO HORIZONTE

20°

TONGA

Nuku'alofa

FRENCH POLYNESIA

TUBUAI

EASTER IS. (Ch.)

S. FÉLIX (Ch.)

S. AMBROSIO (Ch.)

CHILE

PARAGUAY

Asunción

RIO DE JANEIRO

SÃO PAULO

PITCAIRN (U.K.)

Córdoba

Valparaíso

SANTIAGO

Rosario

PÔRTO ALEGRE

ARCH. JUAN FERNÁNDEZ (Ch.)

Concepción

URUGUAY

MONTEVIDEO

BUENOS AIRES

ARGENTINA

Bahía Blanca

(U.K.)

40°

FALKLAND

Stanley

SOUTH GEORGIA

SOUTH SANDW

Cape Horn

Drake Strait

SOUTH ORKNEYS (U.K.)

ANTARC

60°

ARCH. PALMER

ANTARCTIC PEN.

BELLINGHAUSEN SEA

WEDDELL SEA

ANTARCTI

80°

A

BARD NOVAYA KARA LAPTEV SEA NEW SIBERIAN IS. 80°

BARENTS SEA ZEMLYA SEA
North C.

Yenisey Yakutsk

U. S. S. R. 60° BERING

LENINGRAD SVERDLOVSK NOVOSIBIRSK Irkutsk HOTSK SEA SEA

KIEV L. BAIKAL SAKHALIN PACIFIC

Volgograd ARAL SEA Ulan-Bator Vladivostok

CHEC HUN MONGOLIA Shenyang SEA OF JAPAN 40°

ROM BUCAREST TASHKENT PEKIN NORTH KOREA JAPAN

BULG BLACK SEA Pyongyang SOUTH Tokyo

SOFIA ANKARA LANCHOU KOREA

TURKEY AFGANISTAN CHINA Hwang Ho RYU-KYU

TEH-RAN Kabul WUHAN SHANGHAI VULCANO

TUNISIA MALTA SYRIA IRAQ IRAN Islamabad Yangtse

ISRAEL BAGDAD PAKISTAN New Delhi NEPAL BHUTAN TAIPEI 20°

JORD KUWAIT Ganges Dacca HONG-KONG TAIWAN

LIBYA BAHR. Riyadh QATAR CALCUTA BANG. CANTON MACAO

EGYPT Cairo Mascate INDIA BURMA (Port.)

SAUDI ARABIA OMAN BOMBAY Gulf of SOUTH

NIGER CHAD Khartoum YEMEN Bengal RANGUN LAOS CHINA PHILIPPINES MARIANAS WAKE (U.S.A.)

San Shaʼab THAILAND VIETNAM Hanoi

SUDAN YEMEN SOCOTRA MADRAS BANGKOK Vientiane MANILA PHILIPPINES GUAM (F.S.A.) MARSHALL

NIGERIA N'Djamena DJIBOUTI ANDAMAN Phnom Penh CAMB. PACIFIC IS. TER.

CENT. ADDIS ABEBA (Ind.) HO CHI MINH (Saigon)

AFRICAN REP. ETHIOPIA NICOBAR SEA CAROLINE IS.

CAMEROUN Yaounde (Ind.) SRI LANKA MALAYSIA BRUNEI

Libreville UGANDA KENYA SOMALI REP. Mogadiscio MALDIVAS Colombo Kuala Lumpur EQUATOR KIRIBATI (U.K.) 0°

GABON L. VICTORIA FALKLAND IS. Male SINGAPUR BORNEO

CONGO Nairobi 100° 120° 140° 180°

Brazzaville ZAIRE BURUNDI Mombasa SEYCHELLES SUMATRA CELEBES PAPUA N GUINEA NAURU

CABINDA KINSHASA TANZANIA INDIAN YAKARTA INDONESIA SALOMON TUVALU (U.K.)

Luanda Dar es Salaam JAVA ARAFURA Port Moresby IS. SALOMON SLICE

ANGOLA MALAWI L. MALAWI TIMOR SEA Honiara

Lilongwe CHRISTMAS (Aust.) Darwin NEW HEBRIDES FIJI

ZAMBIA Lusaka OCEAN CORAL Vila VANUATU Suva

ZIMBABWE MADAGASCAR NEW 20°

SOUTH WEST MOZAMBIQUE Harare SEA CALEDONIA

AFRICA Antananarivo (Fr.)

Windhoek REUNION TROPIC OF CAPRICORN AUSTRALIA

Walvis Bay BOTSWANA Maputo MAURICIO (Fr.)

JOHANNESBURGO SWAZILAND Brisbane NORFOLK (Aust.)

REP. Pretoria LESOTHO Perth Great SYDNEY

SOUTH AFRICA Australian Bight Canberra TASMAN SEA Auckland

Capetown Cape of Good Hope MELBOURNE NEW NORTH ISLAND 40°

PRINCE EDWARD IS. CROZET (Fr.) KERGUELEN TASMANIA ZEALAND Wellington

(Fr.) SOUTH ISLAND

HEARD (Aust.)

MACQUARIE (Aust.)

OCEAN 60°

IVET OCEAN

RCLE

Queen Maud Land Wilkes Land

ARCTICA 80°

1:75,000,000 1" = 1200 mi
1 cm = 750 km

ARC

80°
QUEEN ELIZABETH IS.
ELLESMERE
JAN Ma
BEAUFORT SEA
BANKS
Baffin Bay
GREENLAND
ICELAND
Str. of Denmark
FAEROES
Str. of Bering
VICTORIA
GREAT BEAR LAKE
Mackenzie
BAFFIN
Davis Strait
Hudson Strait
60°
BERING SEA
Yukon • McKinley 6.194
Alaska
Mts. Mackenzie
GREAT SLAVE LAKE
Hudson Bay
C. Farvel
REYKJANES RIDGE
FAEROERNE PLATFORM
BRI
ISL
ALEUTIAN IS.
ALEUTIAN 8.100
TRENCH
ROCKY MOUNTAINS
Athabasca
Nelson
L. WINNIPEG
LABRADOR
NEWFOUNDLAND
Ba
Bis
VANCOUVER
NORTH
L. SUPERIOR
Missouri
St. Lawrence R.
NOVA SCOTIA
PACIFIC
NORTH BASIN
L. HURON
L. MICHIGAN
L. ONTARIO
L. ERIE
C. Race
40°
C. Mendocino
Colorado
AMERICA
NORTH WESTERN ATLANTIC BASIN
BERMUDA
AZORES
C. Roca
SP
MENDOCINO SEASCARP
S. Nevada
Rio Grande
NORTH ATLANTIC
MADEIRA
4.165
To
SIERRA MADRE
FLORIDA
RIDGE
CANARY IS.
TROPIC OF CANCER
Str. of Florida
CUBA
Gulf of México
SAHA
C. San Lucas
V. Pico de Orizaba 5.747
YUCATAN
HISPANIOLA
CANARY
20°
MARCUS-NECKER RISE
HAWAII
JAMAICA 9.219 PUERTO RICO TRENCH
GREATER ANTILLES
CAPE VERDE IS. C. Verde
A
Hawaii
REVILLAGIGEDO
MIDDLE AMERICA TRENCH
L. NICARAGUA
CARIBBEAN SEA
LESSER ANTILLES
CAPE VERDE BASIN
MID-PACIFIC BASIN
PALMYRA
CLIPPERTON
ALBATROSS PLATEAU
6.669
ISTHMUS OF PANAMA
Orinoco
P O L Y N E S I A
0°
OCEANÍA
EQUATOR
160°
140°
120°
100°
ARCH. DE COLON
80°
CORDILLERA
60°
Chimborazo 6.267
Negro
Amazon
40°
FERNANDO DE NORONHA
20°
PHOENIX IS.
Llanura Amazónica
Madeira
C. San Roque
TOKELAU IS.
I S L A N D
Ucayali
ASCENSION
CAROLINE
MARQUESAS ARCH.
SOUTH
BRAZILIAN
SAMOA IS.
SOCIETY IS.
TUAMOTU
PERU
S. do Roncador
Tocantins
AMERICA
BASIN
ST. HELENA
MICRONESIA
TITICACA
S. do Espinhaço
São Francisco
20°
TONGA IS.
COOK IS.
TUBUAI
GAMBIER
BASIN
Illampú 6.550
AMERICA
TONGA TRENCH 10.882
PITCAIRN
PACIFIC
S. Felix 8.064
Paraná
Aconcagua 6.959
KERMADEC TRENCH
RAPA
EASTER IS.
SALA Y GÓMEZ
S. Ambrosio
ARGENTINE
TRISTAN DA CUNHA
RIDGE
10.047
KERMADEC IS.
SOUTH-WESTERN PACIFIC
PACIFIC RIDGE
ARCH. JUAN FERNANDEZ
CHILE BASIN
Bahía Blanca
BASIN
N.
40°
BASIN
GRANDE DE CHILOÉ
O C E A N
G. San Jorge
ANTIPODES IS.
C. Tres Puntas
P A C I F I C
C
Str. of Magellan
IS. FALKLAND
SOUTH GEORGIA IS.
TIERRA DEL FUEGO
C. Horn
Drake Passage
SCOTIA SEA
SOUTH SANDWICH IS.
60°
PACIFIC-ANTARCTIC
SOUTH SHETLAND
SOUTH ORKNEYS IS.
ANTAR
Ant.
BASIN
ANTARCTIC PEN.
BELLINGHAUSEN SEA
ALEXANDER PETER 1
WEDDELL SEA
AMUNDSEN SEA
Byrd Land
Mt. Vinson 5.140
80°
A N T A
A

EUROPE

ASIA

SPITZBERGEN
FRANZ JOSEF LAND
NOVAYA ZEMLYA
BARENTS SEA
Nordkapp
SCANDINAVIAN PEN.
GIAN
L. ONEGA
L. LADOGA
KARA SEA
POL YAMAL PEN.
TAYMYR PEN.
LAPTEV SEA
NEW SIBERIAN IS.
NORTH LAND
ARCTIC POLAR CIRCLE
Yenisón
Shrene
Lena
Verchoyanskiy Khrebet
80°
Zapadno Sibirskaya Nizmennost
Ploskogorye
Sibirskoye
URALSKY KHREBET
Obi
Obi
Irtysh
60°
Volga
Mts. Sayán
OZ. BAYKAL
ALTAI
Amur
BERING SEA
SEA OF KAMCHATKA
OKHOTSK
Lopatka
SAKHALIN
KURIL IS.
KURIL TRENCH 10.542
mt Blanc
CARPATHIAN MTS.
Danubio
BLACK SEA
CASPIAN SEA
CAUCASUS
Elbrus 5.633
Syr Daria
Aral Sea
TIAN SHAN
OZ. BALKHASH
Amu Darya
HINDU KUSH
Pamir
K 2 8.611
KARAKORUM
GOBI
Mts. Gran Jingán
Huang-ho
Mts. Sayán
SEA OF JAPAN
HOKKAIDO
HONSHU
Kinsiu
JAPAN TRENCH 10.375
ITALY
BALKANIC PEN.
SICILY
CRETA
CYPRUS
Taurus
MEDITERRANEAN SEA
Mts. Zagros
Iranian Plateau
KUEN LUN
Tibet
HIMALAYA
Everest 8.848
Brahmaputra
Yangtse
Korea
YELLOW SEA
EAST CHINA SEA
KYUSHU
RYU-KYU
FORMOSA
PACIFIC
40°
20°
RICA
RED SEA
ARABIC PEN.
Tibesti 3.415
Emi Koussi
Nile
RUB AL JALI
DES. THAR
INDIA
Bay of Bengal
Ganges
Mekong
HAINAN
SOUTH CHINA SEA
LUZON
PHILIPPINE SEA
GUAM
MARIAN
MARIANAS TRENCH
MARIANAS BASIN
MARSHALL
WAKE
ARABIAN SEA
Ethiopian Mt.
G. of Adén
SOCOTORA
C. Guardafui
SOMALI REP.
LACCADIVE IS.
C. Comorin
ANDAMAN
NICOBAR
ISTHMUS OF KRA
INDOCHINA PEN.
MALACCA PEN.
MINDANAO
PHILIPPINE TRENCH 10.497
CAROLINE IS.
PALAOS
MICRONESIA
OCEAN
Adamaua
Ubangui
SOMALI BASIN
MALDIVE IS.
CEYLON
SULU SEA
CELEBES SEA
MOLUCCAS
SUMATRA
BORNEO
CELEBES
BISMARCK
GILBERT
0°
P. Margherita 5.109
L. RUDOLF
KENYA 5.199
L. VICTORIA
Kilimanjaro 5.895
ZANZIBAR
SEYCHELLES
AMIRANTE IS.
SUNDA IS.
JAVA SEA
P. Djaja 6.030
NEW GUINEA
SOLOMON
STA. CRUZ
ELLICE
Congo
Kasai
TANGANYKA
L. MALAWI
COMORES
CHAGOS
MALDIVE RIDGE
JAVA 7.450
JAVA TRENCH
FLORES
SUMBA
TIMOR
ARAFURA SEA
Str. of Torres
C. York
Arnhem Land
G. of Carpentaria
CORAL SEA
NEW HEBRIDES
20°
A BASIN
NAMIB DESERT
Zambs.
MADAGASCAR
MOZAMBIQUE CHANNEL
MASCARENE IS.
CHRISTMAS
COCOS
TIMOR SEA
GREAT SANDY DESERT
FIJI
KALAHARI DES.
Orange
Mts. Drakensberg
OCEAN
MID INDIAN RIDGE
EAST INDIAN RIDGE
WEST AUSTRALIAN BASIN
TROPIC OF CAPRICORN
AUSTRALIA
GREAT VICTORIA DESERT
C. Byron
NEW CALEDONIA
FIJI SEA
Cape of Good Hope
C. Agulhas
CAPE BASIN
MADAGASCAR BASIN
S.W. INDIAN RIDGE
C. Steep
C. Leewin
Great Australian Bight
Kosciusko 2.230
Murray
Darling
GREAT DIVIDING RANGE
TASMAN SEA
NORTH ISLAND
40°
MTIC-ANTARCTIC BASIN
CROZET
KERGUELEN
HEARD
AUSTRAL-ANTARCTIC RIDGE
ATLANTIC-INDIAN-ANTARCTIC BASIN
TASMANIA
SOUTH IS.
AUCKLAND
MACQUARIE
Stewart
NEW ZEALAND
EAN
CIRCLE
Maud Land
Enderby Land
Wilkes Land
Victoria Land
ROSS SEA
M. Kirkpatrick 4.530
60°
80°
CTICA

1:48,000,000 1" = 760 mi / 1 cm = 480 km

ICELAND

GREENLAND SEA · NORWEGIAN SEA · JAN MAYEN · Nordkapp · BARENTS SEA · NOVAYA ZEMLYA · KARA SEA · YAMAL POL. · LAPTEV SEA · NEW SIBERIAN I.

Murmansk · Arkhangelsk · KOLA PEN. · OZ. KOLGUYEV

Shrene Sibirskoye Ploskogorye · Tunguska · TAYMYR

STOCKHOLM · Oslo · Trondheim · Helsinki · LENINGRAD · Zapadno Sibirskaya Nizmennost

COPENHAGUE · Riga · MOSCOW · Novosibirsk · Omsk · U.S.S.R. · SEA OF KAMCHATKA

HAMBURG · BERLIN · WARSAW · Minsk · Mts. Sayan · Irkutsk · Baikal · Petropavlovsk · OKHOTSK · SAKHALIN

GERMANY · POLAND · PRAGUE · KIEV · Rostov · OZ. BALKHASH · MONGOLIA · Altai Mongol · HARBIN · Vladivostok · HOKKAIDO · Hakodate

VIENNA · BUDAPEST · Belgrade · Odesa · Astrakhan · TASHKENT · Alma-Ata · GOBI DESERT · PEKIN · NORTH KOREA · SEA OF JAPAN · JAPAN · HONDO

YUGOS. · ROMANIA · BLACK SEA · CRIMEA · CAUCASUS · BAKU · Ashkhabad · TIEN SHAN · TAKLA MAKLAN · TIENTSIN · Pyongyang · SOUTH KOREA · TOKYO · NAGOYA

BULG. · ISTANBUL · TURKEY · Ankara · Erevan · Tabriz · Hindu Kush · Astin tagh · Mer de Kuen Lun · TSINGTAO · YELLOW SEA · PUSAN · OSAKA · SHIKOKU

GREECE · ATHENS · CYPRUS · CRETE · RHODES · SYRIA · Damascus · Mosul · IRAN · Iranian Plateau · AFGHANISTAN · Kabul · CHINA · NANKING · SHANGHAI · Str. or Korea · Nagasaki · KYUSHU

MEDITERRANEAN SEA · Beirut · Bagdad · IRAQ · Jerusalem · PAKISTAN · NEW DELHI · LAHORE · NEPAL · Katmandu · CHUNGKING · WUHAN · Hoang-Ho · IZU TRENCH

ALEXANDRIA · CAIRO · LIBYA · EGYPT · NEFUD · SAUDI ARABIA · Riyadh · KARACHI · CALCUTTA · BANGL. · Dacca · CANTON · Victoria · HONG KONG · Macao · TAIPEH · TAIWAN · BORODINO · IWO SIMA · MARCUS

EASTERN DES. · Mecca · QATAR · RUB AL KHALI DES. · OMAN · INDIA · HYDERABAD · Chittagong · BURMA · Mandalay · Hanoi · Foochow · RASA · BONIN

Port Sudan · RED SEA · YEMEN · SOUTH YEMEN · KURIA MURIA IS. · ARABIAN SEA · BOMBAY · MADRAS · Bay of Bengal · Rangoon · LAOS · Vientiane · VIETNAM · Hué · LUZON · PHILIPPINES · MARIANAS IS. · PAGAN · SAIPAN

SUDAN · Khartoum · Asmara · SOCOTRA · LACCADIVES IS. · ANDAMAN · THAILD · BANGKOK · CAMB. · Pnom Penh · MANILLA · MINDORO · MICRO

ETHIOPIA · Addis Abeba · Ethiopian Plateau · SRI LANKA · CEYLON · Colombo · C. Comorin · NICOBAR · INDOCHINA · CHIHMIN (Saigon) · SOUTH CHINA SEA · NEGROS · MINDANAO · DAVAO · PALAOS · YAP · SONSOROL

SOMALI REP. · Mogadisho · MALDIVE IS. · MALAYSIA · SINGAPORE · Kuala Lumpur · Medan · CELEBES SEA · TALAUD · NUKUORO · KAPINGAMARANGI

UG. · KENYA · Nairobi · Kilimanjaro · SOMALI BASIN · SEYCHELLES · CHAGOS · Diego Garcia · Padang · BORNEO · CELEBES · MOLUCCAS · MELA

TANZANIA · Mombasa · ZANZIBAR · MAFIA · AMIRANTES IS. · MID INDIAN BASIN · SUMATRA · INDONESIA · JAVA · Macasar · WAIGEO · NEW GUINEA · PAPUA · BISMARCK · NEW IRELAND · BOUGAINVILLE

ZAMBIA · MALAWI · ALDABRA · COSMOLEDO · PROVIDENCE · FARQUHAR · CHRISTMAS · YAKARTA · SURABAYA · JAVA SEA · TIMOR · FREDERIK HENDRIK · Port Moresby · SOLOMON IS.

ZIMBABWE · Harare (Salisbury) · COMORES · Antseranana · TROMELIN · CARGADOS · JAVA TRENCH · SUNDA ISLANDS · TIMOR SEA · ARNHEM LAND · MELVILLE · BATHURST · Darwin · LOUISIADE ARCH · CORAL SEA

MOZAMBIQUE · Beira · MADAGASCAR · MASCARENE IS. · MAURICIO · RODRIGUEZ · REUNION · WEST AUSTRALIAN BASIN · Northwest Cape · Barrow · GROOTE · GREAT BARRIER REEF

BOTS · Maputo · Pretoria · MOZAMBIQUE CHANNEL · Antananarivo · Toamasina · TROPIC OF CAPRICORN · GREAT SANDY DESERT · AUSTRALIA · Townsville

SOUTH AFRICA · Maseru · Durban · INDIAN OCEAN · Geraldton · GREAT VICTORIA DESERT · L. EYRE · L. TORRENS · Brisban

Capte Town · East London · Port Elizabeth · SW INDIAN RIDGE · S. STA. MARIA · CROZET BASIN · S. PABLO · Perth · C. Naturaliste · Great Australian Bight · ESPERANCE · Adelaide · SIDNEY · Newcastl

Cape of Good Hope · C. Agulhas · MADAGASCAR BASIN · SOUTH AUSTRALIAN BASIN · CANGURO · MELBOURNE · TASMAN SEA

PRINCE EDWARD IS. · APOSTLES IS. · CROZET · MARION · HOG I. · POSSESSION · KERGUELEN · INDIAN RIDGE · KING I. · FURNEAUX · TASMANIA · Hobart · South East C.

AFRICAN ANTARCTIC BASIN · HEARD · ANTARCTIC RIDGE · AUSTRAL ANTARCTIC BASIN · ANTARCTIC OCEAN · DAVIS SEA

ANTARCTIC OCEAN · Enderby Land · ANTARCTICA · Wilkes · Larus

GREENLAND

EAST SIBERIAN SEA
CHUKOTSKIY POL.
CHUKCHI SEA
Kolyma
Anadyr
Ahadyrskiy Zaliv
St. Lawrence
NUNIVAK
ST. MATTHEW
PRIBILOF
BERING SEA
COMMANDER
KARAGUINSK
ATTU
KISKA
ALEUTIAN IS.
ALEUTIAN TRENCH
Dutch Harbor
UNALASKA
KODIAK
Anchorage
Mt. Mc Kinley 6.194
Alaska
Juneau
ALEXANDER ARCH.
QUEEN CHARLOTTE IS.
Mts. Brooks
Mackenzie
BEAUFORT SEA
Banks
Melville Strait
McClure Strait
Magnetic North Pole
Victoria
GREAT BEAR LAKE
GREAT SLAVE LAKE
ATHABASCA
REINDEER LAKE
Churchill
Hudson Bay
James Bay
Belcher
Port Harrison
Southampton
MELVILLE
BOOTHIA PEN.
BAFFIN
Baffin Bay
Davis Strait
Godthaab
Brewster
Farewell
Denmark Str.

CANADA
ROCKY MOUNTAINS
Coast Range
Cascade Range
UNITED STATES
Vancouver
Seattle
Mt. Rainier 4.391
Missouri
WINNIPEG
L. SUPERIOR
L. MICHIGAN
L. HURON
ONTARIO
ERIE
NIAGARA FALLS
CHICAGO
Great GREAT SALT LAKE Basin
Mt. Elbert 4.399
Colorado
Plateau
Mt. Whitney 4.418
Sierra Nevada
S. FRANCISCO
LOS ANGELES
S. Diego
Colorado
Pecos
Red
Mississippi
S. MADRE OCC.
S. MADRE OR.
NEW ORLEANS
QUEBEC
Otawa
MONTREAL
BOSTON
N. YORK
WASHINGTON
Chesapeake B.
Mt. Mitchell 2.037
C. Hatteras
Charleston
Jacksonville
Miami
G. of FLORIDA
G. of Mexico
Tampico
Veracruz
LA HABANA
CUBA
BAHAMA
REP. DOM.
Santo Domingo
HAITI
Puerto Principe
LA ESPANOLA
JAMAICA
Kingston
MEXICO
CIUDAD DE MEXICO 5.747
V. Pico de Orizaba
V. Popocatepetl
Acapulco
ISTMO DE TEHUANTEPEC
G. de Guatemala
BELIZE
GUATEMALA
HONDURAS
EL SALVADOR
San Salvador
NICARAGUA
Tegucigalpa
Managua
L. NICARAGUA
COSTA RICA
San José
CLIPPERTON
REVILLAGIGEDO
C. S. Lucas
GUADALUPE
CUMBRE ERBEN

TROPIC OF CANCER

PACIFIC OCEAN
NORTH PACIFIC BASIN
MENDOCINO SEASCARP
Mendocino
MURRAY SEASCARP
HAWAIIAN BASIN
MIDWAY
LISIANSKI
LAYSAN
GARDNER PINACLES
NIIHAU
OAHU
Honolulu
MAUI
HAWAII Mauna Kea 4.205
KAUAI
HAWAIIAN
JOHNSTON
WAKE
TAONGI
BIKINI
WETAR
MARSHALL
RATAK
RALIK
KWAJALEIN
MID-PACIFIC
YYAVIN
NGELAP
EBON
KUSAIE
MAKIN
BARAKEI
NONOUTI
TABITEVEA
GILBERT
ARORAE
KINGSMILL GR.
NANUMEA
NUKUFETAU
FUNAFUTI
ELLICE
PALMIRA
WASHINGTON
FANNING
CHRISTMAS
LINE ISLANDS
JARVIS
MALDEN
STARBUCK
PHILIP'S REEF
CAROLINA
FLINT
VOSTOK
NUKU HIVA
HIVA OA
MARQUESAS IS.
FATUHIVA
TONGAREVA
MANIHIKI
PUKAPUKA
NASSAU
(UNION)
POLYNESIA
RAIATEA
Tahiti
Papeete
SOCIETY IS.
FAKARAVA
TUAMOTU
DISAPPOINTMENT
ARRECIFE ANTIOPE
PALMERSTON
ARRECIFE BEVERIDGE
COOK
RAROTONGA
MANGAIA
RURUTU
TUBUAI
VAITU
MANGAREVA
GAMBIER
RENO
ACTAEON
MORUROA
HENDERSON
DUCIE
PITCAIRN
EASTER I.

EQUATOR

EAST PACIFIC BASIN
ALBATROSS PLATEAU
CUENCA DE GUATEMALA
COCOS RIDGE
I. DEL COCO
ARCH. DE COLÓN GALAPAGOS IS.
STA. CRUZ
ISABELA
SAN CRISTOBAL
CARNEGIE RIDGE
CENTRAL AMERICA
PANAMA
Cartagena
Maracaibo
VEN.
BOGOTÁ
COLOMBIA
Buenaventura
Quito
ECUADOR
Chimborazo 6.267
Guayaquil
MILNE EDWARDS DEPTH
BR.
PERU
LIMA
TITICACA
Arequipa
BOL.
PERU BASIN
NAZCA RIDGE
PERU CHILE TRENCH
Antofagasta
S. FELIX
S. AMBROSIO
CHILE BASIN
ANDES CORDILLERA
Aconcagua 6.959
SANTIAGO
Valparaíso
BUENOS AIRES
MONTEVIDEO
Concepción
Mar del Plata
Bahía Blanca
Valdivia
Pto. Montt
GRANDE DE CHILOE
ARCH. CHONOS
PEN. VALDES
Comodoro Rivadavia
S. Cruz
Magellan
MALVINAS IS.
Stanley
TIERRA DEL FUEGO
Strait of Magellan
Pta. Arenas
ATLANTIC OCEAN
SCOTIA SEA
SOUTH ORKNEYS
SOUTH SHETLANDS
WEDDELL SEA

TROPIC OF CAPRICORN

NEW HEBRIDES
GAVA
ESPIRITU SANTO
MALEKULA
EFATE
SANTA CRUZ
MALAITA
GUADALCANAL
W. CALEDONIA
Noumea
MATTHEW
FIJI
VITI LEVU
Suva
LAU
SEA
NIUE
Nakualofa
TONGA
TONGATABU
WESTERN SAMOA
SAVAII
UPOLU
TUTUILA
HORN
KERMADEC
WACHUSETT SHOAL
ERNEST LEGOUVE REEF
MARIA THERESA REEF
THREE KINGS IS.
C. Norte
NEW ZEALAND
NORTH ISLAND
Auckland
Wellington
SOUTH ISLAND
Mt. Cook 3.764
Christchurch
UMBRAL DE CHATHAM
CHATHAM
STEWART
NEW ZEALAND PLATEAU
ANTIPODES IS.
BOUNTY
AUCKLAND
CAMPBELL
MACQUARIE
SOUTH PACIFIC BASIN
PACIFIC ANTARCTIC RIDGE
BELLINGHAUSEN SEA
PACIFIC ANTARCTIC BASIN
ANTARCTIC CIRCLE
BALLENY IS.
SCOTT
ADELAIDE
PETER I
ALEXANDER I.
ANTARCTIC

31

SEA OF OKHOTSK

U. S.

ARCTIC OCEAN

NORTH POLE

BERING SEA

East Siberian Sea

LAPTEV

Beaufort Sea

CHUKCHI SEA

WRANGEL

CANADA

GREENLAND

Baffin Bay

Davis Strait

FINLAND

SWEDEN

NORWAY

U R S S

Hudson Bay

CIRCLE

Godthab

UNITED STATES

QUEBEC

NORTH SEA

GLASGOW
Dublin
London
Hamburg
Amsterdam
Paris
Berlin

CASPIAN SEA

IRAN

MINNEAPOLIS
KANSAS CITY
CHICAGO
DETROIT
CLEVELAND

NEWFOUNDLAND
BANK

NORTH EASTERN
ATLANTIC BASIN

Bay of Biscay

BLACK SEA

MEXICO

Bay of Fundy
New York
Boston
Washington

MADRID
Lisbon
PORTUGAL
BARCELONA

MEDITERRANEAN SEA

SAUDI

CIUDAD DE MEXICO

Houston

BERMUDA
NORTHWESTERN

AZORES IS.

MADEIRA

CASABLANCA
MOROCCO

EGYPT

LIBYA

Florida
CUBA
LA HABANA

ATLANTIC BASIN

CANARIAS

SAHARA

ALGERIA

GUATEMALA
San Salvador
GREATER
ANTILLES

PUERTO RICO TRENCH

SAHARA

NIGER

CHAD

SUDAN

Khartoum

CARIBBEAN SEA

MAURITANIA

MALI

Dakar
SENEGAL

NIGERIA

Kano

VENEZUELA
CARACAS
COLOMBIA
BOGOTA

Georgetown
Paramaribo

CAPE VERDE

CAPE VERDE
BASIN

GUINEA
SIERRA LEONE
LIBERIA
IVORY COAST
Accra
GHANA

CAMEROUN

CONGO

ECUADOR

GALAPAGOS
CARNEGIE RIDGE

Guayaquil

GUINEA

GUINEA
BASIN

GABON
ZAIRE
KINSHASA

PERU
LIMA
BRAZIL

EQUATOR

ROMANCHE
GAP

Luanda

ANGOLA

Belém

RECIFE
Natal
Salvador

BRAZILIAN
BASIN

ASCENSION

ZAMBIA

BOLIVIA
LA PAZ
BELO HORIZONTE
Brasília

ST. HELENA

ZIMBABWE

NAMIBIA

BOTSWANA

PARAGUAY
Asunción
RIO DE JANEIRO
SÃO PAULO

ATLANTIC OCEAN

JOHANNESBURG

Pretoria

CHILE
ARGENTINA
Porto Alegre

WALVIS RIDGE

SOUTH AFRICA

Durban

Valparaíso
Rosario
SANTIAGO
BUENOS AIRES
MONTEVIDEO
URUGUAY

Cape of Good Hope

Port Elizabeth

Mar del Plata
Bahía Blanca

ARGENTINE
BASIN

CAPE BASIN

AGULHAS BASIN

CHILE
BASIN

GRANDE DE CHILOE

Comodoro Rivadavia

METEOR SEAMOUNT

AFRICAN
ANTARCTIC
RIDGE

MALVINAS
Strait of Magellan

SOUTH GEORGIA

SOUTH SANDWICH
TRENCH

SCOTIA SEA

SOUTH ORKNEYS

Drake Passage

PACIFIC ANTARCTIC BASIN

SOUTH SHETLAND

ANTARCTIC

WEDDELL SEA

ATLANTIC
ANTARCTIC
BASIN

MAUD
SEAM

BELLINGHAUSEN

ANTARCTIC OCEAN

Coats Land

Princess
Martha
Land

Queen Maud Land

ANTARCTICA

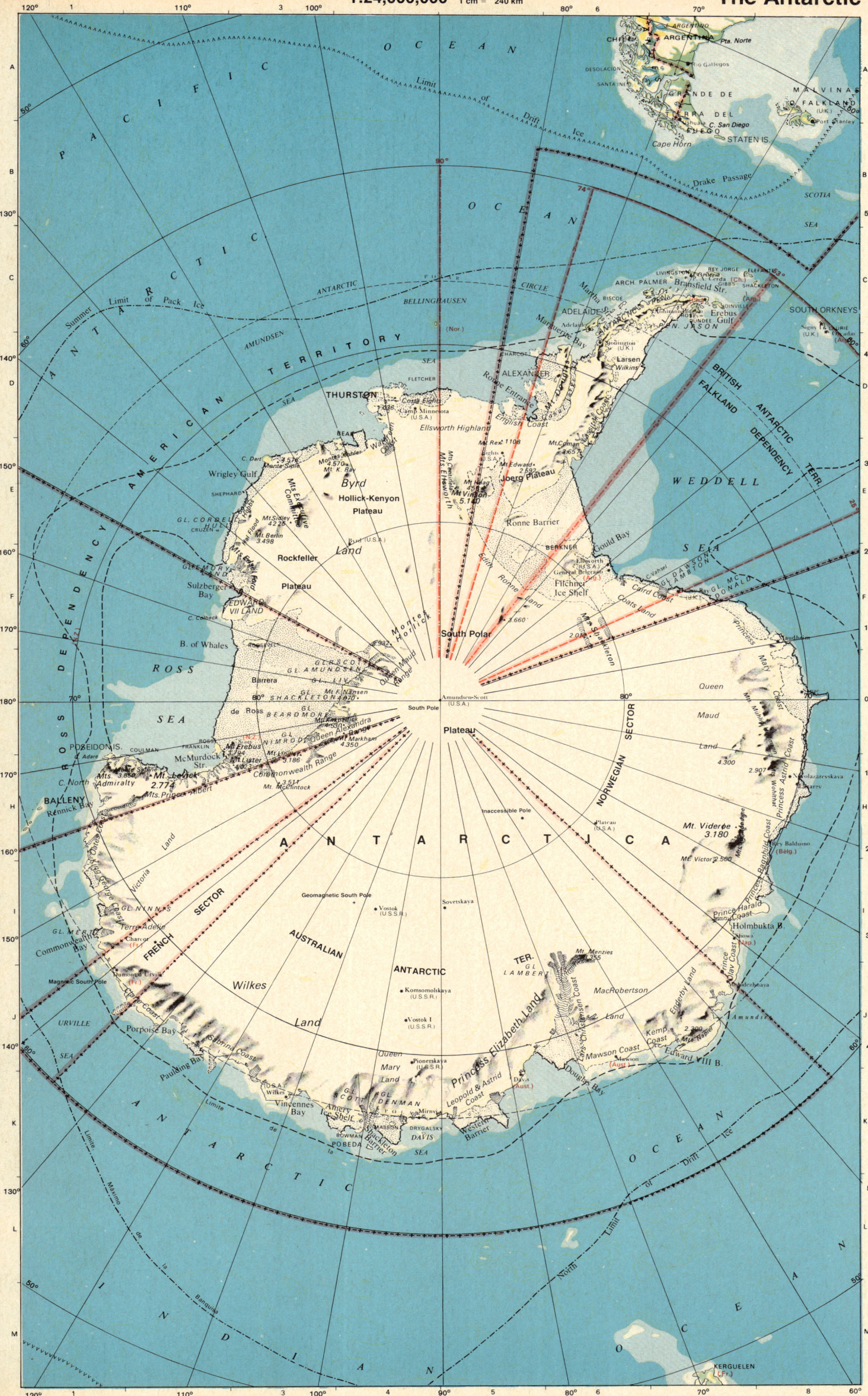

1:24,000,000 1" = 380 mi 1 cm = 240 km

PACIFIC OCEAN

Limit of Drift Ice

ANTARCTIC CIRCLE

BELLINGHAUSEN SEA

AMUNDSEN SEA

AMERICAN TERRITORY

Summer Limit of Pack Ice

THURSTON

Ellsworth Highland

Camp Eight (U.S.A.)
Camp Minnesota (U.S.A.)

Byrd
Hollick-Kenyon
Plateau

Byrd Land

Rockefeller
Plateau

Wrigley Gulf

Sulzberger Bay

EDWARD VII LAND

B. of Whales

ROSS SEA

Barriera

Ross

Amundsen-Scott (U.S.A.)
South Pole

South Polar Plateau

Geomagnetic South Pole

Vostok (U.S.S.R.)

Sovetskaya

Inaccessible Pole

ANTARCTICA

ANTARCTIC SECTOR

AUSTRALIAN SECTOR

FRENCH SECTOR

NORWEGIAN SECTOR

Magnetic South Pole

Wilkes Land

Komsomolskaya (U.S.S.R.)

Vostok I (U.S.S.R.)

Pionerskaya (U.S.S.R.)

Queen Mary Land

Princess Elizabeth Land

Leopold & Astrid Coast

Mawson (Aust.)

TER. GL LAMBERT

Mt Menzies 3,355

MacRobertson Land

Kemp Coast

Edward VIII B.

Enderby Land

Queen Maud Land

Mt Vider¢e 3,180
Mt. Victor 2,560
Plateau (U.S.A.)

Roi Balduino (Belg.)

Novolazarevskaya

Holmbukta B.

Prince Harald Coast
Prince Olav Coast

WEDDELL SEA

Ronne Barrier

Filchner Ice Shelf

Berkner

Coats Land

Gould Bay

General Belgrano (Arg.)

Ellsworth (U.S.A.)

Ronne Entrance

ALEXANDER

Joerg Plateau

Mt Vinson 5,140

Larsen

Wilkins

BRITISH FALKLAND DEPENDENCY TERR.

ADELAIDE I.

ARCH PALMER

Bransfield Str.
Erebus Gulf

SOUTH ORKNEYS

SCOTIA SEA

Drake Passage

Cape Horn

TIERRA DEL FUEGO

GRANDE DE

STATEN IS.

C. San Diego

ARGENTINA

CHILE

MALVINAS o FALKLAND (UK)

Port Stanley

KERGUELEN (Fr.)

INDIAN OCEAN

Balleny

Rennick Bay

Victoria Land

Terre Adelie

Commonwealth B.

Porpoise Bay

URVILLE

SEA

Paulding Bay

Vincennes Bay

DAVIS SEA

Mirnyy

McMurdock Str.

Mt Erebus
Mt Lister

Queen Alexandra Range
Commonwealth Range

ROSS SEA

POSEIDON IS.

C. Adare
C. North
Mts Prince Albert
Mts Admiralty

1:12,000,000   1" = 190 mi   1 cm = 120 km

ARCTIC

NORWEGIAN SEA

ICELAND

Reyjanes Ridge

Iceland Platform

ARCTIC CIRCLE

ATLANTIC OCEAN

FAERÖES (FAERDES)

SHETLAND

ORKNEY IS.

IRELAND (EIRE)

UNITED KINGDOM

GREAT BRITAIN

SCOTLAND

GLASGOW
Edinburgh
Newcastle-upon-Tyne
Teesside
LIVERPOOL
MANCHESTER
Sheffield
BIRMINGHAM
ENGLAND
Cardiff
Bristol
Southampton
Portsmouth
Plymouth
LONDRES

Dublin

NORTH SEA

NORWAY

STOCKHOLM
GOTEBURG
COPENHAGEN
DENMARK
Malmö

SWEDEN

BALTIC

NETHERLANDS
AMSTERDAM
The Hague
Rotterdam
BELGIUM
BRUSSELS

GERMANY
HAMBURG
HANNOVER
Dortmund
Duisburg
Düsseldorf
Frankfurt
WEST GERMANY
EAST GERMANY
Mannheim
Stuttgart
MÜNICH

POLAND
Szczecin
Poznań
Wrocław

PRAGUE
CZECHOSLOVAKIA
Bratislava

AUSTRIA
Salzbourg

HUNGARY
BUDAPEST

SWITZERLAND
Bern

FRANCE
PARIS
Rouen
Reims
Strasbourg
Nancy
BRETAGNE
Rennes
Nantes
NORMANDIE
Le Havre
Cherbourg
Brest
Orléans
Tours
La Rochelle
Bordeaux
Toulouse
Clermont-Ferrand
MASSIF CENTRAL
Lyon
Marseille
NICE
MONACO
Nîmes
Montpellier
Perpignan
Toulon
Golfe du Lion

ITALY
TURIN
MILANO
Genova
Bologna
Firenze
Livorno
VENEZIA
Verona
LA SPEZIA
ROMA
VATICAN CITY
NAPOLES
Ancona
Pescara
Bari
Foggia
Salerno
Taranto
Brindisi

SAN MARINO

CORSE (CORSICA)
SARDEGNA (SARDINIA)
Cagliari

ADRIATIC SEA

YUGOSLAVIA
Zagreb
Rijeka
Split

LIGURIAN SEA

TYRRHENIAN SEA

SICILIA
Palermo
Messina
Catania
Siracusa
Reggio di Calabria

IONIAN SEA

MALTA
Valletta

PORTUGAL
LISBOA
Oporto
Coimbra
Badajoz

SPAIN
MADRID
BARCELONA
Valencia
Sevilla
Córdoba
Málaga
Murcia
Alicante
Cartagena
Almería
Granada
Cádiz
Zaragoza
Bilbao
San Sebastián
Santander
Gijón
Oviedo
La Coruña
Vigo
Tarragona
Castellón de la Plana

Bay of Biscay

CORDILLERA CANTÁBRICA
SIERRA MORENA
SISTEMAS BÉTICOS

BALEARES
MALLORCA
MENORCA
IBIZA
Palma

MEDITERRANEAN SEA

MOROCCO
CASABLANCA
Rabat
Kenitra
Fès
Meknès
Marrakech
Safi
Tánger
Tetuán
Gibraltar

ALGERIA
Alger
Oran
Constantine
Annaba
Bizerta

TUNISIA
Tunis

PELAGE

KERKENNA

25° 30° 35° 40° 45° 50° 12 55° 60° 13 65° 14 70° 15 75°

OCEAN BARENTS SEA

North C.
Murmansk
KOLGUEV
C. Kanin Nos
PEN. RIBACI

FINLAND
KOL'SKIY
L. IMANDRA
BELOYE MORE

Helsinki
Turku
Tampere
Gulf of Bothnia
Gulf of Finland

Arkhangelsk
Onega
Severodvinsk
U R A L

Tallinn
Tartu
LENINGRAD
PEIPUS
LADOGA
Petrozavodsk

Riga
Daugavpils
Pskov

RUSSIAN

Kalinin
MOSCOW
Yaroslavl
Kostroma
Ivanovo
Vladimir
GORKI
Cheboksari
Kazan
Ioshkar-Ola
Izhevsk
Perm
Nizhni Taguil
SVERDLOVSK
Chelyabinsk
Magnitogorsk

SOVIET   FED.   SOC.

KLAIPEDA
LITHUANIA
Kaunas
Vilnius
MINSK
Grodno
Bialystok

WARSAW
Brest
Lublin

Smolensk
Moguilov
Briansk
Orel
Kaluga
Tula
Riazan
Tambov
Penza
Sizran
KUYBYSHEV
Uljanovsk
Saratov
Orenburg
Ufa
Sterlitamak
Aktyubinsk
Ural'sk

Kursk
Voronezh
Lipetsk
Borisoglebsk
Volgograd Res.
Saratov

ROMANIA
Oradea
Cluj-Napoca
Timgu-Mures
Bacau
Iasi
Kishinev
Arad
Timisoara
Sibiu
Brasov
Buzau
Galati
Braila
BUCAREST
Ploiesti
Constanta

Chernovtsi
Vinnitsa
Kirovograd
Krivoi Rog
Nikopol
Zaporozhie
DNEPROPETROVSK
DONETSK
Rostov-na-Donu
Zhdanov
Voroshilovgrad
Kremenchug
Poltava
JARKOV
KIEV
Cherkass
Zhitomir
Rovno
Lvov

Sumi
Belgorod
Chernigov

Volgodonsk
Tsimliansk Res.
Astrakhan
Gurjev
PEN. BUZACHI
Shevchenko

Kherson
Nikolaiev
ODESSA
CRIMEA
Simferopol
Sebastopol
Kerch
AZOV SEA
Novorossiisk
Krasnodar
Stavropol
Maikop
Armavir
Grozni
Majachkala
Ordzhonikidze
Nalchik

BULGARIA
SOFIYA
Plovdiv
Haskovo
Stara Zagora
Nis
Pristina
Skopje

Varna
Burgas

BLACK SEA

C. Ince
Sinop
Samsun
Trabzon
Zonguldak
ISTANBUL
Izmit
Adapazari
ANKARA
Eskisehir
Bursa

Sochi
Sujumi
Batumi
Kutaisi
TBILISI
Kirovabad
BAKU
CASPIAN SEA
Lenkoran

Armenia Plateau
EREVAN
L. Van
Tabriz
URMIA

GREECE
AEGEAN SEA
ATENAS
Salonica
Larisa
Volos
Patras
SPORADHES
RHODES

Izmir
Smyrna
Konya
TURKEY
ASIA MINOR
Kayseri
Malatya
Diyarbakir
Gaziantep
Adana
Mersin
Antalya
Antakya
Latakia

Mosul
KIRKUK
Hamadan
Kermanshah
IRAN
Iranian Plateau

CYPRUS
Nicosia
Tripoli
Beirut
LEBANON
Damascus
Homs
Aleppo
SYRIA
SYRIAN DESERT
Deir-ez-Zor
IRAQ
BAGDAD
An Najaf
Basra
Euphrates

35

1" = 47,5 mi
1 cm = 30 km

ATLANTIC OCEAN

PORTUGAL

La Coruña
El Ferrol del Caudillo
Gijón
Oviedo
Santander
CORDILLERA CANTÁBRICA
Lugo
Pontevedra
Vigo
Orense
Burgos
Ponferrada
Astorga
León
Valladolid
Porto (Oporto)
Vila Nova de Gaia
Douro
VIANA DO CASTELO
VILA REAL
BRAGANÇA
Salamanca
MADRID
VISEU
GUARDA
Ciudad Rodrigo
Segovia
Alcalá de Henares
COIMBRA
Tajo
Plasencia
Cáceres
Toledo
MONTES DE TOLEDO
LEIRIA
CASTELO BRANCO
S.ᵗᵃ DE GUADALUPE
LISBOA
SANTARÉM
PORTALEGRE
Mérida
Badajoz
EXTREMADURA
Sierra de Guadalupe
ÉVORA
SETÚBAL
BEJA
SIERRA MORENA
Guadalquivir
Córdoba
FARO
Huelva
Sevilla
S.ᵃ DE ARACENA
C. de S. Vicente
Sagres
Jerez de la Frontera
Cádiz
San Fernando
Algeciras
Gibraltar
La Línea
Málaga
Granada
Sierra Nevada
Ronda
Strait of Gibraltar
Ceuta
Tanger (Tánger)
Titt'aouen (Tetuán)

MOROCCO

Bay of Biscay

S. Sebastián

Pamplona

Logroño

Zaragoza

Toulouse

Montpellier

Marseille

La Ciotat

Aix-en-Provence

PYRENEES

Perpignan

Golfe du Lion
(Gulf of Lions)

Gerona

BARCELONA
Hospitalet

Tarrasa
Sabadell
Badalona

Tarragona

Castellón de la Plana
Villarreal de los Infantes

ISLAS COLUMBRETES

MENORCA

MALLORCA

Palma

Valencia

IBIZA
Ibiza

FORMENTERA

(BALEARIC ISLANDS)

Albacete

Alicante
Elche

Murcia

Lorca

Cartagena

Almería

El Djeza'ir
Alger

ALGERIA

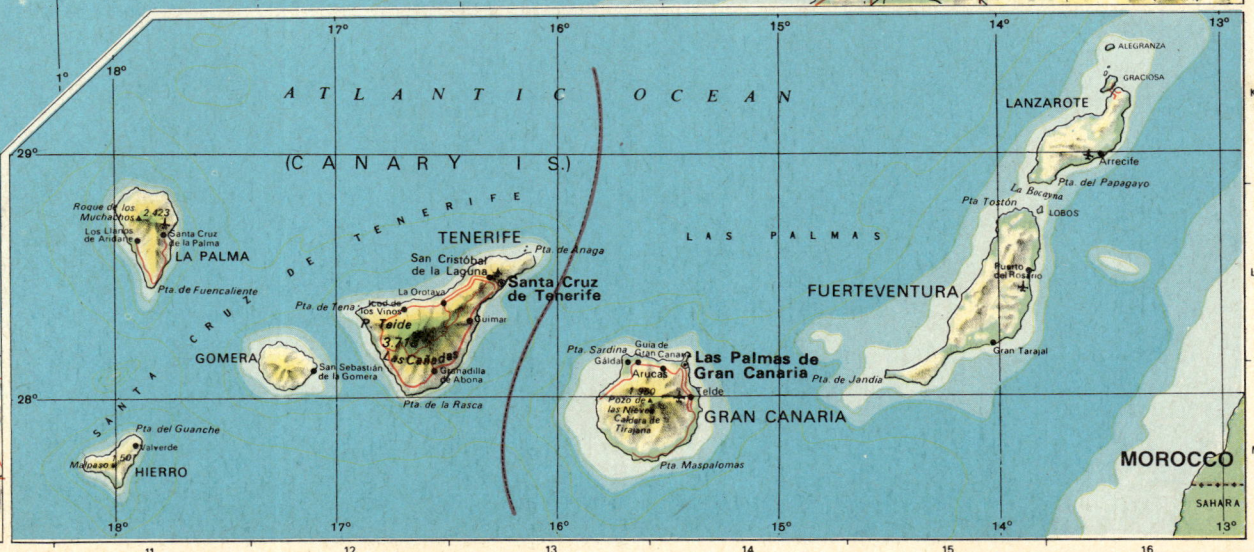

ATLANTIC OCEAN

(CANARY IS.)

LANZAROTE

Arrecife

TENERIFE

Santa Cruz
de Tenerife

FUERTEVENTURA

LA PALMA

GOMERA

Las Palmas de
Gran Canaria

GRAN CANARIA

HIERRO

ALGERIA

Oujda

MOROCCO

SAHARA

37

1 : 3,000,000 1" = 47,5 mi 1 cm = 30 km

BUDAPEST

HUNGARY

YUGOSLAVIA

SWITZERLAND

FRANCE

Salzburg

Graz

Zagreb

Ljubljana

Rijeka (Fiume)

Trieste

Venezia (Venice)

Padova

Verona

Milano (Milan)

Torino (Turin)

Genova

San Remo

Bologna

Ravenna

Rimini

Ancona

Pescara

Firenze (Florence)

Siena

Perugia

Livorno (Leghorn)

Piombino

Zürich

Luzern

Basel

Lausanne

CORSE (CORSICA)

Ajaccio

Canale di Corsica

Golfo di Genova

Riviera di Levante

Golfo di Venezia

APPENNINO

ALPI

DINARA

VATICANO

MEDITERRANEAN SEA
TYRRHENIAN SEA
IONIAN SEA

SARDEGNA / SARDINIA
SICILY
MALTA
TUNISIA
ALGERIA

Bari
Brindisi
Taranto
Napoli / Naples
Salerno
Cosenza
Reggio Calabria
Messina
Catania
Siracusa
Palermo
Trapani
Marsala
Cagliari
Sassari
Tunis
Bizerte
Annaba (Bône)

PUGLIA
BASILICATA
CALABRIA
CAMPANIA
APPENNINO LUCANO
IE LIPARI (IE EOLIE)
IE PONZIANE
IE EGADI
IE PELAGIE
PANTELLERIA

1 : 3,000,000

47,5 mi
30 km
1 cm

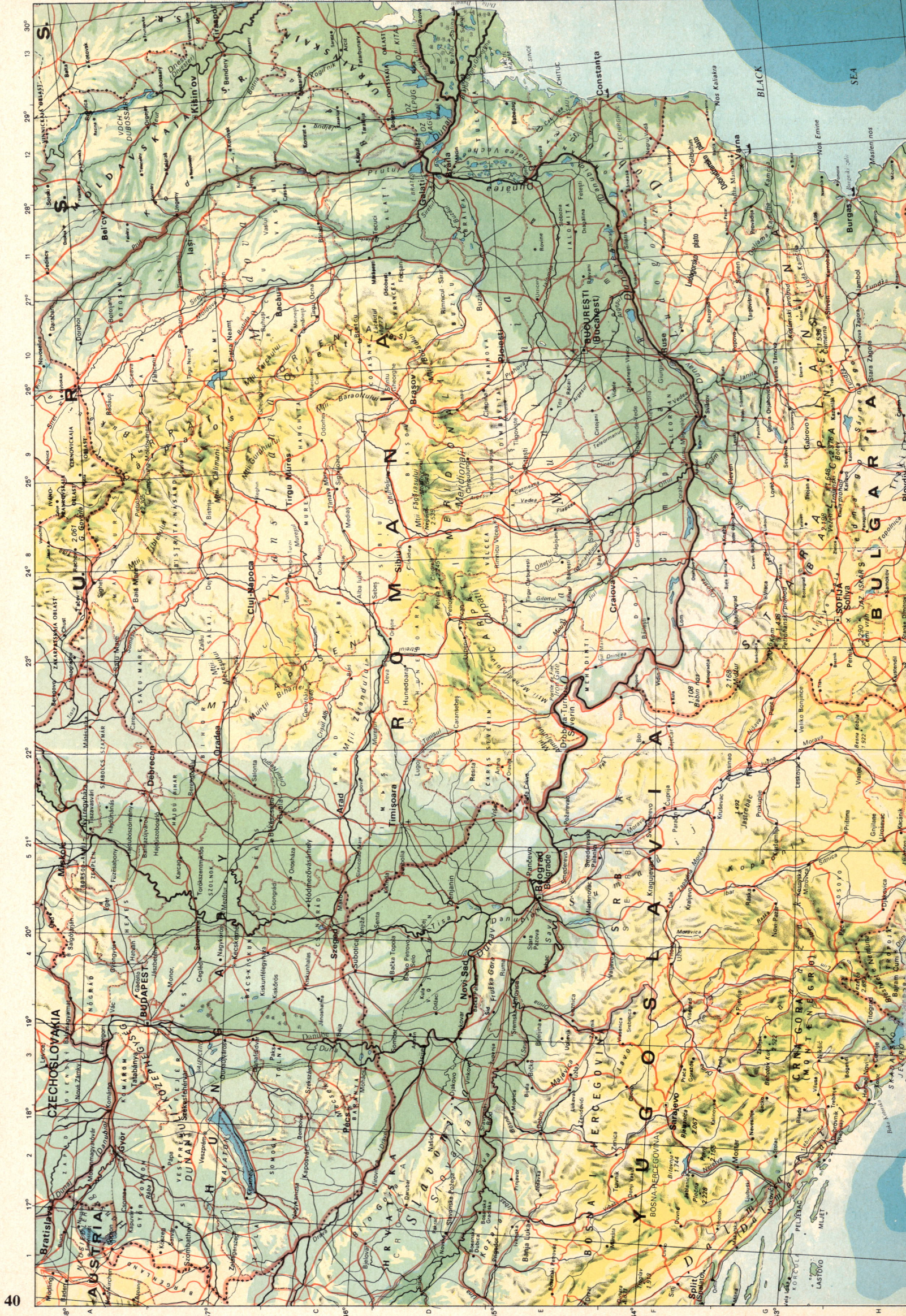

BLACK SEA

CZECHOSLOVAKIA

AUSTRIA

HUNGARY

ROMÎNIA

BULGARIA

YUGOSLAVIA

SRBIJA

BOSNA-HERCEGOVINA

HRVATSKA

CRNA GORA

BUCUREȘTI (Bochest)

SOFIJA · Sofiya

BEOGRAD · Belgrade

BUDAPEST

BRATISLAVA

Constanța

Varna

Burgas

Ploydiy

Sarajevo

Novi Sad

Cluj-Napoca

Tirgu Mureș

Brașov

Craiova

Arad

Timișoara

Oradea

Debrecen

Split

Mostar

Iași

Bacău

Kišinov

Belcy

Ruse

TURKEY

İSTANBUL

SEA OF MARMARA

MARMARA DENIZI

AEGEAN SEA

THRAKIKÓN PELAGOS

LESVOS (LESBOS)

KHIOS

SAMOS

IKARÍA

SPORADHES

DHODHEKÁNISOS
DODECANESE

RODHOS (RHODES)

KARPATHOS

KIKLÁDHES
CYCLADES

NÁXOS

ANDROS

EVVOIA (EUBOEA)

ATHÍNAI
Athens

Piraiévs
(Piraeus)

SKÍROS

VÓRIOI SPORÁDHES

THESSALÍA

Thessaloniki
Salonica

MAKEDONÍA

ALBANIA
Tiranë

GREECE

STEREÁ ELLÁS

PELOPÓNNISOS

IÓNIOI NÍSOI
IONIAN ISLANDS

KÉRKIRA (CORFU)

KEFALLINÍA (CEPHALONIA)

ZÁKINTHOS

LÉVKAS (S. MÁURA)

ADRIATIC SEA

ITALY

Bari

Brindisi

Taranto

Golfo di Taranto

Canale d'Otranto

IONIAN SEA
IÓNION PÉLAGOS

M E D I T E R R A N E A N   S E A

KRITIKÓN PÉLAGOS
SEA OF CRETE

KRITI

1 : 3,000,000

1" = 47.5 mi
1 cm = 30 km

NOORDZEE
(NORDSEE)
NORTH SEA

DENMARK

NETHERLANDS

AMSTERDAM
's Gravenhage
Haarlem
Rotterdam
Dordrecht

HAMBURG
Bremen
Bremerhaven
Oldenburg
Hannover
Braunschweig
Magdeburg
EAST GERMANY
Potsdam
BERLIN
Schwerin
Rostock
Lübeck
Kiel

WEST GERMANY

BRUXELLES
Brussels
Antwerpen
Maastricht
Aachen
Köln
Cologne
Bonn
Düsseldorf
Dortmund
Essen
Wuppertal
Münster
Bielefeld
Kassel
Erfurt
Weimar
Halle
Leipzig
Karl-Marx-Stadt (Chemnitz)
Gera
Zwickau
Plauen

Koblenz
Wiesbaden
Mainz
Frankfurt am Main
Offenbach
Darmstadt
Worms
Mannheim
Ludwigshafen
Heidelberg
Würzburg
Bamberg
Nürnberg (Nuremberg)
Regensburg

Luxembourg
Trier
Saarbrücken
Metz
Nancy
Karlsruhe
Stuttgart
Strasbourg
Freiburg
Ulm
Augsburg
München (Munich)
Ingolstadt
Landshut
Salzburg

Reims
Châlons-sur-Marne
Verdun
FRANCE
Dijon
Besançon
Mulhouse
Belfort
Basel
Zürich
Sankt Gallen
BODENSEE
Bregenz
Innsbruck
TIROL

SWITZERLAND
Bern
Lausanne
Genève Geneva
Lyon
Saint-Étienne

Bolzano
Trento
ALPI

42

BORNHOLM

O S T S E E
(MORZE BAŁTYCKIE)
BALTIC SEA

RÜGEN

Zatoka
Pommersche Bucht
USEDOM

Kaliningrad
(Königsberg)

R.S.F.S.R. U.S.S.R. RUSSIA

Gvardejsk

L I T V A
LITHUANIA

Gdynia
Gdańsk

Grodno

Szczecin

Olsztyn

Białystok

Bydgoszcz

Toruń

WARSZAWA
Warsaw

B E L .
S.S.R.
BYELORUSSIA

Brest

Poznań

P O L A N D

Łódź

Lublin

CHEŁM

Wrocław

Radom

Kielce

TARNOBRZEG

ZAMOŚĆ

Dresden

Częstochowa

GÓRY ŚWIETOKRZYSKIE

Ruda Śląska
Gliwice Zabrze
Katowice
Chorzów Bytom
Sosnowiec

Kraków
Cracov

TARNÓW

RZESZÓW

PRZEMYŚL
Przemyśl

PRAHA
Prague

Ostrava

Bielsko-Biała

KROSNO

C Z E C H O S L O V A K I A

B E S K I D

K A R P A T Y

Gerlachovský
2.654

U K R .
S.S.R.
UKRAINE

Brno

SLOVAKIA

Košice

Užgorod
Mukačevo

WIEN
Vienna

Bratislava

Miskolc

Debrecen

Győr

BUDAPEST

A U S T R I A

H U N G A R Y

Graz

Balaton

Oradea

Szeged

Arad

R O M A N I A

Timişoara

Ljubljana

Maribor

HRVATSKA
CROATIA

Y U G O S L A V I A

Zagreb

Pécs

43

France

1 : 3,000,000   1" = 47.5 mi / 1 cm = 30 km

UNITED KINGDOM

Southampton · Portsmouth · Brighton · Worthing · Hastings · Eastbourne · Lewes

Plymouth · Torquay · Paignton · Exeter · Bournemouth · Weymouth · Portland

Penzance · Falmouth · Camborne · Land's End · Isles of Scilly · Lizard Pt.

CHANNEL ISLANDS (ILES NORMANDES) · ALDERNEY · GUERNSEY · St Peter Port · JERSEY

Le Havre · Rouen · Fécamp · Dieppe · SEINE-MARITIME · Caen · Bayeux · PRESQU'ÎLE DU COTENTIN · Cherbourg

Brest · Quimper · Lorient · Vannes · Saint Brieuc · Saint-Malo · Rennes · FINISTÈRE · CÔTES-DU-NORD · MORBIHAN · ILLE-ET-VILAINE

Le Mans · Angers · Tours · Vendôme

Saint Nazaire · Nantes · LOIRE-ATLANTIQUE · VENDÉE · Yeu · Les Sables d'Olonne · La Roche-sur-Yon

La Rochelle · Rochefort · OLÉRON · RÉ · Niort · Poitiers · Châtellerault · DEUX-SÈVRES · VIENNE

Limoges · HAUTE-VIENNE · Angoulême · CHARENTE · Cognac · Royan · Pte. de Grave

Bordeaux · Arcachon · Bassin d'Arcachon · GIRONDE · Libourne · Bergerac · DORDOGNE · LANDES · Mimizan

ATLANTIC OCEAN

Golfe de Gascogne · Bay of Biscay · Gouf de Capbreton

Bayonne · Biarritz · St Sébastien · Pau · Tarbes · Lourdes · Toulouse · PYRÉNÉES-ATLANTIQUES · HAUTES-PYRÉNÉES · HAUTE-GARONNE

Santander · Bilbao · Gijón · Oviedo · León · Burgos · Valladolid · Zaragoza · Logroño · Pamplona · Vitoria

CORDILLERA CANTABRICA · Costa Verde · C. de Peñas · C. Vidio

PYRÉNÉES · PIRINEOS

44

BELGIUM

GERMANY

EAST GERMANY

BRUXELLES

Aachen · Bonn · Bad-Godesberg

Koblenz

LUXEMBOURG · Luxembourg

Wiesbaden · Frankfurt am Main · Offenbach

Mainz · Darmstad

GERMANY · Würzburg

Saarbrücken · Ludwigshafen am Rhein · Mannheim · Heidelberg

Nürnberg

Metz · Nancy

Karlsruhe

BAYERN

Strasbourg · Stuttgart

Augsburg

Reims

Colmar · WÜRTTEMBERG · SCHWAB

Mulhouse · Freiburg

Belfort

Basel · Zürich

Dijon · Besançon

AUSTRIA

SUISSE · Bern

SWITZERLAND

Lausanne

Genève · Annemasse · ALPES BERNOISES · ALPI · TRENTINO-ALTO ADIGE

Lyon · Villeurbanne · Annecy · Chambéry · ALPI BERGAMASCHE · Bergamo · Brescia · Verona

Saint-Étienne · Grenoble · MILANO · Milano

Clermont-Ferrand · TORINO · Alessandria · Genova · La Spezia

Valence · ALPES DAUPHINÉ · Cuneo · Savona · Livorno

MASSIF CENTRAL · ALPES MARITIMES · Nice · LIGURIAN SEA

Montpellier · Marseille · Toulon · Côte d'Azur · TOSCANA

Narbonne · Golfe du Lion · Gulf of Lions · ÎLES D'HYÈRES · ELBA

Perpignan

CORSE · CORSICA · HAUTE-CORSE

MEDITERRANEAN SEA

CORSE-DU-SUD · Bonifacio

1 : 3,000,000   1″ = 47.5 mi   1 cm = 30 km

NORWAY

SOTRA

SOUND

NORTH SEA

VIKING BANK

BERGEN OR OLD

VIKING BANK

GREAT FISHER BANK

LONG FORTIES

Herma Ness

UNST

FETLAR

SHETLAND ISLAND AREA
ZETLAND ISLANDS
(SHETLAND)
MAINLAND

Sumburgh Hd

FOULA

Fair Isle

Orkney

N. RONALDSAY

SANDAY

STRONSAY

WESTRAY

ROUSAY

MAINLAND
ORKNEY ISLAND AREA
ORKNEY ISLANDS
(ORCADAS)
Stromness

S. RONALDSAY

Duncansby Hd

Dunnet Hd

Wick

SULE SKERRY

STACK SKERRY

NORTH RONA

S-A SGEIR

Cape Wrath

UNITED KINGDOM

OF

GREAT BRITAIN

Kinnairds Hd
Fraserburgh
Peterhead

GRAMPIAN

Aberdeen
Stonehaven

Montrose
Arbroath

TAYSIDE
Dundee

FIFE

Fife Ness

St. Abb's Hd

Edinburgh
LOTHIAN

HIGHLAND

SCOTLAND

CENTRAL

STRATHCLYDE

SPEY HIGHLANDS

Moray Firth

Butt of Lewis
Stornoway

LEWIS

WESTERN
ISLES

SKYE

North Minch

Little Minch

NORTH UIST

BENBECULA

SOUTH UIST

FLANNAN IS.

St. KILDA

OUTER HEBRIDES

BARRA

RUM

EIGG

COLL

TIREE

MULL

IONA

COLONSAY

JURA

FAERØERNE
(FAEROES)

STRØMØ
SYDERØ

FAERØE BANK

N

C

O

R

W

E

G

BILL BAYLEY'S BANK

ROSEMARY BANK

OUTER BAILEY
OR LOUSY BANK

NORTHERN IRELAND

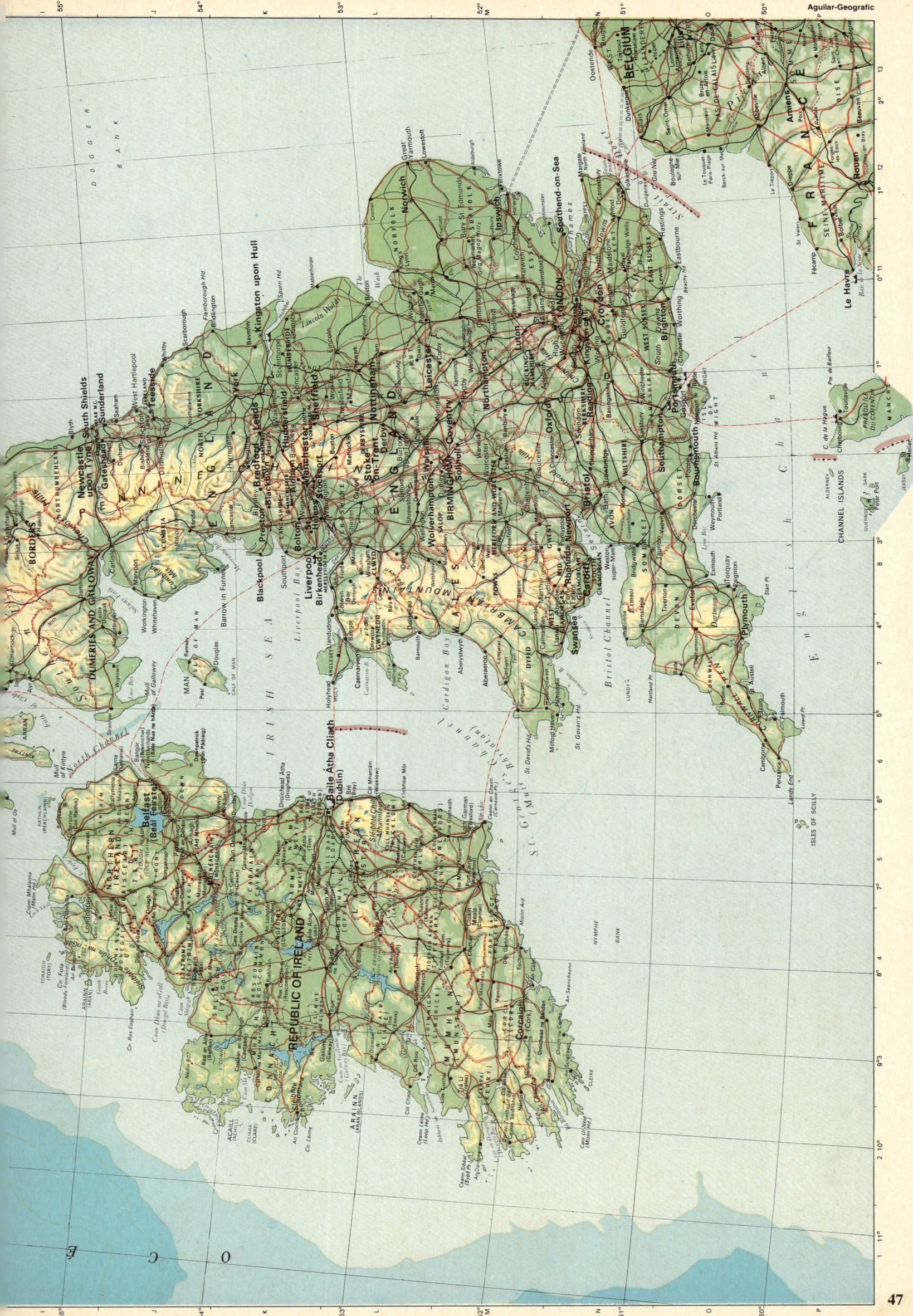

BELGIUM

FRANCE

ENGLAND

REPUBLIC OF IRELAND

Kingston upon Hull
Norwich
Great Yarmouth
Lowestoft
Ipswich
Bury St Edmunds
Southend-on-Sea
LONDON
Croydon
Brighton
Eastbourne
Worthing
Portsmouth
Southampton
Bournemouth
ISLE OF WIGHT

Le Havre
Rouen
Amiens
Boulogne-sur-Mer

Newcastle upon Tyne
South Shields
Sunderland
Gateshead
Middlesbrough
West Hartlepool
Scarborough
York
Leeds
Bradford
Huddersfield
Manchester
Liverpool
Birkenhead
Blackpool
Southport
Preston
Blackburn
Bolton
Stoke
Nottingham
Leicester
Coventry
Birmingham
Wolverhampton
Solihull
Hereford and Worcester
Northampton
Oxford
Bristol
Swansea
Cardiff
Exeter
Plymouth
Torquay
Penzance
Lands End
ISLES OF SCILLY
CHANNEL ISLANDS

CAMBRIAN MOUNTAINS

Belfast (Béal Feirste)
NORTHERN IRELAND
Baile Átha Cliath (Dublin)
Corcaigh (Cork)

IRISH SEA
ISLE OF MAN
Douglas
Peel
Ramsey

Liverpool Bay
Cardigan Bay
Bristol Channel
St George's Channel
English Channel
DOGGER BANK
NORTH SEA

47

1:1,000,000 1" = 158 mi 1 cm = 100 km

ATLANTIC OCEAN

COLONSAY
ORONSAY
IONA
SKERRYVORE

ISLAY
Portnahaven
Rhinns Pt.
Laggan B.
Mull of Oa
Port Ellen
Ardbeg
Bridgend
Port Charlotte
Machir B.

JURA
GIGHA
CARA
Campbeltown
Machrihanish
Mull of Kintyre
Southend
SANDA

ARRAN
Goat Fell 874
Brodick
Lamlash
Blackwater Foot
HOLY

KNAPDALE
Tarbert
Lochgilphead
COWALL
Dunoon
Gourock
Greenock
Port Glasgow
Helensburgh

BUTE
Rothesay
GREAT CUMBRAE
LITTLE CUMBRAE
Largs
Fairlie
Kilwinning
Irvine
Ardrossan
Saltcoats
Stevenston

Ailsa Craig
Girvan
Ballantrae

North Channel

RATHLIN (REACHLAINN)
Benbane Head
Bushmills
Portrush
Portstewart
Portballintrae
Ballycastle Bay
Fair Head
Runabay Hd.
Cushendun
Cushendall

INIS EOGHAIN
Malin Hd.
Cionn Mhálanna (Malin Hd.)
Carndonagh
Moville
Buncrana
Lough Foyle
Coleraine (Cúil Raithin)
Limavady
Ballymoney
Ballymena (An Baile Meánach)
Ballyclare
Larne (Latharna)

DONEGAL (DÚN NA nGALL)
Letterkenny (Leitir Ceanainn)
Lifford
Strabane
Londonderry (Doire)
Dungannon
Sperrin Mts.
Draperstown
Magherafelt
Cookstown
Antrim (Aontroim)
Randalstown
Carrickfergus (Carraig Fhearghais)
Whiteabbey
Newtownabbey

LOUGH NEAGH (LOCH NEATHACH)

TYRONE (TÍR EOGHAIN)
Omagh (An Ómaigh)
Dromore
Coalisland
Lurgan (An Lorgain)
Portadown (Port an Dúnáin)
BELFAST (Béal Feirste)
Lisburn
Holywood
Bangor (Beannchar)
Newtownards (Baile Nua na hÁrda)
Comber
Donaghadee
Groomsport

FERMANAGH (FEAR MANACH)
Enniskillen (Inis Ceithleann)
UPPER LOUGH ERNE
LOWER LOUGH ERNE
Armagh (Ard Mhacha)
ARMAGH
DOWN
Banbridge
Dromore
Hillsborough
Crossgar
Downpatrick (Dún Pádraig)
Killyleagh
Strangford

NORTHERN IRELAND

MONAGHAN (MUINEACHÁN)
Clones (Cluain Eois)
Castleblayney
Newry (An tÚr)
Warrenpoint
Rostrevor
Kilkeel (Cill Chaoil)
Newcastle
Mourne Mts.
Dundrum B.

CAVAN (AN CABHÁN)
Cootehill
Carrickmacross (Carraig Mhachaire Rois)
Dundalk (Dún Dealgan)
Dundalk Bay (Cuan Dhún Dealgan)

LEITRIM
Carrick-on-Shannon (Carraig Dhroma Ruisc)

LONGFORD (AN LONGFORT)

CAVAN
Kells (Ceanannas)
Navan (An Uaimh)
Drogheda (Droichead Átha)

ROSCOMMON
WESTMEATH (AN IARMHÍ)
Mullingar (An Muileann gCearr)
Trim (Baile Átha Troim)
Athboy
MEATH (AN MHÍ)
Balbriggan (Baile Brigín)
Skerries (Na Sceirí)

GALWAY
OFFALY (UÍBH FHAILÍ)
Tullamore
KILDARE (CILL DARA)
Naas
Maynooth
Leixlip
Lucan
BAILE ÁTHA CLIATH (Dublin)
Dún Laoghaire
Malahide
Howth
Dublin Bay (Cuan Baile Átha Cliath)

Holyhead

IRISH SEA

ISLE OF MAN
Peel
Port Erin
Calf of Man
Chicken Rock
Castletown

Luce Bay
Mull of Galloway
Portpatrick
Stranraer
Port Logan

NORTH SEA

FIFE
St. Andrews Bay
St. Andrews

Edinburgh
GLASGOW
Paisley
LOTHIAN
Musselburgh
Dalkeith

Firth of Forth

Berwick Upon Tweed
Spittal

BORDERS
SCOTLAND
ENGLAND

Galashiels
Melrose
Hawick
Jedburgh

CHEVIOT HILLS

Moorfoot Hills
Lammermuir Hills

NORTHUMBERLAND

Kielder
Forest

DUMFRIES AND GALLOWAY

Dumfries
Lockerbie

Carlisle

Newcastle upon Tyne
Gateshead
South Shields
Tynemouth
Sunderland
Washington
Durham
Hartlepool
West Hartlepool

Wigtown
Bay

Solway
Firth

Maryport
Workington
Whitehaven

Lake District
CUMBRIAN
MOUNTAINS
CUMBRIA

Bishop Auckland
Stockton on Tees
Teesside
Thornaby on Tees
Darlington

MAN
Douglas

Barrow
in Furness
Morecambe
Heysham
Lancaster

NORTH YORKSHIRE

Richmond

Ripon
Harrogate
Knaresborough
York

Morecambe
Bay

Liverpool Bay

Fleetwood

Blackpool
Lytham
St. Annes
Southport

Preston
Blackburn
Burnley

Bradford
Leeds
WEST YORKSHIRE M.C.
Halifax
Huddersfield
Wakefield
Pontefract

Formby
Crosby
Bootle
Wallasey
Birkenhead
Liverpool
St. Helens

GREATER MANCHESTER M.C.
Bolton
Oldham
Manchester
Stockport
Rochdale

SOUTH YORKSHIRE
Barnsley
Doncaster
Rotherham
Sheffield

ANGLESEY
GWYNEDD
Llandudno
Colwyn Bay
Rhyl

CHESHIRE
DERBYSHIRE
Chesterfield

I R I S H   S E A

49

1: 1,000,000   1 cm = 16 mi / 10 km

NORTH SEA

The Wash

The Marsh

LINCOLNSHIRE

NOTTINGHAMSHIRE

LEICESTERSHIRE
Leicester

NORFOLK
Norwich
Great Yarmouth
Lowestoft

The Broads

Breckland

NORTHAMPTONSHIRE
Northampton

CAMBRIDGESHIRE
Cambridge
Peterborough

The Fens

SUFFOLK
Ipswich
Bury St. Edmunds
Newmarket

Nottingham
Grantham
Boston
King's Lynn
Wisbech
March
Ely

BEDFORDSHIRE
Bedford
Luton

BUCKINGHAMSHIRE

HERTFORDSHIRE
St. Albans
Watford

Oxford
OXFORDSHIRE

ESSEX
Chelmsford
Colchester
Clacton on Sea
Southend on Sea
Harwich
Felixstowe
Brentwood
Basildon

LONDON
Harrow Hendon Enfield Romford Ilford Dagenham
Wembley Willesden Barking
Ealing Twickenham Turrock
Kingston Wimbledon Croydon
Sutton Bromley

BERKSHIRE
Reading
Newbury

Windsor
Slough
Maidenhead

SURREY
Guildford
Woking
Epsom
Leatherhead
Dorking
Reigate

KENT
Maidstone
Canterbury
Rochester
Chatham
Gillingham
Sittingbourne
Faversham
Ashford
Folkestone
Dover
Deal
Ramsgate
Margate
Broadstairs
Herne Bay
Whitstable
THANET
SHEPPEY
Gravesend
Dartford
Sevenoaks
Tonbridge
Royal Tunbridge Wells
Cranbrook
Hythe
New Romney

Vale of Kent

HAMPSHIRE
Winchester
Southampton
Portsmouth
Gosport
Havant
Basingstoke
Andover

WEST SUSSEX
Chichester
Worthing
Bognor Regis
Littlehampton
Arundel
Horsham
Crawley
East Grinstead
Haywards Heath
Burgess Hill
Cuckfield

EAST SUSSEX
Lewes
Brighton
Hove
Eastbourne
Hastings
Bexhill-on-Sea
Battle
Uckfield
Rye

ISLE OF WIGHT
Newport
Cowes
Ryde
Sandown
Shanklin
Ventnor
St. Catherine's Point

Strait of Dover

FRANCE
Boulogne-sur-Mer
le-Portel
le-Touquet Paris-Plage
Berck
Neufchâtel-Hardelot
Baie de la Somme
Cayeux-sur-M
St. Valery-sur-Somme

Beachy Hd.

St. Mary's Bay
Dungeness

53°

52°

51°

51

1 : 3,000,000
1" = 47,5 mi
1 cm = 30 km

BARENTS SEA

ICELAND

ATLANTIC OCEAN

NORWEGIAN SEA

GREENLAND SEA

ARCTIC CIRCLE

Reykjavík

Murmansk

S.

OZ NUK
KARELSKAJA

SUOMI

FINLAND

Pielinen
NÄSIVESI
PÄIJÄNNE
SAIMAA

Helsinki
Espoo
Tallinn

E E S T I · S. S. V.

LENINGRAD
Sestroreck
KOTLIN
ZAPADNYJ
BER BOLSOJ
BER OZOVYJ
Mys Kolgompa
Mys Kurgal'ski

Gulf of Finland

NAISSAAR

B o t t e n v i k e n

Gulf of Bothnia

BALTIC SEA

AHVENANMAA

Ö v e r k e n

HEMSON
ALNON
BRAMON
RAPPALUOTO

STOCKHOLM
UPPSALA
Västerås
Falun

VÄNERN

V Ä S T E R N

NORRLAND
VÄSTERNORRLAND
DALARNE
VÄRMLAND
JÄMTLAND

S V E R I G E

TRÖNDELAG

53

FINLAND

Gulf of Bothnia

AHVENANMAA

Uppsala

B o t t e n v i k e n

NORRBOTTEN

VÄSTERNORRLAND

JÄMTLAND

GÄVLEBORG

KOPPARBERG

HEDMARK

NORDTRØNDELAG

SØRTRØNDELAG

Trondheim

NORSKEHAVET

NORWEGIAN SEA

MALTENBANKEN

SKLINNABANKEN

HITRA

SMØLA

MØRE OG ROMSDAL

OPLAND

BUSKERUD

SOGN OG FJORDANE

HORDALAND

Bergen

N O R W A Y

FÆRØERNE

STRØMØ   VÅGØ   SUDERØ   SKUØ   SYDERØ

STORE DIMON   LILLE DIMON   HESTØ

A T L A N T I C   O C E A N

STOCKHOLM

Västerås SÖDERMAN Norrköping Linköping

Örebro Karlstad Kristinehamn

VÄNERN VÄTTERN Jönköping Borås

GÖTEBORG BOHUS ORUST TJÖRN

OSLOFJORD Kristiansand LISTA STAVANGER Haugesund KARMØY

ANHOLT LÆSØ Skagen Frederikshavn

Ålborg NORDJYLLAND Århus Randers

DENMARK Odense FYN LANGELAND

KØBENHAVN / Copenhagen AMAGER SJÆLLAND MØN FALSTER LOLLAND

Helsingborg Malmö MALMÖHUS KALMAR BLEKINGE Karlskrona

ÖLAND GOTLAND

BORNHOLM

FREDERIKSHAVN

Rostock EAST GERMANY RÜGEN Stralsund

Lübeck HAMBURG WEST GERMANY Kiel HOLSTEIN SCHLESWIG

Bremerhaven Wilhelmshaven JADE OSTFRIESISCHE INSELN NORDFRIESISCHE INSELN SYLT FÖHR AMRUM HELGOLAND

NETHERLANDS AMELAND

NOORDZEE NORTH SEA VESTERHAVET

Ringkøbing Fjord Blåvands Huk Esbjerg RIBE RØMØ

SKAGERRAK KATTEGAT

LATVIA Ventspils Liepāja

LITHUANIA Klaipeda

U.S.S.R. Kaliningrad Königsberg

POLAND Gdynia Gdansk Gulf of Danzig Słupsk Koszalin Kołobrzeg

Gotland (Visby) Boda

HANÖ-BUKTEN Ystad Kristianstad

Göteborg–Copenhagen

55

ARCTIC OCEAN

NORWAY

SWEDEN

FINLAND

GERMANY

POLAND

CZECHOSL.

ROMANIA

TURKEY

IRAQ

IRAN

AFGHANISTAN

PAKISTAN

OMAN

INDIA

BLACK SEA

CASPIAN SEA

BARENTS SEA

KARSKOJE SEA

ZEMLJA FRANCA IOSIFA (FRANZ-JOSEF LAND)

NOVAJA ZEMLJA (NOVAYA ZEMLYA)

SVALBARD

ROSSIJSKAJA SOVETSKAJA SOCIALISTICESKAJA FEDERAT

RUSSIAN SOVIET FEDERAT

Leningrad

Moskva

Gorkiy

Kazan'

Kirov

Kuybyshev

Saratov

Volgograd

Rostov-na-Donu

Sverdlovsk

Cel'abinsk

Kurgan

Omsk

Tomsk

NOVOSIBIRSK

Kemerovo

Prokop'evsk

Novokuzneck

Barnaul

ALTAJ

Zapadno-Sibirskaja

KAZACHSKAJA S.S.R.

KAZAKHSKAYA S.S.R.

KAZAKHSTAN

Celinograd

Karaganda

Semipalatinsk

Ust'-Kamenogorsk

TURKMENSKAJA S.S.R.

TURKMENISTAN

KARAKUMY

KYZYLKUM

Aschabad

UZBEKSKAJA S.S.R.

(UZBEKISTAN)

TASHKENT

Samarkand

KIRGIZSKAJA S.S.R.

KIRGIZIJA

Frunze

Alma-Ata

TIAN SHAN

XINJIANGWEIWUERZIZH (SINKIANG)

TAKELAMAGANSHAMO

KUNLUN MTS.

XIZANGZIZHIQU (TIBET)

C H I N A

Tehran

Esfahan

Mashhad

DASHT-E KAVIR

DASHT-E LUT

Kabul

HINDU KUSH

PAMIR

Kashi (Kashgar)

Peshawar

Rawalpindi

Lahore

Amritsar

Multan

ARABIAN SEA

TROPIC OF CANCER

ARCTIC OCEAN

SEVERNAJA ZEML'A (NORTH LAND)

ARHIPELAG NORDENŠEL'DA

MORE NOVOSIBIRSKIJE OSTROVA NEW SIBERIAN ISLANDS

KOTEL'NYJ

VOSTOČNO-SIBIRSKOJE MORE EAST SIBERIAN SEA

WRANGELL I.

CHUCHI SEA

LAPTEVYCH MORE LAPTEV SEA

ARCTIC CIRCLE

BERING STR.

SEWARD PEN.

A L A S K A (U.S.A.)

BERING SEA

Gory Byrranga

POLUOSTROV TAJMYR

OZERO TAJMYR

Plato Putorana

Momskij Chrebet

Kolymskoje Nagorje

Korjakskoje Nagorje

VERCHOJANSKIJ CHREBET

Srednesibirskoje

Nizhnjaja Tunguska

Ploskogorje

SREDNESIBIRSKOJE

Aldan

Jakutsk

Lena

Aldanskoje Nagorje

SEA OF OKHOTSK

SAKHALIN

Patomskoje Nagorje

Severo Bajkalskoje Nagorje

RESPUBLIKA

STANOVOJ

POLUOSTROV KAMČATKA

KURIL'SKIJE OSTROVA

Angara

ROSSIJSKAJA SOVETSKAJA

Bratskoje VODOCHRANILIŠČE

VOSTOČNYJ SAJAN

Čeremchovo

Angarsk

Irkutsk

Ulan-Ude

Čita

Komsomol'sk na-Amure

SICHOTE ALIN'

HOKKAIDO

SAPPORO

CHÖVSGÖL NUUR

Moron

CHANGAJN NURUU

HANGAJN NURUU

DAXINGANLING Ta Hingan Ling

Xiaoxing'anling Xiao Hingan Ling

Qiqihar (Tsitsihar)

HAERBIN (Harbin)

Mudanjiang

Jilin

Vladivostok

SEA OF JAPAN

HONSHŪ

Niigata

Kanazawa

NAGOYA

TOKYO

M O N G O L I A

Ulaanbaatar (Ulan-Bator)

Baruun Urt

CHANGCHUN

SHENYANG

FUSHUN

NORTH KOREA

PYÖNGYANG

KYŌTO

ŌSAKA

KOBE

Hiroshima

FUKUOKA

KYŪSHŪ

Kagoshima

Nagasaki

Kumamoto

SHIKOKU

Sendai

NURUU

GOVIALTAJN NURUU

G O B I

NEI MENGGU Mongolia Interior

Yinshan

Zhangjiakou (Kalgan)

BOHAI (POHAI)

Dairen (Port Arthur)

Lushun

SOUTH KOREA

Seoul

Inchon

Taegu

Pusan

YELLOW SEA

CHEJU

SHAN

ALANSHANSHAMO

NINGXIA

Baotou (Paotow)

Huhehaote (Hoh-Hote)

BEIJING

Datong

TIANJIN

Tangshan

Yantai

Qingdao

C H I N A

Yinchuan

TAIYUAN

Shijiazhuang

Jinan

Xuzhou

Lianyungang

QINGHAI

Xining

LANZHOU Lanchow

Baoji

XIAN

Zhengzhou

Luoyang

Kaifeng

NANJING

SHANGHAI

Suzhou

Hangzhou

Ningbo

EAST CHINA SEA

Hofei

ZHALINGHU

ELINGHU

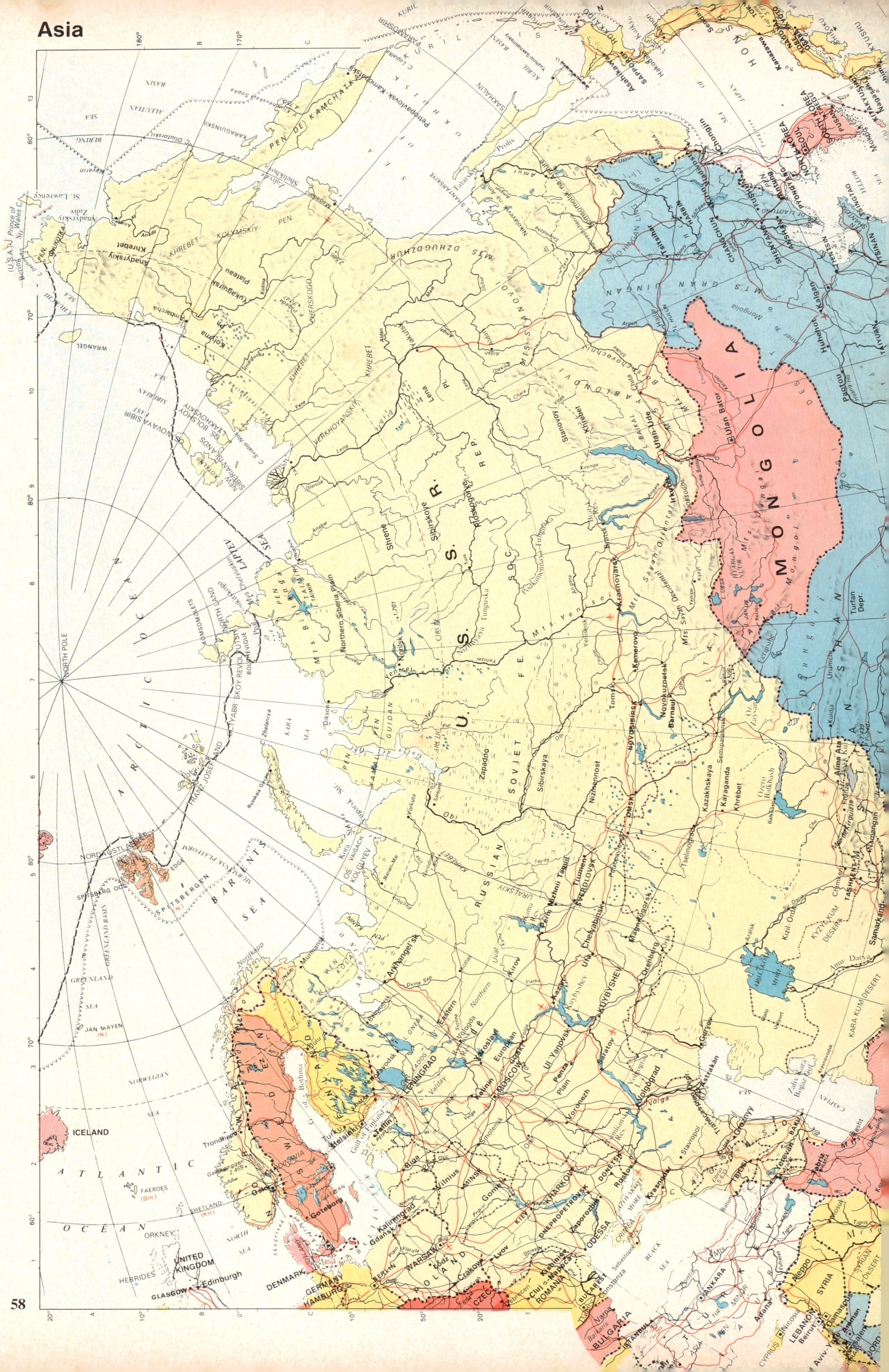

ARCTIC OCEAN

NORTH POLE

U.S.S.R.

RUSSIAN SOVIET F. E. SOC. REP.

MONGOLIA

Ulan Bator

BARENTS SEA

KARA SEA

LAPTEV SEA

NEW SIBERIAN ISLANDS

WRANGEL

BERING SEA

SEA OF OKHOTSK

PEN. DE KAMCHATKA

KURIL IS.

SAKHALIN

KHREBET KOLYMSKIY

KHREBET

MTS DZHUGDZHUR

Northern Siberia Plain

Central Siberian Plateau

West Siberian Plain

Lena

Ob

Yenisei

Amur

JAPAN

KOREA

SPITSBERGEN

FRANZ JOSEF LAND

NORDAUSTLANDET

EDGE

GREENLAND SEA

JAN MAYEN

NORWEGIAN SEA

ICELAND

ATLANTIC OCEAN

FAEROES

SHETLAND

ORKNEY

HEBRIDES

UNITED KINGDOM

GLASGOW Edinburgh

DENMARK

GERMANY

HAMBURG

BERLIN

POLAND

WARSAW

CZECH.

ROMANIA

BULGARIA

SYRIA

LEBANON

CYPRUS

NORWAY

SWEDEN

FINLAND

Oslo

Trondheim

Göteborg

Helsinki

LENINGRAD

MOSCOW

KIEV

ODESSA

KHARKOV

DNEPROPETROVSK

DONETSK

Rostov

Volgograd

Astrakhan

KUYBYSHEV

Saratov

SVERDLOVSK

CHELYABINSK

Omsk

NOVOSIBIRSK

Barnaul

Novokuznetsk

Tomsk

Kemerovo

Karaganda

TASHKENT

ALMA ATA

Samarkand

CASPIAN SEA

ARAL SEA

KARA KUM DESERT

KYZYL KUM

Ural

Volga

Don

BLACK SEA

Urumchi

Turfan Depr.

Gobi Desert

58

1:12,000,000 1" = 190 mi 1 cm = 120 km

KAZACHSKAJA S.S.R.

Bet-Pak-Dala

PESKI MUJUNKUM

Karaganda

Celinograd

Pavlodar

Barnaul

Semipalinsk

Ust Kamenogorsk

TASKENT

UZB

KIRGIZSKAJA S.S.R.

Alma-Ata

Frunze

TADZIKSKAJA S.S.R.

AFGHANISTAN

PAKISTAN

Kašgar

XINJIANGWEIWUERZIZHIQU (SINKIANG)

TAKELAMA GANSHAMO

Wulumuqi (Urumchi)

BEISHAN

MONGOL

CHOVSGOL NUUR

MON

BAJAN ULGIJ

GOV ALTAJ

CHONGOR

KUNLUN

QINGHAI (TSINGHAI)

BUERHANBUDASHAN

BAYANKALASHAN

HUANGHE

XIZANGZIZHIQU

HIMALAYA

Lhasa

NEPAL

Katmandu

BHUTAN

ARUNACHAL PRADESH

ASSAM

NAGALAND

MEGHALAYA
Shillong

Dispur

BANGLADESH

DACCA

Narayanganj

Chittagong

TRIPURA

MIZ

BURMA

Mandalay

THAILAND

INDIA

New Delhi
DILLI

RAJASTHAN

Jaypur

Agra

KANPUR

Lakhnau

MADHYA PRADESH

Nagpur

Raypur

ORISSA

MAHARASHTRA

HAIDARABAD

Vijayawada

Vishakhapatnam

Bay of Bengal

Calcutta

WEST BENGAL

Bhubaneswar

Cox's Bazar

Akyab

PAKISTAN

Rawalpindi

Faisalabad

Amritsar

Jalandhar

Ludhiana

Patiala

Saharanpur

Muzaffarnagar

Meerut

Bikaner

PAMIR

YUNNAN

OZ ZAJSAN

OZ BALKAŠ

ISSYK KUL

Bangal Ki Khadi

Bay of Bengal

105° 110° 115° 11 120° 12 125° 13 130° 14 135° 15

S.   S.   R.

BRATSKOJE
VDCHR
Ceremchovo
Usť Ordynsk
sk Irkutsk
Ulan-Ude
Petrovsk
Zabajkalskij
Cita

Komsomolsk-na-Amure
Sovetskaja Gavan
Tatarskij Proliv

MONGOLIA
Ulaambaatar
(Ulan-Bator)
Mandalgov
Chuld
Sajnsand
Dalandzadgad

Manzhouli
(MANCHOULI)
Hailar
(Hailar)

Qiqihaer
(Tsitsihar)
HAERBIN
(Harbin)
Mudanjiang
Vladivostok
Nachodka
Ussurijsk

HEILUNGKIANG
CHANGCHUNG
Jilin (Kirin)
Siping
Chongjin

Baotou
Huhehaote
Datong
Zhangjiakou
BEIJING
Tangshan
Peking
TIANJIN (Tientsin)

SHENYANG
FUSHUN
ANSHAN
Jinzhou
Antong

NORTH KOREA
PYONGYANG
Sinuiju
Hungnam
Wonsan

Yinchuan
TAIYUAN
Shijiazhuang
Baoding (Paoting)
Yangquan

JINAN (Tsinan)
Zibo
QINGDAO (Tsingtao)
Weifang

SOUTH KOREA
Seoul
Inchon
Suwon
TAEGU
PUSAN
Kwangju
Mokpo
CHEJU

SEA OF JAPAN
NIPPON-KAI

HONSHU (HONDO)
Hiroshima
FUKUOKA
Nagasaki
Kagoshima
Miyazaki

Sian
ZHENGZHOU
Luoyang
HENANSHENG
Shangqiu
Xuzhou
Huaiyin
Bengbu

Lianyungang

YELLOW SEA
HUANGHAI

Wuhan
Nanchang
HEFEI (Hofei)
NANJING (Nankin)
Suzhou
Wuxi
SHANGHAI
Hangzhou
Ningbo

DONGHAI
EAST CHINA SEA

Chongqing (Chungking)
Guiyang (Kueiyang)
Changsha
Xiangtan
Hengyang
Nanchang
Wenzhou

Guilin
Ganzhou
Fuzhou
TAIPEI
Chi-lung (Tailung)
Tai-chung
TAIWAN (FORMOSA)
T'ai-nan
Kao-hsiung
P'ing-tung

Guiyang
Liuzhou
Wuzhou
Shantou
Xiamen (Amoy)

NANHAI
Canton
Foshan
HONG KONG (U.K.)
Macau

Nanning
Zhanjiang
Haikou
HAINANDAO

VIETNAM
HANOI
Haiphong

SOUTH CHINA SEA

TAIPING YANG
PACIFIC OCEAN

LUZON
PHILIPPINES

NANSEI-SHOTO
OKINAWA-SHOTO
AMAMI-SHOTO

50°   45°   40°   35°   30°   25°   20°   15°

61

Japan

1 : 3,000,000   1" = 47,5 mi / 1 cm = 30 km

HOKKAIDŌ (YESO)

HOK-KAI (OKHOTSK SEA)

HOKKAIDŌ (YESO)

U. S. S. R.

KUNASHIR

SHIKOTAN

SEA OF JAPAN

NIHON-KAI

SEA OF JAPAN

PACIFIC OCEAN

JAPAN

SADO

SAPPORO

Asahikawa

Kushiro

Obihiro

Muroran

Hakodate

Otaru

HONSHŪ

OSHIMA-HANTŌ

SHIMOKITA-HANTŌ

TSUGARU-KAIKYŌ

Sōya-Kaikyō

RISHIRI

REBUN

OKUSHIRI

Hidaka Sanmyaku

ISHIKARI Sanchi

Hakodate

Muroran

Aomori

Hirosaki

Akita

Niigata

Sendai

Fukushima

Iwaki

Hachinohe

Morioka

Sendai Wan

Ishikari-Wan

Uchiura-Wan

Uchiura-Wan

Funka-Wan

ULLUNGDO

HONSHŪ
(HONDO)

IZU-SHOTŌ

TAI-HEI-YŌ

PACIFIC OCEAN

SHIKOKU

KYŪSHŪ

YOKOHAMA
KAWASAKI
Ashikaga
Maebashi
Takasaki
Kasugai
Kiryu
Hachiōji
Odawara
Fuji
Numazu
Shimizu
Shizuoka
Hamamatsu
NAGOYA
Okazaki
Toyohashi
Gifu
Ōgaki
Suzuka
Matsuzaka
Fukui
Takefu
Tsuruga
KYOTO
Toyonaka
KŌBE OSAKA
Sakai
Kishiwada
Akashi
Himeji
HYŌGO
Tottori
Okayama
Kurashiki
Fukuyama
Onomichi
Kure
Hiroshima
Iwakuni
Yamaguchi
Ube
Shimonoseki
KITAKYŪSHŪ
FUKUOKA
Kurume
Saga
Nagasaki
Sasebo
Ōmuta
Kumamoto
Kagoshima
Miyazaki
Nobeoka
Ōita
Beppu
Nakatsu
Takamatsu
Tokushima
Matsuyama
Imabari
Niihama
Kōchi

IZUMO
SHIMANE

OKI-GUNTŌ

AWAJI

Kii-Suidō

Kumano-nada

Suruga Wan

Sagami Wan

Ise Wan

Wakasa Wan

Tosa Wan

Bungo-Suidō

Kammon-Tunnel

ŌSUMI-SHOTŌ
TANEGA
YAKU

AMAKUSA SHOTŌ
KOSHIKI RETTŌ

MIKURA
MIYAKE

HACHIJŌ

63

1 : 9,000,000 1 cm / 142 mi / 90 km

CHINA

BURMA

RANGOON

THAILAND (SIAM)

Bangkok

Thon Buri

LAOS

Vientiane

VIETNAM

Hué

Da Nang

Qui Nhon

Nha Trang

Da Lat

CAMBOYA

Phnom Penh

HO CHI MINH (Saïgon)

Rach Gia

HAINANDAO (HAINAN)

XISHAQUNDAO PARACEL IS (China)

SOUTH CHINA SEA

ANDAMAN SEA

Bay of Bengal

ISTHMUS OF KRA

PHUKET

Alor Star

Kota Bahru

Kuala Trengganu

MALAYSIA

PENINSULAR MALAYSIA

(George Town) PENANG

Ipoh

Kuala Lumpur

Klang

Melaka (Malacca)

Johore Bahru

Singapore

BRUNEI

Bandar Seri Begawan (Brunei)

SARAWAK

Kuching

Banda Atjeh

Medan

Pematangsiantar

Pakanbaru

Padang

SUMATRA

Telanaipura (Djambi)

Palembang

KALIMANTAN (BORNEO)

Pontianak

Bandjarmasin

LAUT DJAWA / JAVA SEA

Tandjungkarang

Telukbetung

DJAKARTA

Bogor

BANDUNG

Tjirebon

Pekalongan

Semarang

Magelang

Surakarta (Solo)

Jogjakarta

SURABAJA

Madura

Malang

Kediri

LAUTAN INDIA / INDIAN OCEAN

DJAWA (JAVA)

COCOS IS. (KEELING IS.) (Aust.) NORTH KEELING / SOUTH KEELING

64

116° 9 10 120° 12 132° 13 14 136° 15

LAUTAN TEDUH

PACIFIC OCEAN

EQUATOR

Laoag
Vigan
Mt. Cleveland
LUZON
Cabanatuan
Tarlac
San Fernando
MANILA
Quezon City
Pasig
San Pablo
PHILIPPINES
Lucena
Batangas
MINDORO
MARINDUQUE
SIBUYAN SEA
MASBATE
PANAY
Iloilo
Cadiz
Bacolod
San Carlos
CEBU
Cebu
NEGROS
BOHOL
PALAWAN

SULU SEA

Calbayog
SAMAR
LEYTE
Butuan
Cagayan de Oro
Iligan
MINDANAO

Zamboanga
Basilan
SULU ARCHIPELAGO
Davao
General Santos

SEM DORERI
(VOGELKOP)
JAPEN
IRIAN JAYA
Pegunungan Van Rees
Jayapura
NEW GUINEA
NEW GUINEA
PAPUA

KEP. MAPIA
(ST. DAVID ISLANDS)

KEP. ARU
WOKAM
KOBROOR

NEW GUINEA

INDONESIA

LAUT ARAFURA

PACIFIC OCEAN

PALAU IS.

SONSOROL IS.

CAROLINE ISLANDS

PACIFIC IS. TER.
(U.S.A.)

LAUT SULAWESI

Samarinda
Balikpapan

Manado
SULAWESI UTARA
Gorontalo
HALMAHERA
MOROTAI
TERNATE
TIDORE

LAUT MALUKU

SULAWESI
SULAWESI TENGAH
SULAWESI TENGGARA
SULAWESI SELATAN

Parepare

KEP. SULA
TALIABU
MANGOLE
BURU
SERAM
AMBON

IRIAN
SEM DORERI
(VOGELKOP)
JAVA
MISOOL
NEW GUINEA
SEM BOMBERAI

Udjungpandang
(Makasar)
SALAYAR

LAUT BANDA

KEP. TUKANGBESI

KEP. ARU

KEP. TANIMBAR
JAMDENA

KEP. BARAT DAJA
WETAR

SUMBAWA
FLORES
KEP. SUNDA KETJIL
SUNDA IS.
TIMOR
SUMBA

LAUT FLORES

LAUT SAWU

LAUT TIMOR

LAUT ARAFURA

1" = 158 mi
1 cm = 100 km

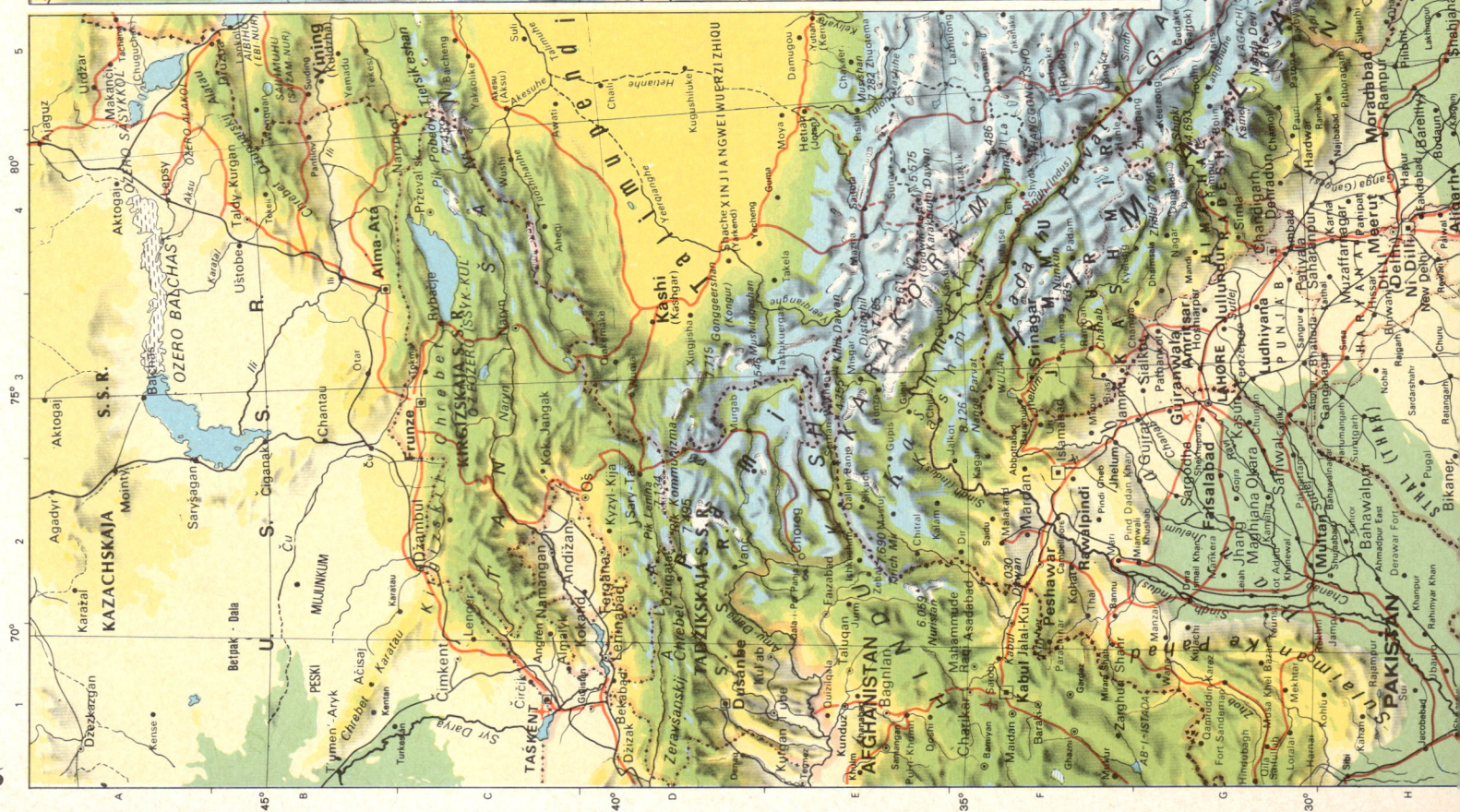

INDIA · BANGLADESH · BURMA · THAILAND · LAOS

SUMATRA
INDONESIA

NICOBAR DVIP (India)
BARA NICOBAR
CHHOTA NICOBAR
KAR NICOBAR
KAMORTA
KATCHALL

ANDAMAN SEA

UTAR ANDAMAN
DAKSHIN ANDAMAN
MADHYA ANDAMAN
ANDAMAN DVIP (India)
CHHOTA ANDAMAN
MYEIK KYUNZU

Bay of Bengal

BANGLADESH
DACCA
Khulna
Chittagong

Calcutta
Howrah
Jamshedpur
Bhubaneswar

Varanasi
KASHMIR
Lucknow
Jodhpur
Ajmer
Jaipur

I N D I A

Bhopal
Indore (Indore)
Nagpur
HAIDARABAD (Hyderabad)
Vishakhapatnam
Machilipatnam (Bandar)
Vijayawada
Guntur

Ahmedabad
Vadodara (Baroda)
Surat
Nasik
PUNE (Poona)
Bombay
Thane (Thana)
Sholapur
Kolhapur

MADRAS
Pondicherry
Nagapatnam

KARNATAKA
BENGALURU
Bangalore
Mangaluru (Mangalore)
Hubli
Dharwar
Belgaum

KERALA
Calicut
Kovampura
TAMIL NADU
Salem
Madura (Madurai)

Ernakulam
Mattancheri
Alleppey
Quilon
Tiruvanantapuram (Trivandrum)
Kanya Kumari Antarip (C. Comorin)

LAKSHADVIP
LACCADIVE IS.
AMIN DVIP
AMINDIVIS.
LAKSHADWEEP
KAVARATTI
ANDROTH

SRI LANKA
CEYLON
Jaffna
Tirukunamalaya (Trincomalee)
Kolamba (Colombo)
Negombo
Galle
Hambantota

ARABIAN SEA

BURMA
Mandalay
RANGOON
MOULMEIN
Pegu Yoma
Arakan Yoma
Gulf of Martaban

THAILAND
LAOS

AMAGANSHAMO
DE (TAKLA-MAKAN)
XINJIANGWEIWUERZIZHIQU

UTTAR PRADESH
DILLI (Delhi)
DILLI
Ni Dilli New Delhi

Kamarhati
Uttar Dum Dum
Dakshin Dum Dum
Bally
Baranagar
Howrah
Garden Reach
KALIKATA Calcutta
Dakshin Suburban
TOLLYGUNGE
SALT WATER LAKE

QINGHAISHENG

CHINA
XIZANGZIZHIQU (TIBET)

Lasa (Lhasa)

QINGHAI SHAN

HIMALAYA SHAN

NEPAL
Katmandu

SIKKIM
Thimbu
BHUTAN
ARUNACHAL PRADESH
Ziro

ASSAM
Gauhati
Shillong
MEGHALAYA
Imphal
MANIPUR
Lumding

Lakhnau (Lucknow)
KANPUR
Gorakhpur
Darbhanga
Muzaffarpur
Dinapore
Patna
Jamalpur Bhagalpur
Bihar
INDIA
Dahabad
Varanasi
Mirzapur
BIHAR
Gaya
Buddh Gaya
Madhupur
DACCA
BANGLADESH
Comilla
TRIPURA
MIZORAM
BURMA

Asansol
Ranchi
Durgapur
Burdwan
Jamshedpur
PASCHIM BANGAL
Chandannagar
Howrah
Baranagar
Khulna
Chittagong
Kharagpur

Raipur
Sambalpur
HIRAKUD RESERVOIR
ORISSA
Cuttack
Bhubaneswar

Mouths of the Ganges
Bangal Ki Khadi
Bay of Bengal

69

The Middle East

1:10,000,000 1" = 158 mi 1 cm = 100 km

Seas and Water Bodies
BLACK SEA
MEDITERRANEAN SEA
AEGEAN SEA
SEA OF MARMARA
CASPIAN SEA
RED SEA
PERSIAN GULF
DEAD SEA

Countries and Regions
GREECE
BULGARIA
TURKEY
CYPRUS
SIRIA (SYRIA)
LEBANON
ISRAEL
JORDAN
IRAQ
IRAN
SAUDI ARABIA
EGYPT
SUDAN
ETHIOPIA
YEMEN
SOUTH YEMEN
KUWAIT
BAHREIN
QATAR
UNITED ARAB EMIRATES
U.S.S.R.
R.S.F.S.R.
GRUZINSKAJA S.S.R.
AZERBAIDZANSKAJA S.S.R.

Selected Cities
ISTANBUL
ANKARA
Izmir (Smyrna)
Bursa
Adana
Konya
Kayseri
Erzurum
TBILISI
JEREVAN
BAKU
Krasnodar
Astrakhan
Machačkala
Tabriz
TEHERAN (Tehran)
Esfahan
Shiraz
Abadan
BAGHDAD
Al Basra
Al Mosul
Karkuk
Aleppo (Haleb)
Homs
Hama
Damascus (Esh Sham)
Beyrouth (Beirut)
Tarabulus (Tripoli)
Tel-Aviv-Yafo
Yerushalayim (Jerusalem)
Amman
EL QAHIRA (Cairo)
El Gizeh
EL ISKANDARIYA (Alexandria)
Bur Said (Port Said)
Suways (Suez)
Asyut
El Minya
Bur Sudan (Port Sudan)
Berber
Asmera (Asmara)
Al Kuwayt
BAHREIN
Al Manaman
Ad Dawha (Doha)
Abu Zabi (Abu Dhabi)
Dubayy
Al Madinah (Medina)
Makkah (Mecca)
Jidda (Jiddah)
Ar Riyad
Al Hufuf
Ad Damman
San'a
Al Hudayda
Al Mukalla
Aden

Physical Features
TOROSLAR
ZAGROS
CAUCASUS MTS. (KAVKAZ)
AN NAFUD
NEFUD
RUB' AL-KHALI
AD-DAHNA
BADIYAT ASH-SHAM
AN NAFUD
NUBIAN DESERT
EASTERN DESERT (ES SAHRA ESH SHARQIYA)
SUDAN
JABAL SHAMMAR
NEJD
HIJAZ
HADRAMAWT
ARABIAN PENINSULA
LAKE NASSER (BUHEIRET EN NASER)
LAKE TANA

70

S.S.R.

KYZYLKUM

TURANSKAJA NIZMENNOST

KARAKUMY

T.S.S.R.

CHINA

Kashi (Kaskgar)

Samarkand

Dušanbe

Mashhad

Khorasan

KAVIR

DASHT-E LUT

AFGHANISTAN

Kabul

Peshawar

Rawalpindi

Srinagar

JAMMU KASHMIR

HIMACHAL PRADESH

Kandahar

Quetta

PAKISTAN

Baluchistan

LAHORE

Faisalabad

Multan

Amritsar

Ludhiyana

Chandigarh

PUNJAB

HARYANA

DILLI DELHI

New Delhi

Meerut

Moradabad

Bareli

Agra

Gwaliyar (Gwalior)

RAJASTHAN

Jaipur

Jodhpur

Bikaner

Ajmer

Kota

MADHYA PRADESH

Bhopal

Indaor (Indore)

Hyderabad

KARACHI

Mouths of the Sindh

ARABIAN SEA

Gulf of Oman

Masqat (Mascate)

TROPIC OF CANCER

GUJARAT

AMDABAD (Ahmedabad)

Vadodara (Baroda)

Rajkot

KATHIYAVAR

Bhavnagar

Surat

MAHARASHTRA

Nasik

Aurangabad

Jalna

MUMBAI (Bombay)

PUNE Poona

Ulhasnagar

Thanen (Thana)

Ahmednagar

SHOLAPUR

HAIDARABAD (Hyderabad)

INDIA

71

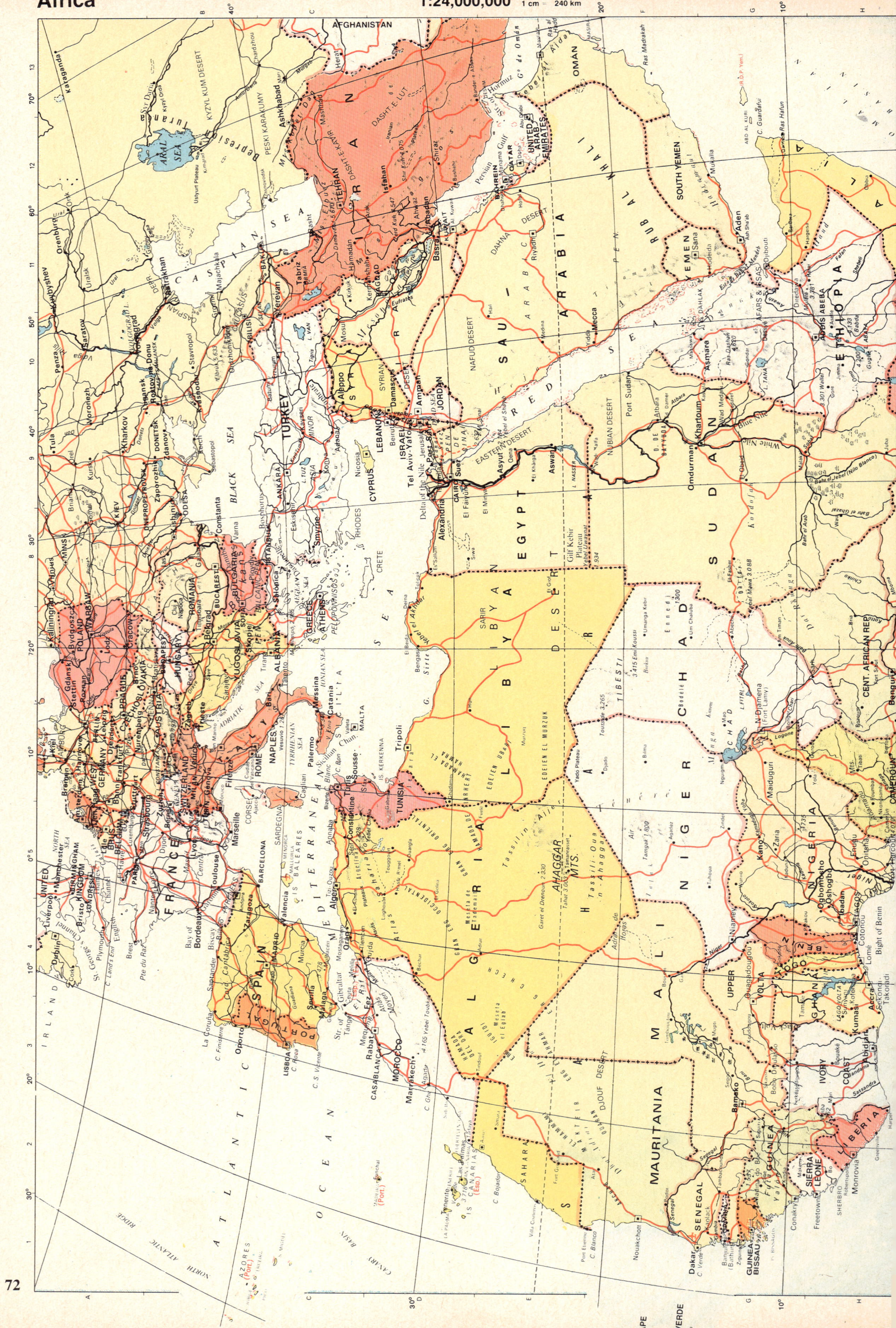

1:24,000,000

1" = 380 mi
1 cm = 240 km

AFGHANISTAN

I R A N

TEHRAN

OMAN

UNITED ARAB EMIRATES

SOUTH YEMEN

YEMEN

Aden

SAUDI ARABIA

RUB' AL KHALI

CASPIAN SEA

ARAL SEA

TURKEY

ANKARA

SYRIA

LEBANON

ISRAEL

JORDAN

CYPRUS

BLACK SEA

EGYPT

LIBYAN DESERT

LIBYA

SUDAN

ETHIOPIA

ADDIS ABEBA

KHARTOUM

RED SEA

NUBIAN DESERT

CHAD

N'Djamena

CENT. AFRICAN REP.

NIGER

MALI

MAURITANIA

AHAGGAR MTS.

TIBESTI

NIGERIA

ALGERIA

TUNISIA

TRIPOLI

MEDITERRANEAN SEA

MALTA

CRETE

GREECE

ATHENS

ALBANIA

YUGOSLAVIA

BULGARIA

ROMANIA

HUNGARY

AUSTRIA

ITALY

ROME

NAPLES

SARDEGNA

CORSE

FRANCE

SPAIN

PORTUGAL

LISBOA

MOROCCO

Rabat

CASABLANCA

Marrakech

ATLANTIC OCEAN

SENEGAL

Dakar

GUINEA

GUINEA BISSAU

SIERRA LEONE

Freetown

LIBERIA

Monrovia

IVORY COAST

GHANA

TOGO

BENIN

UPPER VOLTA

Lagos

POLAND

GERMANY

SWITZERLAND

UNITED KINGDOM

BIRMINGHAM

Liverpool

Manchester

IRELAND

Dublin

Paris

Bordeaux

Marseille

Barcelona

Madrid

BALEARES

NORTH ATLANTIC RIDGE

CAPE VERDE

72

SEYCHELLES
SEYCHELLES
Victoria
ALMIRANTES
AGALEGA (Maur.)
MASCARENE IS.
Port Louis
MAURITIUS I.
Saint Denis
REUNION (Fr.)
TROMELIN (Fr.)

BRITISH TER. OF INDIAN OCEAN
COSMOLEDO
FARQUHAR
ASCENSION
ALDABRA
COMORES
GRANDE COMORE
MAYOTTE (Fr.)
ANJUAN

MALAGASY REP.
Antananarivo (Tananarive)
Majunga
Diégo-Suarez
Tamatave
Fianarantsoa
C. Ste Marie

KENYA
Nairobi
Mombasa

TANZANIA
Dar es Salaam
Zanzibar
Tanga
TANGANYICA

ZAIRE
Kisangani
Kananga
Mbuji-Mayi
Lubumbashi
KINSHASA Léopoldville

CONGO
Brazzaville
Pointe Noire

GABON
EQUATORIAL GUINEA

SAO TOME Y PRINCIPE (Gui. Ec.)
PAGALU (Gui. Ec.)

ANGOLA
Luanda
Benguela
Cabinda (Ang.)

MALAWI (NYASA)
Lilongwe

ZAMBIA
Lusaka
Kitwe

MOZAMBIQUE
Maputo (Lourenço Marques)
Beira
Delagoa
Mozambique

ZIMBABWE (RHODESIA)
Harare (Salisbury)

BOTSWANA
Gaborone
KALAHARI DESERT

SOUTH WEST AFRICA
NAMIBIA
NAMIB DESERT
Mazo de Kaoko

SOUTH AFRICA
JOHANNESBURG
Pretoria
CAPE TOWN
Cape of Good Hope
C. Agulhas
Port Elizabeth
East London
Durban
Bloemfontein
GD KARROO
St. Helena Bay

SWAZILAND
LESOTHO

TROPIC OF CAPRICORN

INDIAN OCEAN
SOUTH ATLANTIC OCEAN
MOZAMBIQUE CHANNEL
MOZAMBIQUE BASIN
WESTERN BASIN
CROZET BASIN
CROZET RIDGE
AGULHAS BASIN
CAPE TOWN BASIN
CAPE TOWN RIDGE
WALVIS RIDGE
PRINCE EDWARD
MASCARENE RIDGE
MALAGASY BASIN

73

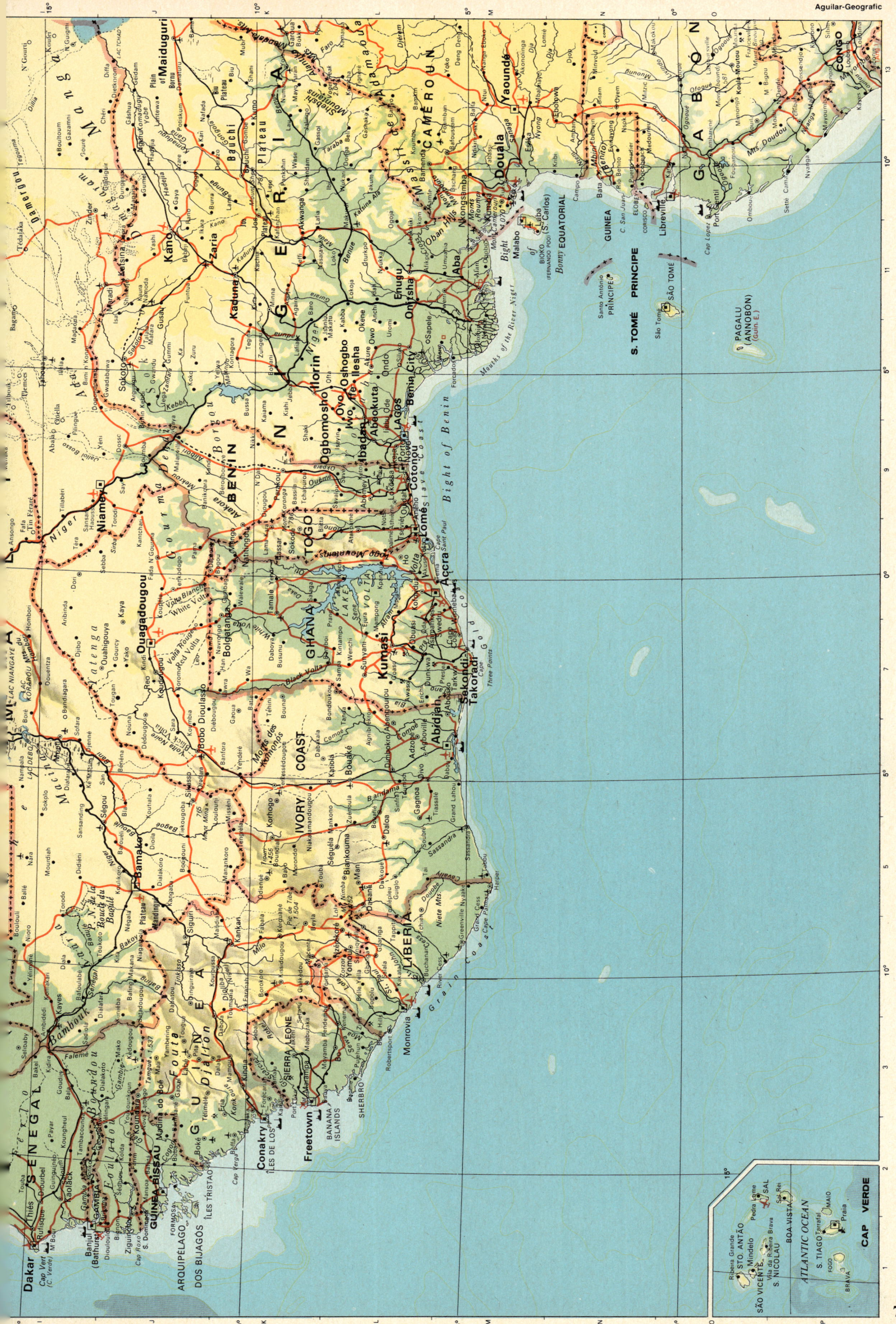

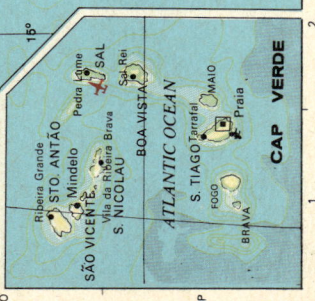

1:10,000,000

1" = 158 mi
1 cm = 100 km

MEDITERRANEAN SEA

Trâbulus (Tripoli)
Benghâzi
EL ISKANDARIYA
Alexandria
Tel Aviv-Yafo
Yerushalayim
Bûr Sa'îd (Port Said)
EL QÂHIRA (El Cairo)
El Suweys (Suez)
El Gîza
El Fayyûm
El Minyâ
Asyût
Sohag
El Khârga
Aswân

Khalij Surt (G. of Sirte)
Baerqa (Cirenaica)
Gardaba

L I B Y A
E G Y P T

Fezzân
EDEIEN EL MURZÛQ
SERIR TIBESTI
T I B E S T I
Pic Toussidé 3.267
Emi Koussi 3415

El Kufra

TROPIC OF CANCER

Wâdi Halfa
ES SAHRÂ E
NUBIAN DESERT

N I G E R
Borkou
Bodele
Ennedi
Erdi

LAC TCHAD
C H A D
S U D A N
N'Djamena (Fort Lamy)

NIGERIA
Maiduguri

Omdurmân
El Obeid

Abéché
Massif de Marfa
Gebel Marra 3088

CAMEROUN
CENT. AFRICAN REP.
Bangui

Parc National de la Bamingui
Massif des Bongos

Z A I R E
Zaïre (Congo)
GABON
CONGO

76

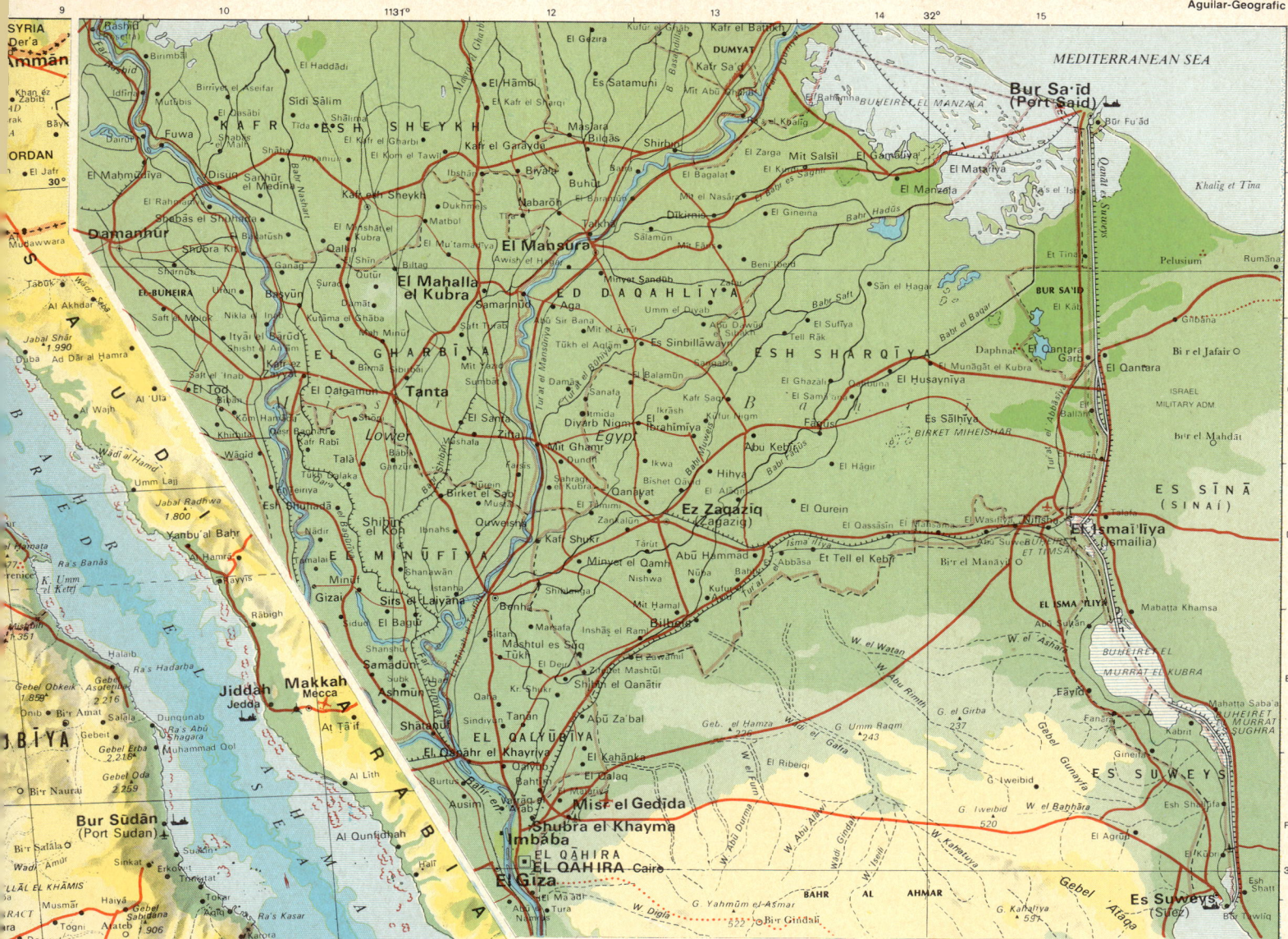

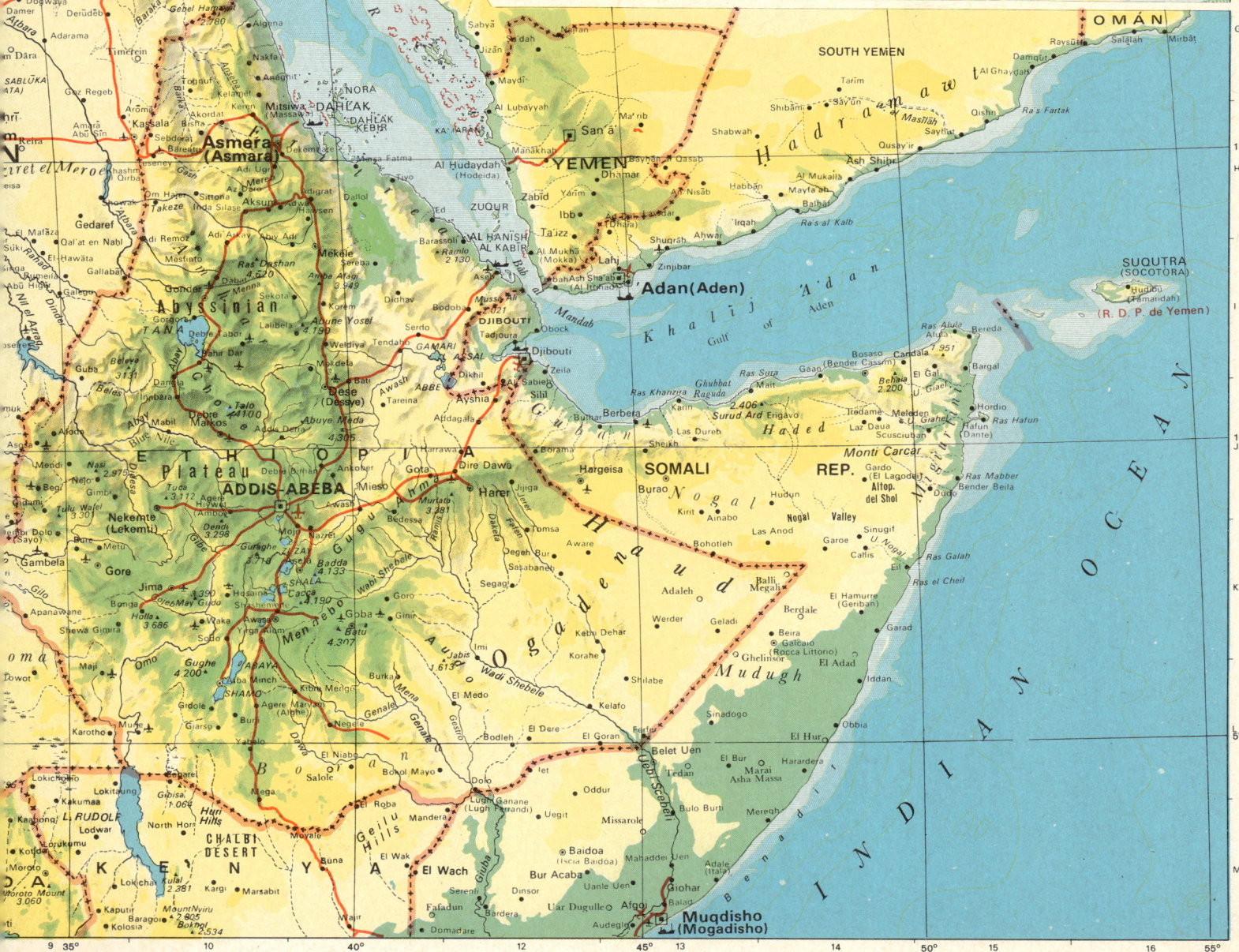

1:10,000,000 1" = 158 mi 1 cm = 100 km

SOMALI REP.

ETHIOPIA

KENYA

UGANDA

SUDAN

RWANDA

BURUNDI

TANZANIA

ZAÏRE

CENTRAL AFRICAN REP.

CAMEROUN

GABON

CONGO

ANGOLA

ZAMBIA

MALAWI

Mozambique

Nairobi

Mombasa

Dar es Salaam

ZANZIBAR
Zanzibar

PEMBA

MAFIA

Kampala

Bukavu (Costermansville)

Kisangani (Stanleyville)

Bangui

Mbandaka (Coquilhatville)

Kinshasa (Léopoldville)

Brazzaville

Matadi

Mbanza-Ngungu

Bandundu (Banningville)

Mbuji-Mayi (Bakwanga)

Kananga (Luluabourg)

Lubumbashi (Elizabethville)

Likasi (Jadotville)

Kolwezi

Ndola

Kitwe

Chingola

Luanshya

Luanda

Lobito

Benguela

LAKE VICTORIA

LAKE TANGANYIKA

LAKE MWERU

LAKE BANGWEULU

LAKE RUDOLF

LAKE EDWARD

LAKE KIVU

CHALBI DESERT

MALAWI (LAKE NYASA)

1:24,000,000   1" = 380 mi
1 cm = 240 km

CHINA
Nanchang
Wenchow
Foochow
Amoy
TAIPEI
TAIWAN
(FORMOSA)
Kaohsiung

Naha
OKINAWA
RETTO
ARCH. DE SAKISHIMA
RYUKYU (Jap.)
IS. DAITO
OKINO DAITO
EAST CHINA SEA

IS OGASAWARA (Jap.)

MINAMI TORI (Jap.)

TROPIC

IS BATAN
IS. BABUYAN
Laoag
LUZON
Quezon City
IS. POLILLO
MANILA
PHILIPPINES
Batangas
MINDORO
Legazpi
CATANDUANES
SAMAR
Cebu
LEYTE
NEGROS
Dumaguete
PANAY
Butuan
PALAWAN
JOLO SEA
Zamboanga
BASILAN
MINDANAO
Davao
2.965

PHILIPPINE SEA

PARECE VELA (Jap.)

WESTERN MARIANAS
MARIANAS BASIN
ASUNCION
AGRIHAN
PAGAN
ALAMAGAN
GUGUAN
SARIGAN
ANATAHAN
Saipan
TINIAN
ROTA
GUAM

IS MAUG
FARALLON DE PAJAROS

EASTERN MARIANAS BASIN

MARIANAS TRENCH

WAKE

ATOLÓN ENIWETOK
ATOLÓN BIKINI
ATOLÓN RONGERIK
ATOLÓN RONGELAP

ATOLÓN ULITHI
IS. NGULU
YAP
FAIS
GAFERUT
ATOLÓN FARAULEP
ATOLÓN NAMONUITO
HALL
ATOLÓN OLIMARAO
ATOLÓN PULAP
TRUK
ATOLÓN OROLUK
ATOLÓN SOROL
ATOLÓN WOLEAI
ATOLÓN IFALIK
LAMOTREK
PONAPE
IS. SENYAVIN
IS. PINGELAP
KUSAIE
ATOLÓN NAMORIK
BABELTHUAP
PALAU
ATOLÓN EAURIPIK
ATOLÓN PULUSUK
IS MORTLOCK
ATOLÓN NUKUORO
ATOLÓN EBON

CAROLINE ISLANDS
PACIFIC ISLAND TER.

IS SONSOROL
PULO ANNA
MERIR
W. CAROLINE
E. CAROLINE
BASIN
BASIN

TOBI
IS. HELENA

CAROLINE BASIN

ATOLÓN KAPINGAMARANGI

NAURU

CELEBES SEA
Manado
HALMAHERA
MOLUCCA SEA
Gorontalo
S. MOLUCCAS
SANGI
PENÍNSULA MINAHASA
WAIGEO
TALAUD
MOROTAI
Ternate
BIAK
Jaya-Pura
C. D'URVILLE
ADMIRALTY IS.
IS SAN MATIAS
NEW HANOVER
IS. TABAR
BISMARCK SEA
ARCH. DE BISMARCK
NEW IRELAND
IS. NUGURIA
IS. NUKUMANU
MELANESIA

YOGIAN
CERAM SEA
BURU
Ambon
CERAM
KAI
Fak-Fak
Cord. Maoke
P. Djaja 5.020
IRIAN JAYA
Mt Wilhelm 4.694
NEW BRITAIN IS.
Rabaul
NISSAN
BUKA
BOUGAINVILLE
IS. ORONG JAVA
CHOISEUL

C E L E B E S
BANDA SEA
IS. KAI
IS. ARU
PAPUA
NEW GUINEA
Lae
NEW GEORGIA
SANTA ISABEL
MALAITA
DUFF

Ujungpandang
Makassar
BANDA
TUKANGBESI
IS. TANIMBAR
C. Valsch
Mt Victoria 4.072
Port Moresby
IS. TROBRIAND
WOODLARK
IS. D'ENTRECASTEAUX
LOUISIADE ARCH.
ROSSEL
TAGULA
GUADALCANAL
Honiara
SAN CRISTOBAL
SALOMON IS
NDENI
UTUPUA
VANIKORO
RENNELL
NUPANI

FLORES
FLORES SEA
SUMBAWA
SUMBA
ALOR
SAWU
ROTI
TIMOR
TIMOR SEA
SUNDA ISLANDS

INDONESIA

ARAFURA SEA

MELVILLE
BATHURST
Darwin
Arnhem Land
GROOTE EYLANDT
Carpentaria
IS. WELLESLEY
CAPE YORK
Mitchell River
PEN.
GREAT BARRIER REEF
SWAIN REEF
WILLIS
CORAL SEA

C. York
Cooktown
Cairns
Townsville

CHESTERFIELD
BELEP
NEW HEBRIDES (U.K.)
ESPIRITU SANTO
VANUATU
MAEWO
PENTECÔTE
MALÉKULA
EPI
EROMANGA
CHUON
Vila
LOYALTY IS
UVÉA
LIFOU
MARE

NEW CALEDONIA (Fr.)
Nouméa

TROPIC

C. Talbot
Joseph Bonaparte Gulf B.
Wyndham
Kimberley Plateau
C. Lévêque
Mt Ord 936
Katherine
Daly
Roper
Victoria
Birdum
Barkly Tableland
Nicholson
Normanton
Mitchell River
Hughenden
Mackay
Rockhampton
Gladstone
Bundaberg
FRASER

Tennant Creek
Mt Isa
Mts Selwyn
Winton
Longreach
Aramac
Capricorn Channel

GREAT SANDY DESERT
Port Hedland
Mts Hamersley
Mt Bruce 1.226
Asbestos
Mts Macdonnell
1.510
Alice Springs
SIMPSON DESERT
GREAT ARTESIAN BASIN
Quilpie
Toowoomba
BRISBANE
C. Byron

GIBSON DESERT
Mt Augustus 1.105
Mts Musgrave 1.515
Mt Woodroffe
Oodnadatta
Cooper
Bourke
Mt Round 1.615
LORD HOWE (Aust.)

GREAT VICTORIA DESERT
Wiluna
Meekatharra
L. EYRE
Marree
Kingoonya
Broken Hill
NORFOLK (Aust.)

Leonora
Llanura de Nullarbor
Ceduna
L. GAIRDNER
L. TORRENS
Mts Flinders
FROME
Wagga Wagga
Newcastle
SYDNEY
Wollongong
CANBERRA
Mt Kosciusko 2.230

Kalgoorlie
Eucla
Great Australian Bight
L. EVERARD
Whyalla
Mts Gawler
EYRE
Adelaide
Murray
Australian Alps
C. Howe

AUSTRALIA

Perth
Narrogin
Esperance
Norseman
C. Catastrophe
Port Lincoln
CANGUROO
KANGAROO
Mount Gambier
Ballarat
MELBOURNE
Geelong
C. Howe

Geographe B.
C. Naturaliste
Mt Bluff 1.109
Bunbury
C. Leeuwin
Albany
C. Jaffa
Nelson
C. Otway
Bass Strait
King
FURNEAUX Bank
FURNEAUX Strait

TASMAN SEA

SOUTH AUSTRALIA BASIN

TASMANIA
Launceston
Hobart
South East C.

I N D I A N   O C E A N

Pt Cascade
C. Providence
Invercargill
STEWART

LORD HOWE RIDGE

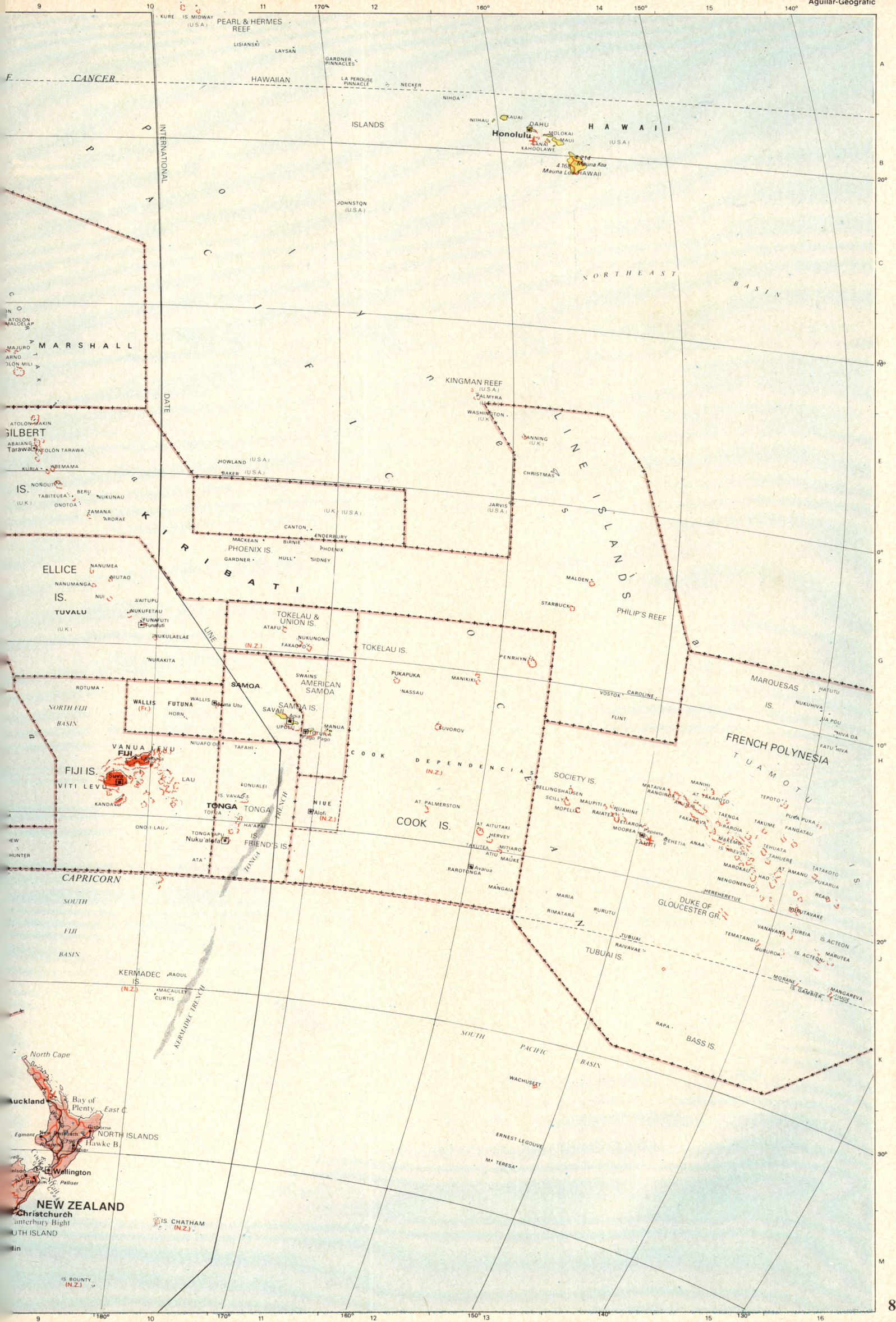

CANCER

PEARL & HERMES REEF

KURE IS. MIDWAY (USA)

LISIANSKI LAYSAN

GARDNER PINNACLES

HAWAIIAN LA PEROUSE PINNACLE NECKER

NIHOA

ISLANDS

NIIHAU KAUAI OAHU MOLOKAI MAUI HAWAII (U.K.)

Honolulu LANAI KAHOOLAWE

4 214 Mauna Kea

4 16 Mauna Loa HAWAII

JOHNSTON (USA)

NORTHEAST BASIN

MARSHALL

ATOLON MALOELAP

MAJURO

ARNO

ATOLON MILI

KINGMAN REEF (U.S.A.)

PALMYRA (U.S.A.)

WASHINGTON (U.K.)

GILBERT

ATOLON MAKIN

ABAIANG Tarawa ATOLON TARAWA

KURIA ABEMAMA

IS.

NONOUTI BANNING (U.K.)

(U.K.) TABITEUEA BERU NUKUNAU

ONOTOA TAMANA

ARORAE

L I N E I S L A N D S

CHRISTMAS

HOWLAND (USA)

BAKER (USA)

KIRIBATI

CANTON JARVIS (U.S.A.)

(U.K.) (U.S.A.)

MACKEAN BIRNIE ENDERBURY

PHOENIX IS. PHOENIX

GARDNER HULL SIDNEY

ELLICE NANUMEA

NIUTAO

NANUMANGA

MALDEN (U.K.)

IS. NUI

VAITUPU

PHILIP'S REEF

TUVALU NUKUFETAU

STARBUCK

FUNAFUTI Funafuti

(U.K.) NUKULAELAE

NURAKITA

TOKELAU & UNION IS.

ATAFU

PENRHYN

NUKUNONO

LINE (N.Z.) FAKAOFO

TOKELAU IS.

SAMOA SWAINS PUKAPUKA MANIHIKI

AMERICAN NASSAU

WALLIS FUTUNA SAMOA MARQUESAS

WALLIS Mata Utu SAMOA IS. IS.

(Fr.) Apia VOSTOK CAROLINE NUKUHIVA

HORN SAVAII UPOLU UA POU

FUTUNA MANUA SUVOROV FLINT HIVA OA

Pago Pago FATU HIVA

VANUA LEVU NIUAFO OU TAFAHI

FIJI COOK DEPENDENCIAS FRENCH POLYNESIA

FIJI IS. LAU (N.Z.) T U A M O T U

VITI LEVU SOCIETY IS. MATAIVA MANIHI

Suva FONUALEI PELLINGHAUSEN RANGIROA AT TAKAPOTO TEPOTO

SCILLY HUAHINE PUKA PUKA

KANDAVU TONGA AT PALMERSTON MOPELIA MAUPITI TAENGA TAKUME FANGATAU

TOFUA HA'APAI NIUE MAURUA RAIATEA FAKARAVA FAROIA

ONO-I-LAU TONGA IS. (N.Z.) HERVEY MOOREA Papeete TAHANEA TEHUATA TAHUERE

TONGATAPU FRIEND'S IS. AT AITUTAKI TAHITI BEHETIA ANAA MAKEMO AMANU TAKATOTO

Nuku'alofa TAKUTEA HIKUERU HAO PUKARUA

ATA MITIARO NENGONENGO REAO

COOK IS. MAUKE MAROKAU HEREHERETUE TATAKOTO

HUNTER Avarua RARATONGA HAO HIKUTAVAKE

CAPRICORN MANGAIA VANAVANA TUREIA IS ACTEON

SOUTH MARIA DUKE OF MARUTEA

FIJI RIMATARA RURUTU GLOUCESTER GR. TEMATANGI IS. ACTEON

BASIN TUBUAI MURUROA

RAIVAVAE

KERMADEC RAOUL TUBUAI IS.

IS. MORANE

(N.Z.) MACAULEY MANGAREVA

CURTIS RAPA IS GAMBIER

BASS IS.

SOUTH PACIFIC BASIN

WACHUSET

North Cape

Auckland Bay of Plenty East C.

Gisborne

NORTH ISLANDS Hawke B. ERNEST LEGOUVE

Napier

Wellington M TERESA

C. Palliser

NEW ZEALAND

Christchurch

Canterbury Bight IS. CHATHAM

SOUTH ISLAND (N.Z.)

IS BOUNTY (N.Z.)

INDONESIA

JAVA   LAUT BALI   LOMBOK   SUMBAWA   SUMBA   LAUT SAWU   SAWU   ROTI

KANGEAN   SUNDA   KEP. PATERNOSTER   KALAD BONERATE   LAUT FLORES   FLORES   SOLOR   PANTAR   ALOR   KALAOTOA   LOMBLEN   WETAR   TIMOR

KEP. LETI   MOA   SELARU   KEP. TANIMBAR

NEW

ARAFURA SEA

TIMOR SEA

CROKER   BATHURST   Fort Dundas   Bathurst Island   MELVILLE   Dundas Str.   GOULBURN IS.   WESSEL IS.   C. Wessel   Tg. Vals   C. Wan

C. Londonderry   Joseph Bonaparte Gulf   Beagle Gulf   Van Diemen Gulf   Arnhem Land   C. Arnhem   Alexander   BICKERTON   GROOTE EYLANDT

INDIAN OCEAN

ARCHIPELAGO BONAPARTE   Admiralty Gulf   King Edward   Port Keats   Katherine   Roper   Roper River   MARIA   Limmen Bight   SIR EDWARD PELLEW GROUP   VANDERLIN

ROWLEY SHOALS

C. Leveque   King Sound   Leopold Ranges   Durack Range   Wyndham   Ord   Ivanhoe   Kimberley Plateau   Antrim Plateau   NORTHERN TERRITORY   BARKLY TABLELAND   Borroloola   McArthur River   Robinson River   Nichols

Pender Bay   Dampier Land   Broome   Yeeda Downs   Fitzroy   Oscar Ra.   Fossil Downs   Fitzroy Crossing   Halls Creek   Gordon Downs   Mt. Winnecke   Tennant Creek   Mt. Samuel   Alexandria   Alroy Downs

Roebuck Bay   Roebuck Downs   Noonkanbah   St. George Ranges   Christmas Creek   Sturt   Balgo Hills   Tanami   TANAMI DESERT   Murchison Ra.   Davenport Ra.   Lake Nash

La Grange   Anna Plains   Eighty Mile Beach   Wells   Kuduarra Well   The Granites   Lander   Singleton   Sandover

GREAT SANDY DESERT

DAMPIER ARCH.   BARROW   Pt. Hedland   De Grey   Warrawagine   WAUKARLICARLY   PERCIVAL LAKES   O Ural Well   Patience Well   Alfred and Marie Rd.   L. MACKAY   Kintore Range   Napperby   Aileron   SIMPSON DESERT

Nickol Bay   Preston   Marble Bar   Nullagine   L. DORA   Argos Hills   Mt. Singleton   Mount Doreen   Central Mt. Stuart

Chichester Range   Hamersley Range   Fortescue   Balfour Downs   Throssel Ra.   L. DISAPPOINTMENT   HOPKINS   L. NEALE   LAKE AMADEUS   Macdonnell Ranges   George Gills Ra.   Hermannsburg   Palmer   Finke   Rodinga   Jervois Ra.

CAPRICORN Ra.   Mt. Bruce   Mt. Vernon   Kunawagga   Mirrindindi   WESTERN   GIBSON DESERT   Mt. Madley   Rawlinson Ra.   Mt. Olga   Erldunda   Finke   Bundooma

SALT LAKE   Teano Ra.   Weld Spring   AUSTRALIA   CARNEGIE   Warburton Range   Mount Davies   Mt. Woodroffe   Everard Ra.   Hamilton   Alberga   Pedirka

Robinson Ranges   Horseshoe Mine   Peak Hill   GREGORY   L. CARNEGIE   L. GILLEN   WELLS   BAKER L.   SERPENTINE LAKES   Mount Willoughby   Coober Pedy   LAKE EYRE

Meekatharra   Nannine   Wiluna   Pt. Lilian   L. WAY   L. YEO   L. RASON   GREAT VICTORIA DESERT   L. DEY-DEY   L. MAURICE   WILKINSON LAKES   SOUTH AUSTRALIA   L. EVERARD   LAKE GAIRDNER   LAKE TORRENS

Roderick   Big Bell   Cue   Mt. Magnet   Sandstone   L. DARLOT   Laverton   MINIGWAL   PLUMRIDGE LAKES   L. NYANGA   Ogilea   Wynbring   Tarcoola   Kingoonya   L. MACFARLANE

WESTERN   AUSTIN   Melrose   Leonora   L. CAREY   L. YINDARLGOODA   Leigh Creek

Geraldton   Morawa   Mt. Singleton   Dalgaranga   Leinster   Menzies   Bardoc   Karonie   Zanthus   Rawlinna   Haig   Cook   Wirrulla   Streaky Bay   Port Augusta

HOUTMAN ROCKS   WESTERN   L. BARLEE   L. BALLARD   LAKE RAESIDE   Malcolm   Kalgoorlie   LEFROY   L. COWAN   Norseman   NUYTS ARCH.   Ceduna   Bookaloo   Iron Knob   Whyalla

MONGER   L. MOORE   Wongan Hills   Dowerin   Southern Cross   Coolgardie   Widgiemooltha   L. DUNDAS   Balladonia   Eucla   Nullarbor   Colona   Penong   Anxious Bay   Elliston   Cowell

Perth   THE JOHNSTON LAKES   Salmon Gums   Nullarbor Plain   Head of Bight   Coffin B.

Mandurah   Pinjarra   Narrogin   Wickepin   Hyden   Lake King   Russell Ra.   Twilight Cove   GREAT AUSTRALIAN BIGHT   C. Catastrophe   Port Lincoln

Bunbury   Collie   Darkan   Katanning   Gnowangerup   Ravensthorpe   Hopetoun   Esperance   Esperance Bay   C. Arid   ARCH. OF THE RECHERCHE

Geographe Bay   Busselton   Margaret River   C. Leeuwin   Manjimup   Stirling Ra.   Albany   Pt. D'Entrecasteaux   Adelaide   Mount Lofty Ranges   KANGAROO   Investigator Strait   Encounter Bay   Lacepe

CAPRICORN   TROPIC   Exmouth Gulf   Onslow   Ashburton   Learmonth   North West C.   Point Cloates   Chabunuardoo Bay   Minilya   Gascoyne Junction   Carnarvon

Geographe Channel   BERNIER   DORRE   Shark Bay   DIRK HARTOG   Steep Point   Denham   Hamelin Pool   Tamala   Gantheaume Bay   Kalbarri   Murchison   Ajana   Northampton   Mullewa

INDIAN OCEAN

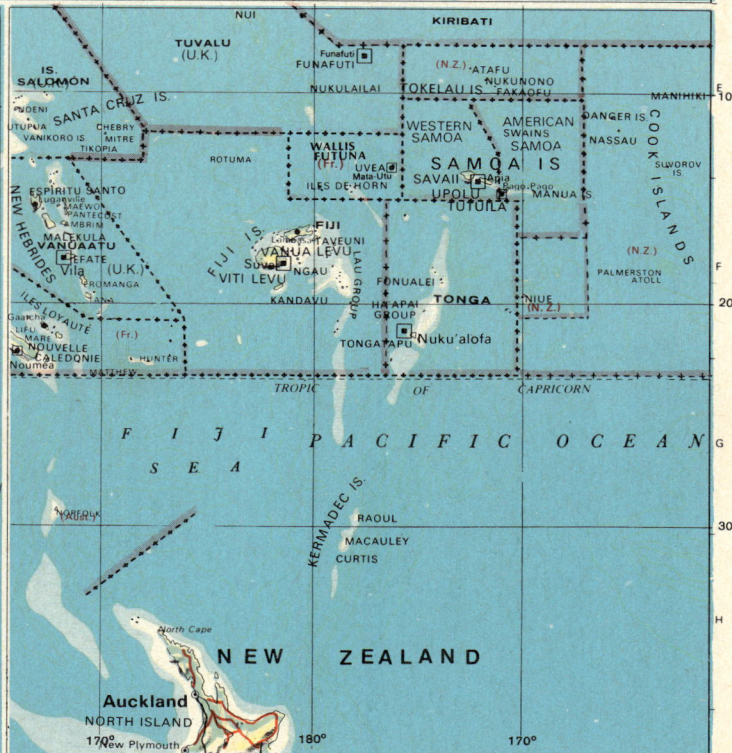
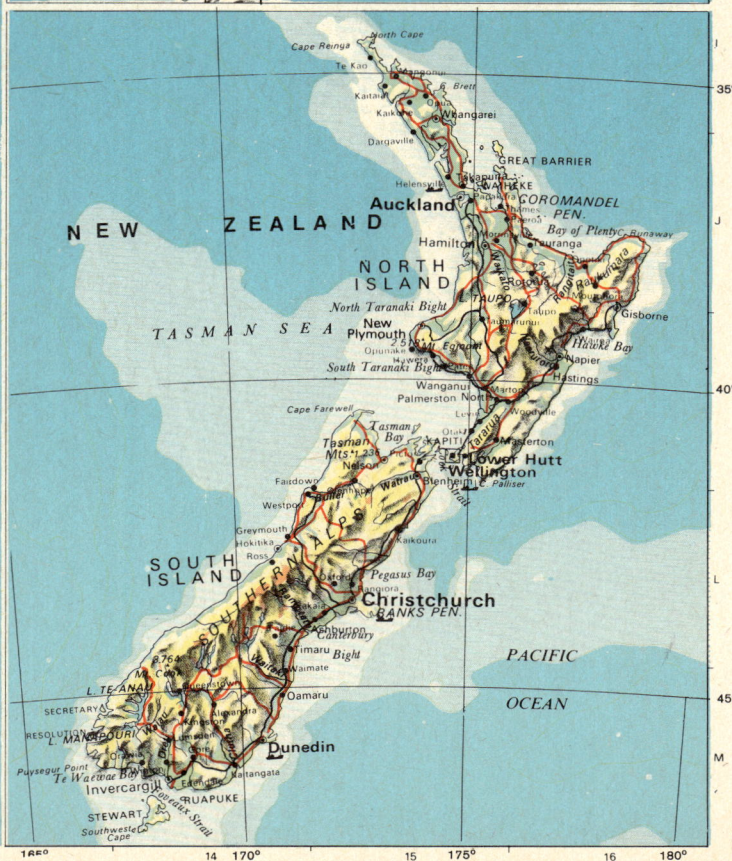

DENMARK
NORWAY
ICELAND
UNITED KINGDOM
OUTER HEBRIDES
ORKNEY IS.
SHETLAND IS.
SKAGERRAK
NORTH SEA
NORWEGIAN SEA
REYKJANES RIDGE

ATLANTIC OCEAN

GREENLAND

Denmark Strait
Kong Christian IX Kyst
Dronning Margrethe

LABRADOR SEA

DAVIS STRAIT
Baffin Bay

ARCTIC OCEAN
North Pole

QUEEN ELIZABETH ISLANDS
PRINCE PATRICK IS.
BANKS
VICTORIA
Gulf of Boothia
BOOTHIA PEN.
McClure Strait
Melville
Prince of Wales
Somerset
Baffin
Foxe Basin
SOUTHAMPTON

HUDSON BAY
BELCHER ISLES
Ungava Bay
QUEBEC
Hudson Strait

CANADA
ONTARIO
Edmonton
Saskatoon
Winnipeg
L. Winnipeg
L. Athabasca
Gt. Bear Lake
Gt. Slave Lake
Caribou Mts.
Mackenzie Mts.
Franklin Mts.
Mackenzie
Brooks Range
Fort Nelson

BEAUFORT SEA
Amundsen G.

Montreal
Ottawa
Quebec
Hamilton
Toronto
L. Superior
L. Huron
L. Michigan
Sudbury
Minneapolis
St. Paul
Duluth
Boston
Providence
NEW YORK
Buffalo
Portland

NOVA SCOTIA
PRINCE EDWARD ISLE
ST. LAWRENCE
NEW BRUNSWICK

Seattle
Vancouver
Vancouver Island
Queen Charlotte Islands
ARCH. ALEXANDER
Queen Charlotte Str.

ALASKA (U.S.A.)
ALASKA RANGE
Mt. McKinley 6193
Mts. Chugach
Mt. St. Elias
Tanana
Yukon
Kuskokwim
Gulf of Alaska
KODIAK IS.
Kotzebue Sd.
Norton Sd.
Bristol B.
Bering Strait
C. of Wales

U. S. S. R.
SIBERIA
EAST SIBERIAN SEA
WRANGEL IS.
BEAR IS.
Long Str.
CHUKCHI SEA
Anadyr Plateau
CHUKOTSKIY POL.
Anadyrskiy Zaliv
Mts. Ahlunay
Koryakskiy Khrebet
Zaliv Shelikhova
ST. LAWRENCE I. (U.S.A.)
ST. MATTHEW I.
NUNIVAK
PRIBILOF
BERING SEA
ALEUTIAN BASIN
KOMANDORSKIYE
Petropavlovsk-Kamchatskii
Klyuchevskaya Sopka

ALEUTIAN IS.
ATTU
IS ANDREANOF

UNITED STATES

MEXICO

BRAZIL

BOLIVIA

VENEZUELA

COLOMBIA

ECUADOR

PERU

CUBA

HAITI

DOMINICAN REP.

JAMAICA

GUATEMALA

EL SALVADOR

HONDURAS

NICARAGUA

COSTA RICA

PANAMA

BELIZE

BERMUDA

BAHAMA IS.

LEEWARDS IS.

BARBADOS

GRENADA

TRINIDAD Y TOBAGO

LESSER ANTILLES

WINDWARD IS.

VIRGIN IS.

CARIBBEAN SEA

GULF OF MEXICO

PACIFIC OCEAN

GALAPAGOS IS.

REVILLA GIGEDO IS. (Méx.)

CANCER

TROPIC

SIERRA MADRE OCCIDENTAL

SIERRA MADRE ORIENTAL

SIERRA MADRE DEL SUR

BAJA CALIFORNIA

G. of California

CORDILLERA OCCIDENTAL

CORDILLERA ORIENTAL

CORDILLERA CENTRAL

San Francisco

Los Angeles

San Diego

Mexicali

Sacramento

Denver

Oliver

Albuquerque

El Paso

Ciudad Juárez

Chihuahua

Phoenix

Wichita

Oklahoma City

Tulsa

Kansas City

Omaha

Lincoln

Little Rock

Shreveport

Fort Worth

Dallas

San Antonio

Corpus Christi

Houston

New Orleans

Memphis

Birmingham

Atlanta

Jacksonville

Savannah

Charleston

Columbus

Cincinnati

Indianapolis

Louisville

Nashville

Pittsburgh

Columbus

Baltimore

Washington

Philadelphia

C. Hatteras

C. Cañaveral

Miami

MONTERREY

Saltillo

Torreón

Durango

Mazatlán

Culiacán

GUADALAJARA

León

Aguascalientes

San Luis Potosí

CIUDAD DE MEXICO

Puebla

Veracruz

Tampico

Mérida

Campeche

YUCATAN

LA HABANA

Santiago

Camagüey

Kingston

Port-au-Prince

Santo Domingo

San Juan

BOGOTA

MEDELLIN

Cali

Barranquilla

Cartagena

Santa Marta

Bucaramanga

Quito

Guayaquil

CARACAS

Maracaibo

Barquisimeto

PANAMA CANAL ZONE

San José

Managua

Tegucigalpa

San Salvador

Guatemala

Belmopan

85

1:12,000,000

1" = 190 mi
1 cm = 120 km

ARCTIC OCEAN

BEAUFORT SEA

QUEEN ELIZABETH ISLANDS

PARRY ISLANDS

MELVILLE

BANKS

VICTORIA

BROOKS RANGE

ALASKA RANGE

UNITED STATES

GULF OF ALASKA

PACIFIC OCEAN

ALEXANDER ARCHIPELAGO

QUEEN CHARLOTTE ISLANDS

NORTHWEST

GREAT BEAR LAKE

GREAT SLAVE LAKE

MACKENZIE MOUNTAINS

ROCKY MOUNTAINS

LAKE ATHABASCA

REINDEER LAKE

WOLLASTON L.

CREE L.

ALBERTA

SASKATCHEWAN

MANITOBA

BRITISH COLUMBIA

YUKON

VANCOUVER

Victoria

SEATTLE

Tacoma

Portland

Edmonton

Calgary

Saskatoon

Regina

Winn.

CASCADE RANGE

COLUMBIA PLATEAU

BLUE MOUNTAINS

IDAHO

MONTANA

WYOMING

Missouri Plateau

NORTH DAKOTA

SOUTH DAKOTA

Great Falls

Spokane

CALIFORNIA

OREGON

86

GREENLAND

ICELAND

Baffin Bay

Baffin Basin

Davis Strait

LABRADOR SEA

ATLANTIC OCEAN

Denmark Strait

ELLESMERE

AXEL HEIBERG

DEVON

SOMERSET

Gulf of Boothia

Hudson Strait

Ungava Bay

Hudson Bay

OTTAWA ISLANDS

BELCHER ISLANDS

James Bay

LABRADOR

QUEBEC PENINSULA

NEW QUEBEC

NEWFOUNDLAND

Gulf of St. Lawrence

PRINCE EDWARD ISLAND

CAPE BRETON

NOVA SCOTIA

NEW BRUNSWICK

Halifax

Québec

MONTREAL-Laval

Ottawa

ONTARIO

LAKE SUPERIOR

LAKE HURON

LAKE MICHIGAN

LAKE ONTARIO

LAKE ERIE

Thunder Bay

Duluth

MINNEAPOLIS-SAINT PAUL

MINNESOTA

WISCONSIN

MICHIGAN

TORONTO

Hamilton

BUFFALO

NEW YORK

BOSTON

Detroit

MAINE

ATLANTIC OCEAN

GEORGES BANK

LAKE WINNIPEG

MANITOBA

Southampton

Foxe Basin

PRINCE CHARLES

CUMBERLAND

Nettilling Lake

Amadjuak Lake

87

1 : 6,000,000   95 mi / 60 km

Map

Coordinate markers (top): 108° 106° 104° 102° 100° 98° 96° 94° 13 92° 14 90° 88° 86° 15 84°

Latitude markers (right): A — 62° — B — 60° — C — 58° — D — E — 56° — F — G — 54° — H — 52° — I — J — 50° — K — L — 48° — M

Major regions and features:

SOUTHAMPTON · COATS · Bell Pen · Fisher Str.

Hudson Bay · Bay of Gods Mercy · Ross Welcome Sd.

D I S T R I C T   O F   K E E W A T I N

N O R T H W E S T   T E R R I T O R I E S

MENZIE TERRITORY

S A S K A T C H E W A N · M A N I T O B A · O N T A R I O

N O R T H   D A K O T A · M I N N E S O T A

Lakes and places:
Baker Lake · Aberdeen L. · Beverly L. · Wharton L. · Dubawnt L. · Tulemalu L. · Yathkyed L. · Angikuni L. · Kaminuriak L. · Kaminak L. · North Henik · South Henik · Chesterfield Inlet · Rankin Inlet · Eskimo Point · Churchill · C. Churchill

Aylmer L. · Clinton Colden L. · Artillery L. · Whitefish L. · Kamilukuak L. · Eileen L. · Lynx L. · Rennie L. · Alcantara L. · Abitau L. · Wignes L. · Selwyn L. · Nueltin · Kasba L. · Wholdaia L. · Flett L. · Snowbird L. · Ennadai L. · Nejanilini L. · Edehon L. · Thlewiaza

Lake Athabasca · Eldorado · Black L. · Stony Rapids · Wollaston L. · Lac Brochet · Tadoule · Reindeer · Cree L. · Southern Indian L. · South Seal · North Knife L. · Big Sand L. · Etawney L. · Gauer L. · Baldock L. · Granville L. · Southend

Lynn Lake · Nelson House · Split L. · Gillam · Sipiwesk L. · Oxford L. · Gods L. · Island L. · Sachigo · Big Trout L. · Wunnummin L. · Shibogama L.

Churchill L. · Île-à-la-Crosse · La Ronge · Deschambault L. · Flin Flon · Creighton · Cranberry Portage · Snow Lake · Herb Lake · Cross L. · Molson L. · Norway House · Cedar Lake · Lake Winnipegosis

LAKE WINNIPEG · Berens · Poplar · Little Grand Rapids · Red L. · Lac Seul · Savant L. · St. Joseph · Sachigo

Saskatoon · Prince Albert · Melfort · Hudson Bay · Swan River · Dauphin · Riding Mountain · Duck Mt. P.P. · Roblin · Ste. Rose du Lac

Regina · Moose Jaw · Yorkton · Melville · Qu'Appelle · The Coteau · Old Wives L. · Last Mountain L. · Quill Lakes · Foam Lake

Winnipeg · Portage la Prairie · Brandon · Steinbach · LAKE OF THE WOODS · LAKE MANITOBA · Lake St. Martin · Kenora · Dryden

Swift Current · Weyburn · Estevan · International Falls · Rainy L. · Fort Peck Res. · Missouri Coteau · Coteau du Missouri

Grand Forks · Devils Lake · Fargo · Moorhead · Jamestown · Bismarck · Duluth · Cloquet · Mille Lacs

CANADA · U.S.A.

Coordinate markers (bottom): 9 108° 10 106° 104° 102° 12 100° 13 98° 14 96° 15 94° 16

1 : 6,000,000 1" = 95 mi 1 cm = 60 km

HUDSON Bay

JAMES Bay

MANITOBA

ONTARIO

QUEBEC

MINNESOTA

WISCONSIN

MICHIGAN

NEW YORK

LAKE SUPERIOR

LAKE HURON

LAKE MICHIGAN

LAKE ERIE

LAKE ONTARIO

GEORGIAN BAY

Belcher Is.

Akimiski

Duluth · Thunder Bay · Sudbury · Timmins · Noranda · Val d'Or

Milwaukee · Chicago · Detroit · Windsor · London · Hamilton · Toronto · Niagara Falls · Buffalo · Rochester · Syracuse · Utica · Albany

Ottawa · Montreal · Kingston · Laval

90

L A B R A D O R   S E A

Ungava Bay

AKPATOK

BUTTON IS.

NORTH AULATSIVIK

SOUTH AULATSIVIK

Coast of Labrador

Groswater Bay

Benedict Mountains

MISTASTIN L.

MICHIKAMAU L.

GRAND FALLS

Mealy Mtns.

Labrador

L A B R A D O R   P E N I N S U L A

N E W F O U N D L A N D

SCARP

Strait of Belle Isle

BELLE ISLE

CANIAPISCAU

MANICOUAGAN L.

Passage de Jacques Cartier

ANTICOSTI

Détroit de Gaspé

Gulf of St. Lawrence

GASPÉ PENINSULA

Notre Dame

NEWFOUNDLAND

GANDER

Cabot Strait

MAGDALEN IS.

CAPE BRETON

PRINCE EDWARD ISLAND

NEW BRUNSWICK

GRAND MIQUELON
SAINT PIERRE ET MIQUELON (Fr.)

SAINT PIERRE BANK

Quebec

Fredericton

Saint John

Bay of Fundy

N O V A   S C O T I A

Halifax

BANQUEREAU

SABLE

SABLE ISLAND BANK

Yarmouth

Gulf of Maine

BROWNS BANK

A T L A N T I C   O C E A N

Portland

Portsmouth

91

1:12,000,000 1" = 190 mi 1 cm = 120 km

PACIFIC OCEAN

CANADA

MEXICO

VANCOUVER
SEATTLE
Tacoma
Spokane
Portland
SAN FRANCISCO
Berkeley
Oakland
Sacramento
San Jose
Stockton
Fresno
Bakersfield
LOS ANGELES
Long Beach
Anaheim
Santa Ana
San Bernardino
SAN DIEGO
Tijuana
Mexicali
Phoenix
Tucson
DENVER
Colorado Springs
Pueblo
Santa Fe
Albuquerque
El Paso
Ciudad Juárez
Amarillo
Lubbock
Oklahoma City
Wichita
Topeka
Omaha
Lincoln
Tulsa
Fort Worth
DALLAS
Waco
Austin
San Antonio
HOUSTON
Corpus Christi
Laredo
Nuevo Laredo
MONTERREY
Matamoros
Chihuahua
Hermosillo
Torreón
Durango
Zacatecas
Aguascalientes
San Luis Potosí
León
GUADALAJARA
Tampico
Ciudad Madero
Querétaro

Edmonton
Regina
Winnipeg
LAKE WINNIPEG
HUDSON BAY

Great Basin
Mojave Desert
Great Salt Lake
Salt Lake City
Yellowstone Lake
BLACK ROCK DESERT
COLUMBIA
ROCKY MOUNTAINS
SIERRA NEVADA
COAST RANGES
CASCADE RANGE
BLUE MOUNTAINS
BAJA CALIFORNIA NORTE
BAJA CALIFORNIA SUR
SIERRA MADRE OCCIDENTAL
SIERRA MADRE ORIENTAL
NORTH DAKOTA
SOUTH DAKOTA
NEBRASKA
NEVADA
UTAH
ARIZONA
NEW MEXICO
COLORADO
OKLAHOMA

TROPIC OF CANCER

GUADALUPE
SANTA MARGARITA
REVILLA GIGEDO IS.
CHARLOTTE ISLANDS
QUEEN CHARLOTTE ISLANDS
CHANNEL ISLANDS
Cabo San Lucas
ALLAIRE BANK

92

Hudson Bay

BELCHER ISLANDS

James Bay

C A N A D A

QUEBEC

LAURENTIDE SCARP

Gulf of St. Lawrence

NEWFOUNDLAND

PRINCE EDWARD ISLAND

CAPE BRETON

LAKE SUPERIOR

LAKE OF THE WOODS

Thunder Bay

Duluth

WISCONSIN

MINNESOTA

Saint Paul

Milwaukee

Rockford

Madison

LAKE MICHIGAN

LAKE HURON

LAKE ERIE

LAKE ONTARIO

Toronto

Hamilton

Rochester

Buffalo

NEW YORK

Montreal

Ottawa

Adirondack Mountains

Gulf of Maine

MOUNT DESERT

Boston

New Bedford

Cape Cod

NANTUCKET

LONG ISLAND

New York

Newark

Paterson

New London

Bridgeport

PHILADELPHIA

Camden

Atlantic City

CHICAGO

Detroit

Windsor

Cleveland

Akron

Youngstown

PITTSBURG

PENNSYLVANIA

Columbus

Dayton

Cincinnati

INDIANAPOLIS

Fort Wayne

INDIANA

OHIO

ILLINOIS

Springfield

Saint Louis

Kansas City

Des Moines

IOWA

MISSOURI

Baltimore

WASHINGTON

WEST VIRGINIA

VIRGINIA

Newport News

Norfolk

Portsmouth

C. Hatteras

Raleigh

Greensboro

Winston Salem

Charlotte

NORTH CAROLINA

SOUTH CAROLINA

Charleston

C. Fear

C. Lookout

KENTUCKY

Louisville

TENNESSEE

Nashville

Knoxville

Chattanooga

Memphis

ARKANSAS

Little Rock

MISSISSIPPI

Jackson

ALABAMA

Birmingham

Montgomery

Mobile

GEORGIA

ATLANTA

Columbus

Albany

Savannah

Jacksonville

FLORIDA

Tampa

Saint Petersburg

LAKE OKEECHOBEE

BIG CYPRESS

Everglades

MIAMI

West Palm Beach

Fort Lauderdale

Cape Canaveral

Cape Sable

FLORIDA KEYS

Key West

DRY TORTUGAS

LOUISIANA

NEW ORLEANS

Baton Rouge

Mississippi River Delta

Port Arthur

TEXAS

Gulf of Mexico

ALACRAN REEF

CAMPECHE BANK

YUCATAN

QUINTANA ROO

COZUMEL

Mérida

Progreso

Canal de Yucatán

TROPIC OF CANCER

LA HABANA

Pinar del Río

C U B A

Santa Clara

Camagüey

Holguín

Santiago de Cuba

BAHAMA

GRAND BAHAMA

GREAT ABACO

LITTLE ABACO

BERRY IS.

NEW PROVIDENCE

Nassau

ANDROS

ELEUTHERA

CAT

SAN SALVADOR

EXUMA

LONG ISLAND

CROOKED ISLAND

ACKLINS

MAYAGUANA

CAICOS ISLANDS

TURKS IS.

GREAT INAGUA

BARTLETT TRENCH

GRAND CAYMAN

LITTLE CAYMAN

CARIBBEAN SEA

HAITI

DOMINICAN REP.

Santo Domingo

Port au Prince

HISPANIOLA

TORTUE

ILE DE LA GONAVE

BERMUDA

Hamilton

BLAKE PLATEAU

BAHAMA TRENCH

A T L A N T I C   O C E A N

GEORGES BANK

GRAND BANKS

Gulf of Maine

Strait of Belle Isle

ANTICOSTI

Gulf of St. Lawrence

MAGDALEN ISLANDS

SABLE

90° 85° 80° 75° 70° 65° 60°

50° 45° 40° 35° 30° 25° 20°

1: 6,000,000   1 cm = 60 km   95 mi

Major labels and features:

PACIFIC OCEAN

VANCOUVER

CANADA
U.S.A.
MEXICO

ALBERTA

Cape Flattery

Cities / places:
- VANCOUVER
- VICTORIA
- New Westminster
- North Vancouver
- Kamloops
- Kelowna
- Penticton
- SEATTLE
- Tacoma
- Olympia
- Bremerton
- Everett
- Bellingham
- Port Angeles
- Hoquiam
- Aberdeen
- Spokane
- Coeur d'Alene
- Wenatchee
- Yakima
- Walla Walla
- Kennewick
- Richland
- Pasco
- Lewiston
- Moscow
- Pullman
- Missoula
- Helena
- Butte
- Anaconda
- PORTLAND
- Vancouver
- Salem
- Corvallis
- Eugene
- Albany
- Newport
- Tillamook
- Astoria
- The Dalles
- Pendleton
- Bend
- Redmond
- Baker
- Boise
- Nampa
- Caldwell
- Twin Falls
- Pocatello
- GREAT SALT LAKE
- GREAT SALT LAKE DESERT
- Roseburg
- Coos Bay
- North Bend
- Bandon
- Cape Blanco
- Gold Beach
- Grants Pass
- Medford
- Ashland
- Klamath Falls
- Crater Lake Nat. Park
- GREAT SANDY DESERT
- HARNEY L.
- MALHEUR L.
- SUMMER L.
- L. ABERT
- GOOSE L.
- UPPER KLAMATH L.
- Crescent City
- Eureka
- Arcata
- Fortuna
- Cape Mendocino
- Redding
- Red Bluff
- SHASTA LAKE
- Mount Shasta
- Weed
- Dunsmuir
- Alturas
- UPPER LAKE
- MIDDLE LAKE
- LOWER L.
- EAGLE L.
- Susanville
- CLEAR LAKE
- Fort Bragg
- Willits
- Ukiah
- Reno
- Sparks
- Carson City
- Winnemucca
- Elko
- Wells
- BLACK ROCK DESERT
- SMOKE CREEK DESERT
- Lovelock
- Fallon
- Battle Mountain
- SACRAMENTO
- Oroville
- Marysville
- Nevada City
- Auburn
- Placerville
- Stockton
- Lodi
- Vacaville
- Fairfield
- WALKER L.
- Hawthorne
- Tonopah
- SAN FRANCISCO
- Oakland
- Berkeley
- Richmond
- San Rafael
- San Mateo
- Redwood City
- SAN JOSE
- Santa Cruz
- Watsonville
- Salinas
- Monterey
- Seaside
- Hollister
- Gilroy
- MONO LAKE
- Fresno
- Madera
- Merced
- Modesto
- Turlock
- Sanger
- Visalia
- Tulare
- Hanford
- Porterville
- Delano
- Mt. Whitney
- Sequoia Nat. Park
- Kings Canyon Nat. Park
- OWENS LAKE
- DEATH VALLEY
- AMARGOSA DESERT
- Las Vegas
- LAKE MEAD
- Boulder City
- Paso Robles
- Atascadero
- San Luis Obispo
- Santa Maria
- Lompoc
- Santa Barbara
- Bakersfield
- Ridgecrest
- MOJAVE DESERT
- Barstow
- Needles
- Lancaster
- SANTA LUCIA RANGE
- Ventura
- Oxnard
- LOS ANGELES
- Santa Monica
- Pasadena
- Glendale
- Burbank
- San Fernando
- SAN BERNARDINO
- Riverside
- Redlands
- Banning
- ANAHEIM
- Santa Ana
- Long Beach
- Redondo Beach
- Santa Catalina
- San Clemente
- Oceanside
- Escondido
- SALTON SEA
- Imperial Valley
- Brawley
- El Centro
- SAN DIEGO
- Chula Vista
- El Cajon
- Mexicali
- Tijuana
- Ensenada
- Yuma
- BAJA CALIFORNIA NORTE
- SONORA
- GRAN DESIERTO
- ARIZONA
- UTAH
- NEVADA
- OREGON
- WASHINGTON
- CALIFORNIA
- IDAHO
- MONTANA
- COLUMBIA RIVER
- SNAKE RIVER
- CASCADE RANGE
- COAST RANGE
- SIERRA NEVADA
- COLORADO RIVER
- CHANNEL ISLANDS
- Santa Rosa
- San Miguel
- Santa Cruz
- Avalon

94

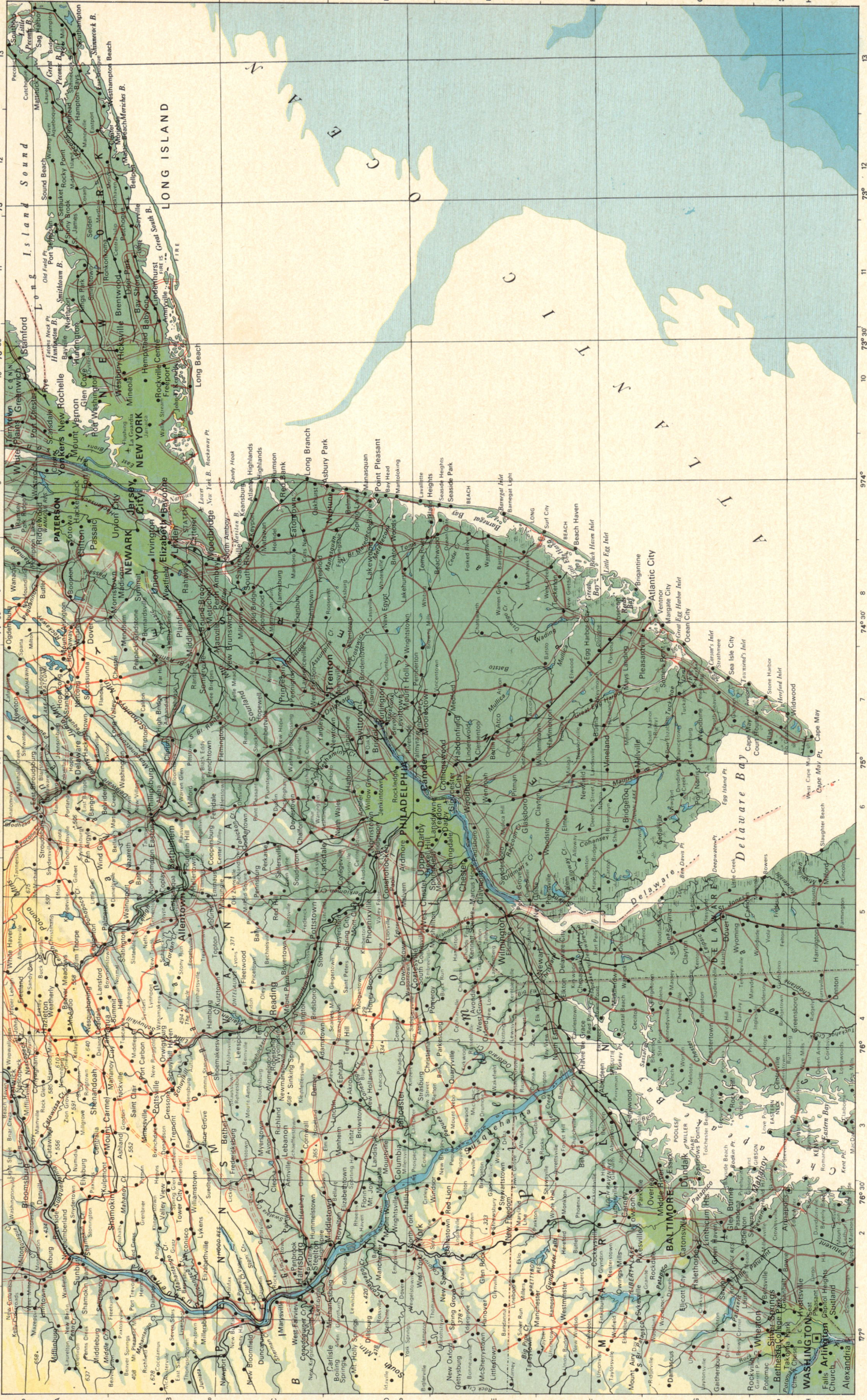

Aguilar-Geografic

LONG ISLAND

Long Island Sound

OCEAN

ATLANTIC

NEW YORK

PATERSON
PASSAIC
NEWARK
JERSEY CITY
ELIZABETH

Stamford
Greenwich
New Rochelle
Yonkers
Mount Vernon
White Plains

Long Beach

Long Branch
Asbury Park
Point Pleasant
Seaside Park
Barnegat Bay

Atlantic City
Margate City
Ocean City
Sea Isle City
Wildwood
Cape May

Delaware Bay

PHILADELPHIA
Camden

Trenton

Reading
Allentown
Bethlehem

Wilmington

DELAWARE

Dover

BALTIMORE

WASHINGTON
Alexandria

95

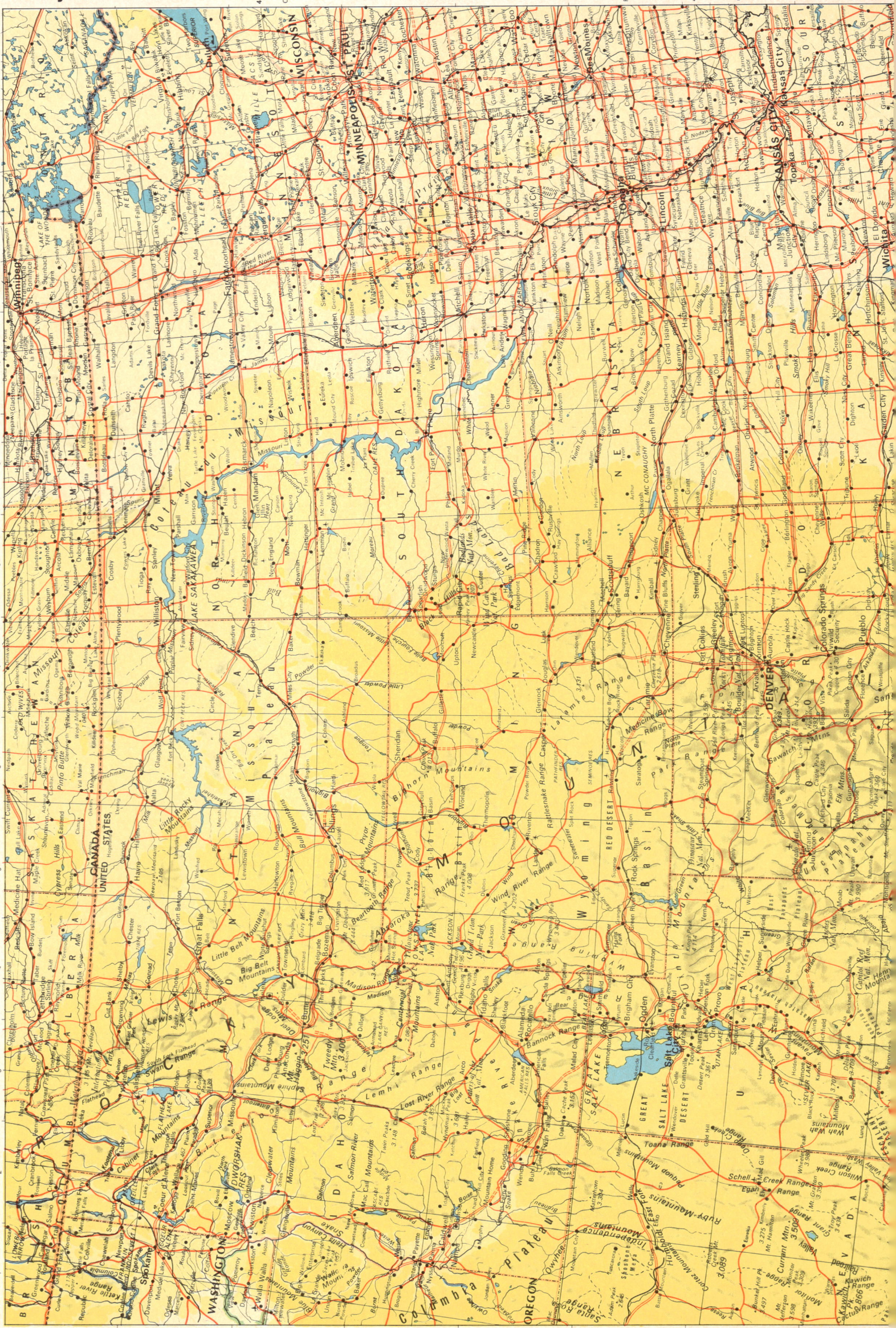

1 : 6,000,000

1" = 95 mi
1 cm = 60 km

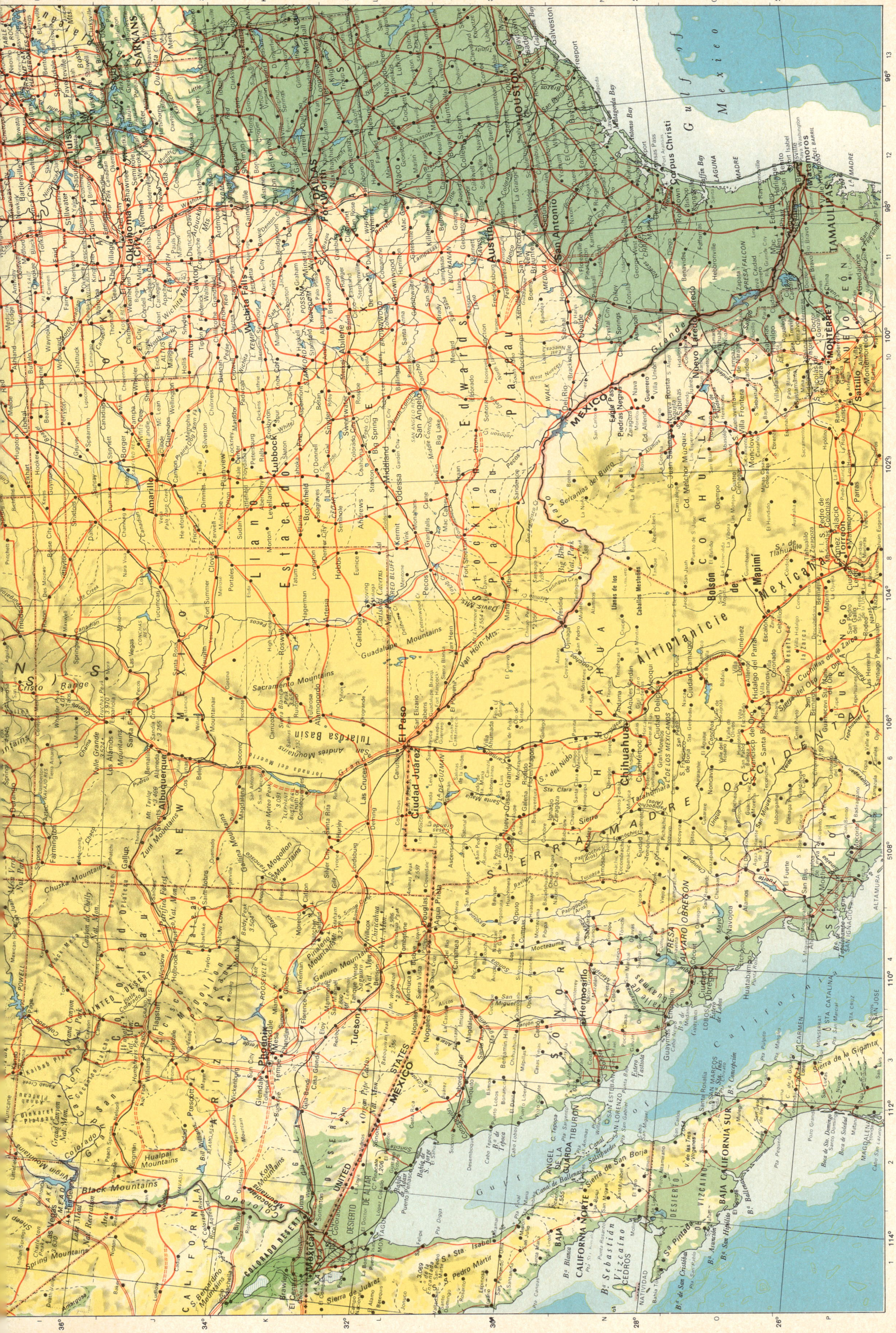

HOUSTON

DALLAS

San Antonio

Austin

Oklahoma

Wichita Falls

Amarillo

Lubbock

Midland

Odessa

San Angelo

Edwards Plateau

Llano Estacado

Stockton Plateau

Davis Mountains

El Paso

Ciudad Juárez

Las Cruces

Guadalupe Mountains

Sacramento Mountains

Roswell

Carlsbad

Tularosa Basin

San Andres Mountains

Albuquerque

Santa Fe

NEW MEXICO

ARIZONA

Phoenix

Tucson

Nogales

Gila Desert

Colorado Plateau

Colorado Desert

Mexicali

Flagstaff

Gallup

Farmington

CALIFORNIA

Black Mountains

Spring Mountains

Colorado

UNITED STATES MEXICO

DESIERTO DE ALTAR

BAJA CALIFORNIA NORTE

BAJA CALIFORNIA SUR

VIZCAINO

Hermosillo

Ciudad Obregón

SONORA

Guaymas

TIBURON

SIERRA MADRE OCCIDENTAL

CHIHUAHUA

Chihuahua

DURANGO

COAHUILA

Bolsón de Mapimí

Altiplanicie Mexicana

MONTERREY

NUEVO LEÓN

Saltillo

TAMAULIPAS

Matamoros

Nuevo Laredo

Piedras Negras

Eagle Pass

Del Rio

Río Grande / Río Bravo

MEXICO

Corpus Christi

Galveston

Gulf of Mexico

Golfo de California

Gulf of Mexico

MONCLOVA

SINALOA

1 : 3,000,000 1" = 47.5 mi 1 cm = 30 km

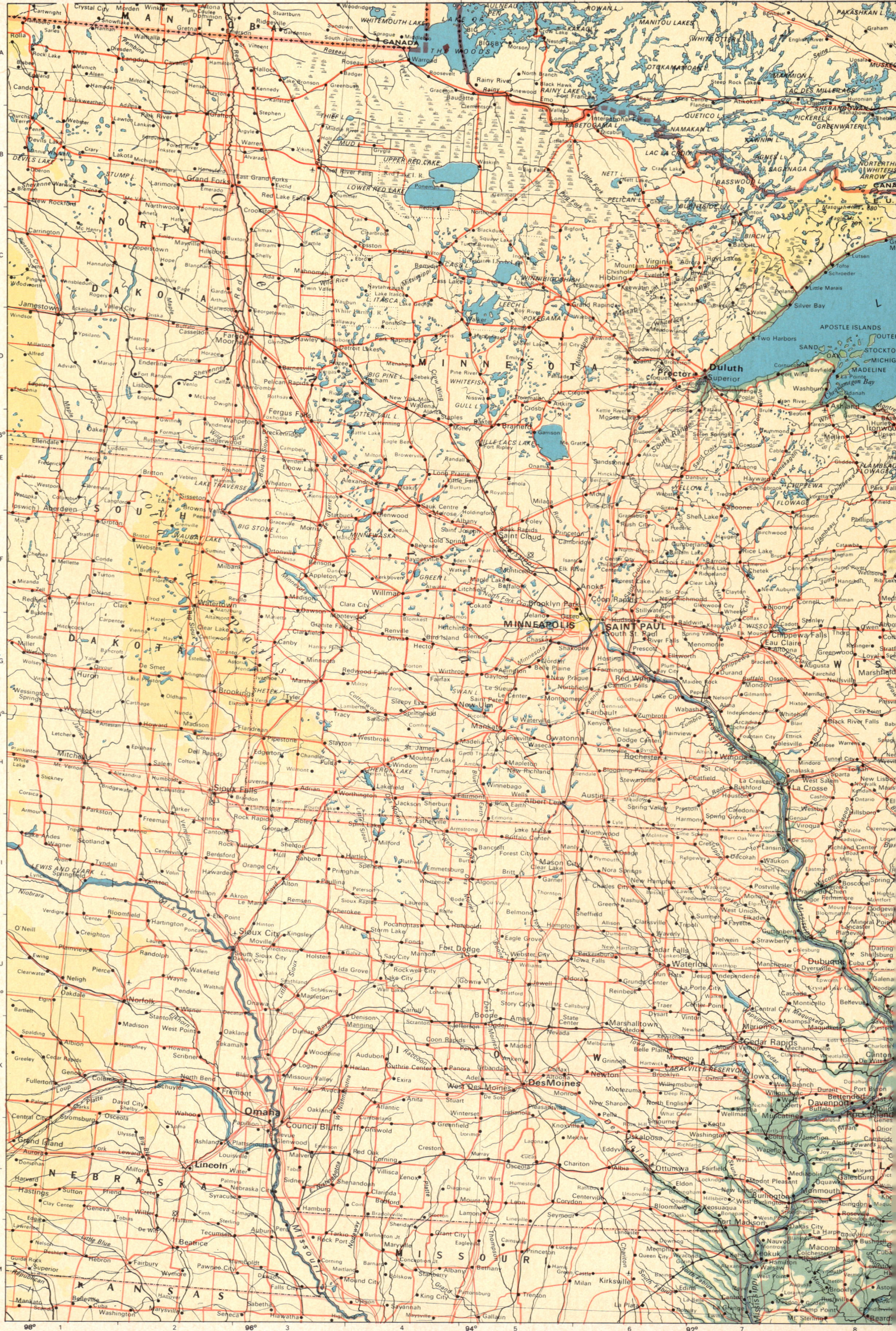

1 : 3,000,000

1 cm 47.5 mi / 30 km

LAKE SUPERIOR

LAKE HURON

Georgian Bay

LAKE ONTARIO

LAKE ERIE

CANADA

U.S.A.

ONTARIO

MANITOULIN

Algonquin Provincial Park

Saginaw Bay

M I C H I G A N

DETROIT
Dearborn
Windsor

TORONTO
Hamilton
Niagara Falls
BUFFALO
Rochester

Lansing

Flint

Saginaw
Bay City

Toledo

CLEVELAND

Akron
Youngstown
Canton

PITTSBURGH

Wheeling

Columbus

Dayton

CINCINNATI

Covington

W E S T   V I R G I N I A

K E N T U C K Y

P E N N S Y L V A N I A

N E W   Y O R K

Erie

Buffalo

Lexington

Huntington

ALLEGHENY MOUNTAINS

BLUE RIDGE MOUNTAINS

WASHINGTON

BALTIMORE

QUÉBEC

Parc Provincial du Mont Tremblant

NEW BRUNSWICK

MONTRÉAL
Laval
Ottawa
Hull

Sherbrooke
Granby

Adirondack Mountains

VERMONT

NEW HAMPSHIRE

MAINE

Augusta
Portland
Biddeford

Gulf of Maine

Watertown
Utica
Syracuse
Oneida
Rome
Herkimer

Saratoga Springs
Glens Falls
Albany
Schenectady
Amsterdam
Gloversville

TROY
Pittsfield
MASSACHUSETTS

Springfield
Worcester
BOSTON
Cambridge
Lynn
Salem

Massachusetts Bay

CONNECTICUT
Hartford
Waterbury
New Haven
Bridgeport
Norwalk
Stamford

New London
Providence
Pawtucket
New Bedford

CAPE COD
Cape Cod Bay
Chatham

Nantucket Sound
MARTHA'S VINEYARD
NANTUCKET

Scranton
Wilkes-Barre

PATERSON
NEWARK
NEW YORK
JERSEY
Elizabeth
Perth Amboy

LONG ISLAND
Long Island Sound

Allentown
Bethlehem
Reading
Trenton
New Brunswick
Lakewood
Point Pleasant

PHILADELPHIA
Camden
Wilmington

Mount Holly
Atlantic City
Ocean City

ATLANTIC OCEAN

Cape May
Rehoboth Beach
Bethany Beach

Salisbury
Cambridge

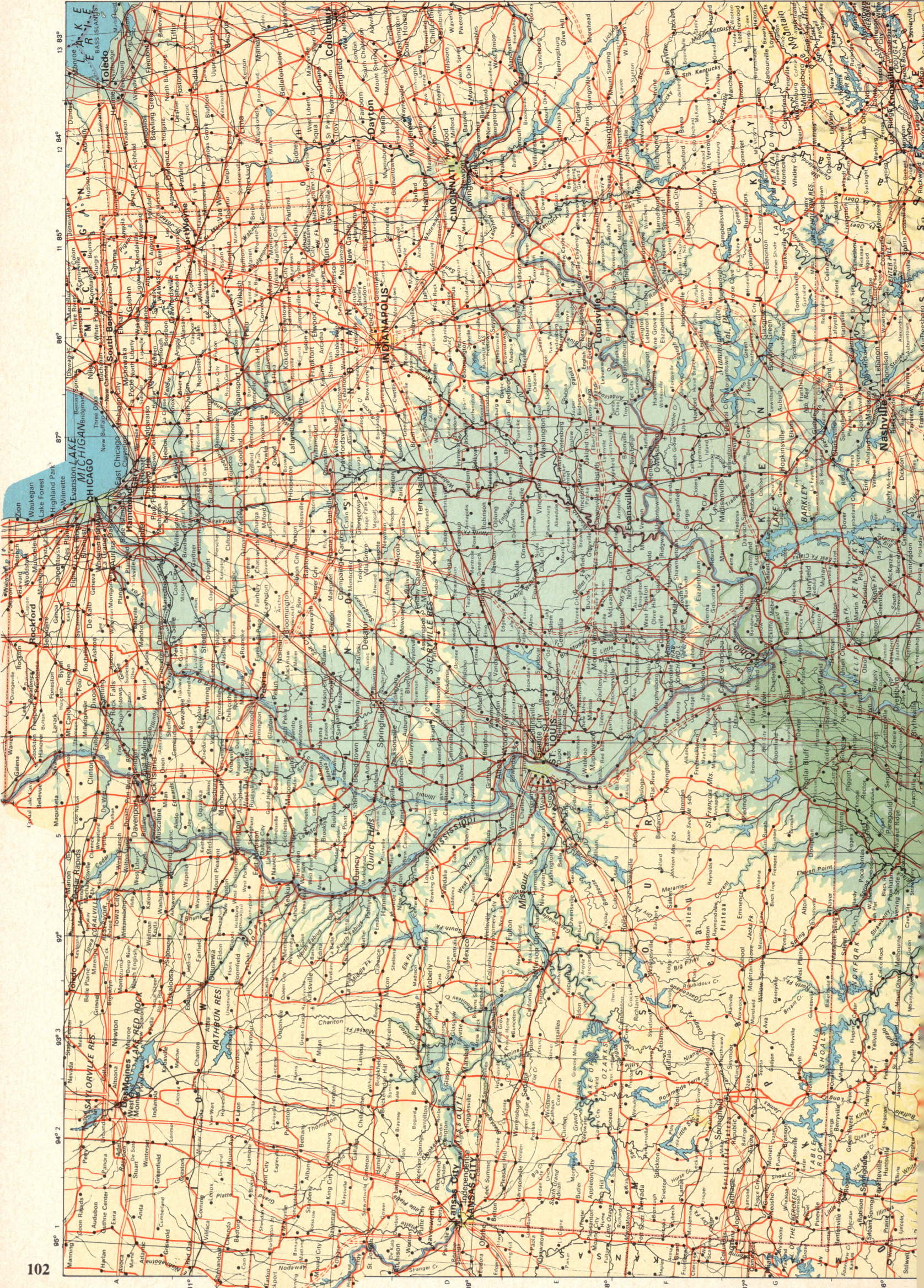

1 : 3,000,000   1 cm = 47,5 mi / 30 km

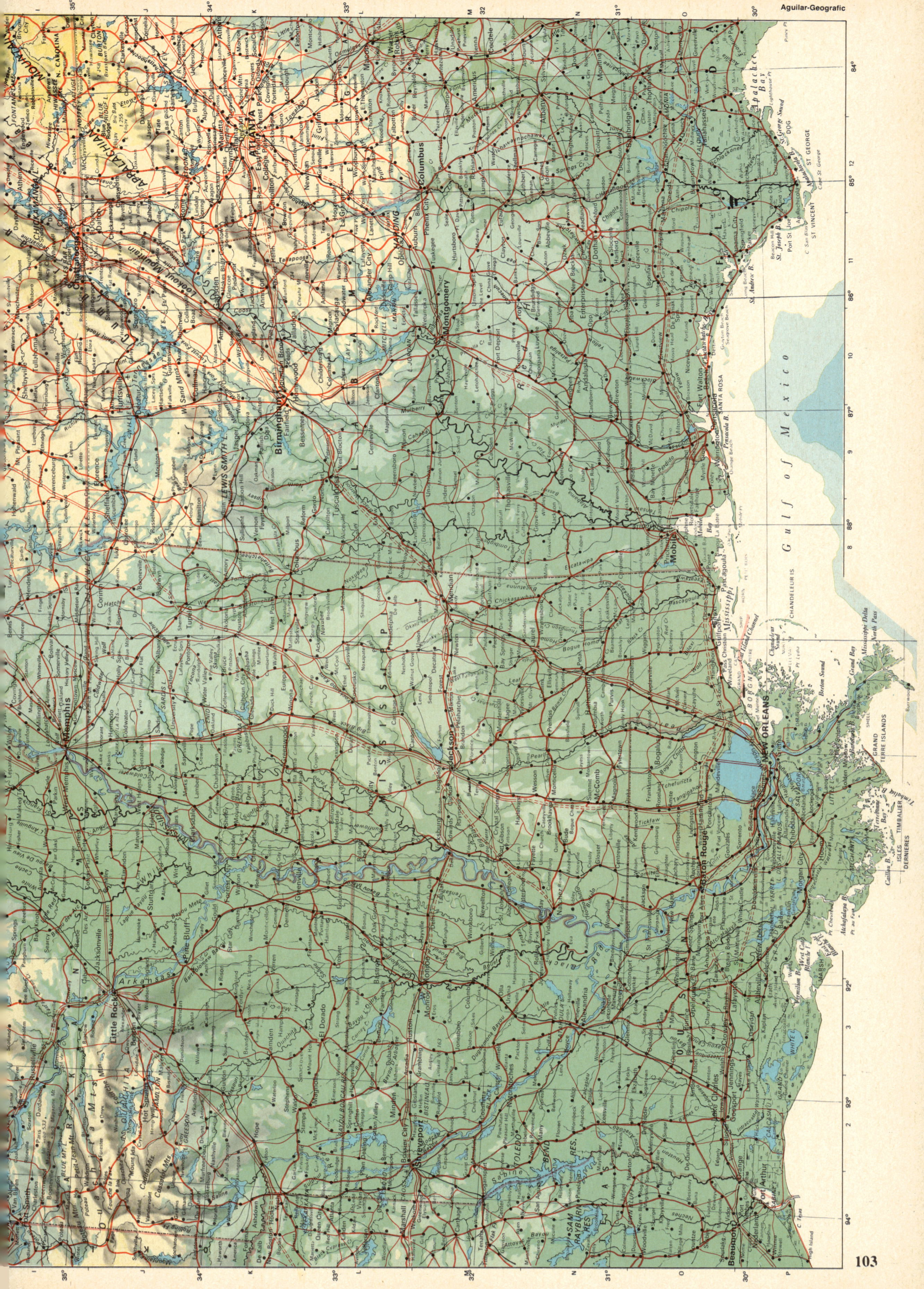

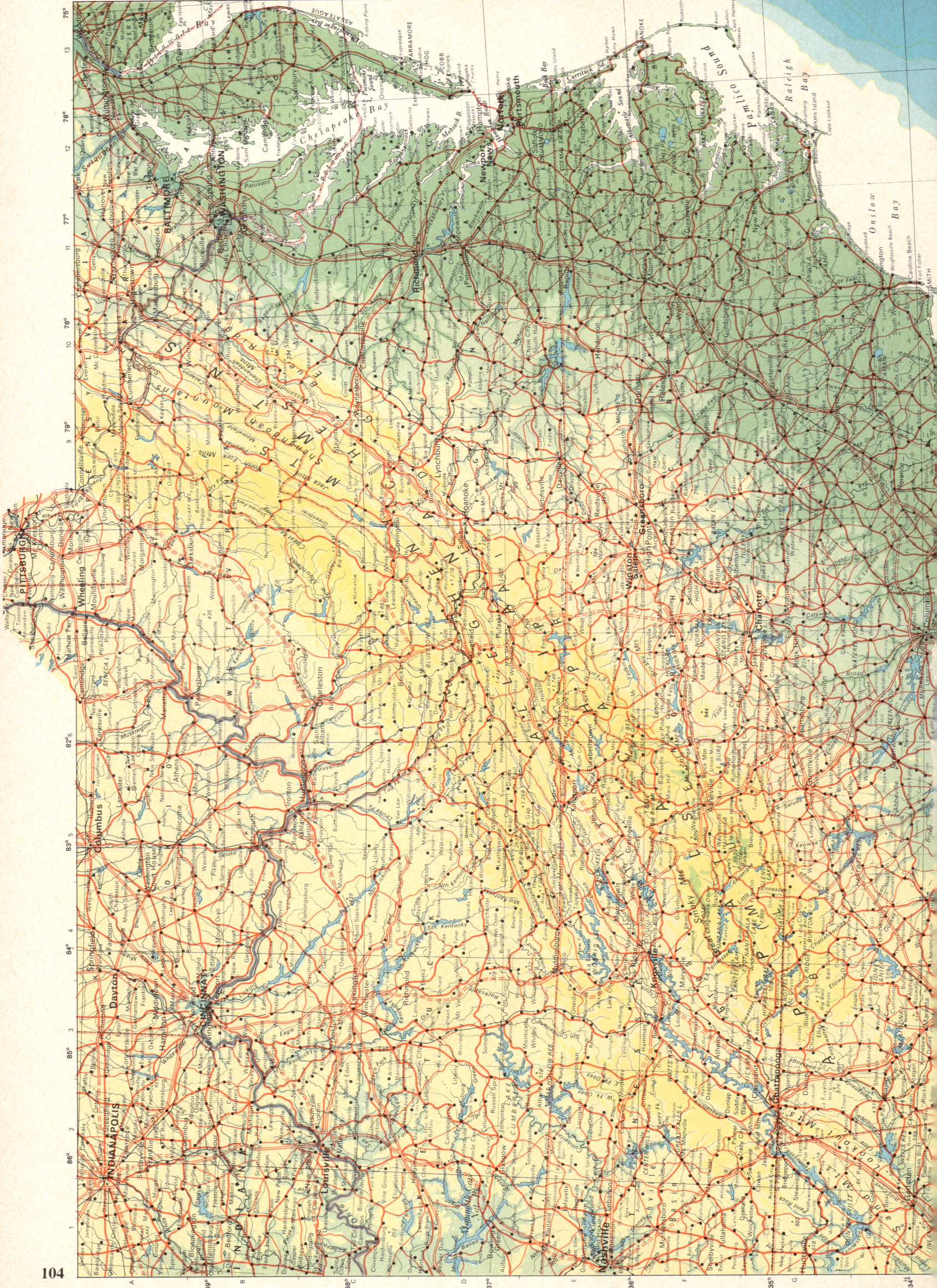

1 : 3,000,000

1" = 47.5 mi
1 cm = 30 km

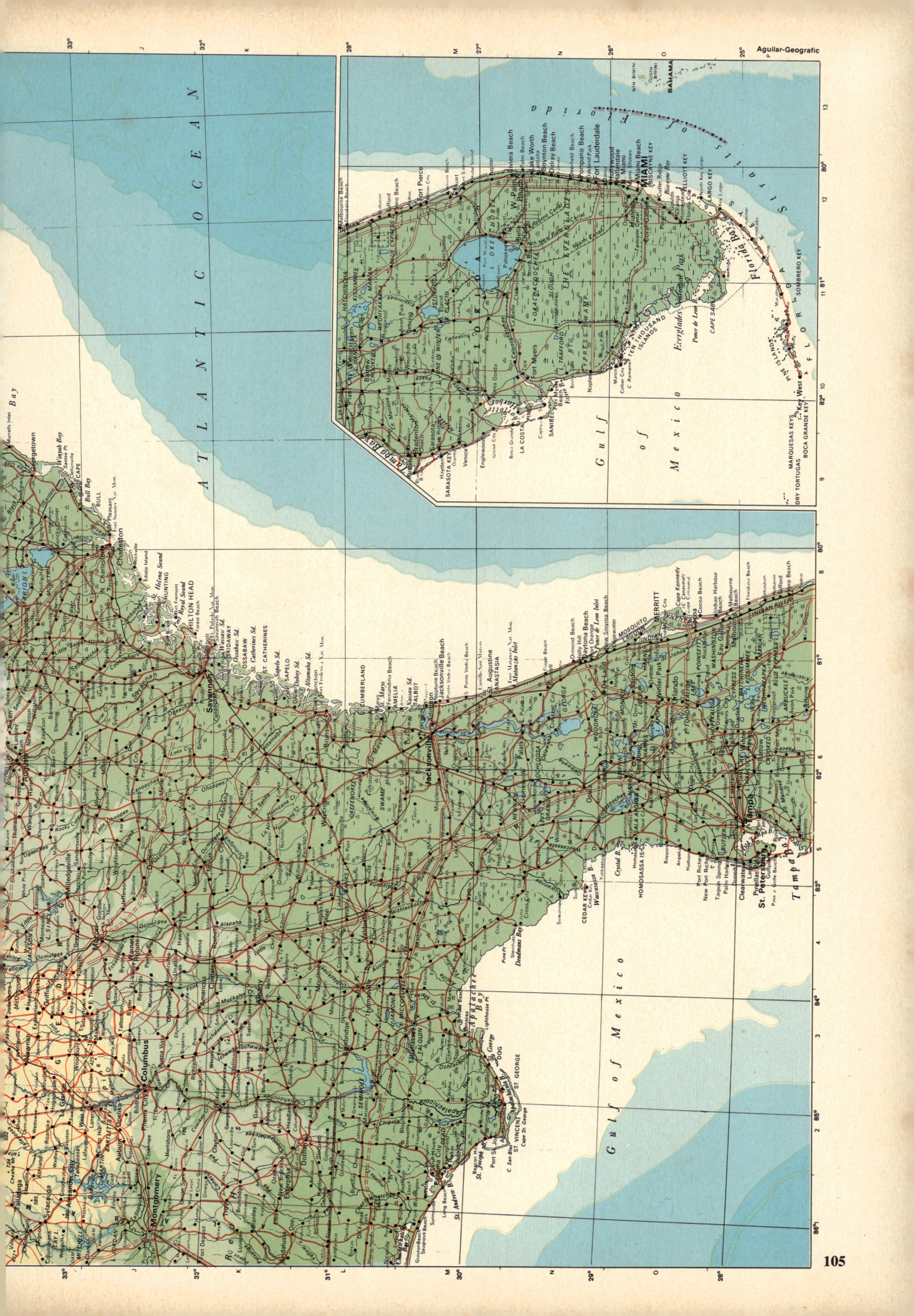

ATLANTIC OCEAN

BAHAMA

NEW BIMINI

Florida

Straits of Florida

Melbourne Beach
Vero Beach
Fort Pierce
W. Palm Beach
Palm Beach
Riviera Beach
Lake Worth
Boynton Beach
Delray Beach
Pompano Beach
Fort Lauderdale
Hollywood
Hialeah
MIAMI
Miami Beach
Coral Gables
KEY BISCAYNE
Cutler Ridge
KEY LARGO
ELLIOTT KEY
Homestead

FLORIDA
THE EVERGLADES
BIG CYPRESS SWAMP
Okeechobee
L. OKEECHOBEE
Kissimmee

Fort Myers
SANIBEL ISLAND
Naples
TEN THOUSAND ISLANDS
CAPE SABLE
Everglades Nat. Park
Flamingo

SOMBRERO KEY
Key West
FLORIDA KEYS
MARQUESAS KEYS
DRY TORTUGAS
BOCA GRANDE KEY

Sarasota
SARASOTA KEY
Venice
LA COSTA
Englewood
BOCA GRANDE

Gulf of Mexico

Winyah Bay
Georgetown
CAPE
BULL
Bull Bay
Charleston
N. Charleston
HUNTING
HILTON HEAD
Savannah
ST. CATHERINES
OSSABAW
St. Catherines Sd.
SAPELO
Sapelo Sd.
Brunswick
CUMBERLAND
St. Marys
AMELIA
Fernandina Beach
Jacksonville Beach
Jacksonville
St. Augustine Beach
St. Augustine
ANASTASIA
Matanzas Inlet
Ormond Beach
Daytona Beach
Port Orange
New Smyrna Beach
MOSQUITO
Titusville
MERRITT
Cape Kennedy
Cocoa Beach
Cocoa
Melbourne
INDIAN RIVER
Vero Beach

Orlando
Winter Park
OKEFENOKEE SWAMP
Lake City
Gainesville
Ocala
Leesburg

CEDAR KEYS
HOMOSASSA ISL.
Crystal B.
Tarpon Springs
Clearwater
St. Petersburg
PINELLAS
Tampa
TAMPA BAY
Bradenton

Tallahassee
St. Marks
St. George
ST. GEORGE
ST. VINCENT
Apalachicola
Apalachee Bay
Apalachicola Bay
Port St. Joe
C. San Blas
Panama City
Cape St. George

Columbus
Americus
Montgomery
Dothan

Gulf of Mexico

1 : 6,000,000 1" = 95 mi 1 cm = 60 km

UNITED STATES

CALIFORNIA

Tijuana
Mexicali
CORONADO

ARIZONA

NEW MEXICO

Tucson

El Paso
Ciudad Juárez

Stockton Plateau

BAJA CALIFORNIA NORTE

MIRAMAR

ANGEL DE LA GUARDA

TIBURON

SONORA

Hermosillo

CHIHUAHUA

Chihuahua

Delicias

Ciudad Camargo

COAHUILA del Norte

Meseta

CEDROS

NATIVIDAD

DESIERTO DE VIZCAÍNO

BAJA CALIFORNIA SUR

Ciudad Obregón

PRESA A. OBREGON

LOBOS

TORTUGA
SAN MARCOS

MONSERRAT
STA. CATALINA
STA. CRUZ

MAGDALENA

SANTA MARGARITA

CRECIENTE

ESPIRITU SANTO

CERRALVO

La Paz

San José del Cabo
Cabo San Lucas

ROCAS ALIJOS

SINALOA

DURANGO

Gómez Palacio
Torreón
Ciudad Lerdo

S. Pedro de las Colonias

ZACATECAS

Culiacán

Durango

Mazatlán

Zacatecas

Fresnillo

Aguascalientes

Meseta Centr

JALISCO

GUADALAJARA

Chapala

Tepic

ISLAS MARIAS
MARIA MADRE
MARIA MAGDALENA
MARIA CLEOFAS

ISLAS MARIETAS
Cabo Corrientes

SIERRA MADRE OCCIDENTAL

COLIMA
Colima

Manzanillo

PRESA DEL INFIERNILLO

MICH

REVILLA GIGEDO IS.

SAN BENEDICTO
ROCA PARTIDA
SOCORRO
CLARION

P A C I F I C   O C E A N

106

114°   112°   110°   108°   106°   104°   102°

DALLAS
Fort Worth
Shreveport
Jackson
Montgomery
MISSISSIPPI
LOUISIANA
ALABAMA
ARKANSAS
Austin
HOUSTON
San Antonio
Baton Rouge
NEW ORLEANS
Mobile
Pensacola
FLORIDA
CHANDELEUR IS.
Galveston
Mississippi Delta

Corpus Christi

Gulf

of

Mexico

Nuevo Laredo
MONTERREY
NUEVO LEÓN
Matamoros
Brownsville
Valle Hermoso
TAMAULIPAS
SIERRA MADRE
PRESA FALCON

TROPIC OF CANCER

Ciudad Victoria
Tampico
Ciudad Madero
VERACRUZ
SAN LUIS POTOSÍ
San Luis Potosí

ALACRAN REEF
DESTERRADA
PEREZ
ARENAS CAY
CAY NUEVO
CAMPECHE BANK
MADAGASCAR REEF
SISAL REEF
Progreso
Mérida
YUCATÁN

WEST TRIANGLE
EAST TRIANGLE

ARCAS CAYS

Poza Rica
Martínez de la Torre
Jalapa Enriquez
Veracruz

Gulf of Campeche

Campeche
CAMPECHE
QUINTANA ROO

CIUDAD DE MÉXICO
Toluca
Puebla
Cuernavaca
MORELOS
HIDALGO
Querétaro
GUANAJUATO
CUMBRES

Córdoba
Orizaba
Alvarado
Coatzacoalcos
Ciudad del Carmen
TABASCO
Villahermosa

BELIZE
Belmopan
TURNEFFE ISLANDS
AMBERGRIS CAY

GUERRERO
Acapulco
Chilpancingo
OAXACA
Oaxaca
ISTMO DE TEHUANTEPEC
Tuxtla Gutiérrez
CHIAPAS
SIERRA MADRE DE CHIAPAS
Las Casas

Golfo de Tehuantepec
MAR MUERTO
Puerto Ángel
Tapachula
Puerto Madero
Ciudad Hidalgo

GUATEMALA
Guatemala
HONDURAS
EL SALVADOR

Central America-the Caribbean

1 : 9,000,000

1" = 142 mi
1 cm = 90 km

UNITED STATES

DALLAS · Fort Worth · Austin · San Antonio · Corpus Christi · Houston · Beaumont · Baton Rouge · NEW ORLEANS · Shreveport · Jackson · Montgomery · Birmingham · ATLANTA · Columbus · Macon · Savannah · Jacksonville · Jacksonville Beach · St. Petersburg · Tampa · MIAMI · Miami Beach

ARKANSAS · LOUISIANA · MISSISSIPPI · ALABAMA · GEORGIA · SOUTH CAROLINA · THE EVERGLADES · FLORIDA KEYS · Key West · DRY TORTUGAS

Gulf of Mexico

MEXICO

Tampico · Ciudad Madero · Poza Rica · Jalapa Enriquez · Veracruz · Puebla · Oaxaca · Villahermosa · Mérida · Progreso · Campeche

YUCATÁN · QUINTANA ROO · CAMPECHE · TABASCO · CHIAPAS · OAXACA

ALACRAN REEF · DESTERRADA · CAYO ARENAS · CAYO NUEVO · WEST TRIANGLE · EAST TRIANGLE · CAYOS ARCAS

Cozumel · Puerto Juárez · Tulum

Sierra Madre de Chiapas · Tehuantepec · Quezaltenango · Tapachula

GUATEMALA · Guatemala · San Salvador · EL SALVADOR

BELIZE · Belmopan · AMBERGRIS CAY · TURNEFFE ISLANDS

HONDURAS · Tegucigalpa · San Pedro Sula · IS. DE LA BAHIA · ROATAN · SWAN ISLANDS (U.S.A.) (Hond.)

NICARAGUA · Managua · León · Puerto Cabezas · Bluefields · CORN ISLANDS

COSTA RICA · San José · Limón

PANAMÁ · Panamá · Colón · CANAL ZONE (U.S.A.) · Canal de Panamá

CUBA · LA HABANA · Matanzas · Santa Clara · Cienfuegos · ARCH. DE LOS CANARREOS · I. DE PINOS · ARCH. DE SABANA · JARDINES DE LA REINA

CAYMAN ISLANDS (U.K.) · GRAND CAYMAN · Georgetown · CAYMAN BRAC · LITTLE CAYMAN

CARIBBEAN SEA · PACIFIC OCEAN

SERRANILLA BANK (Col.) · SERRANA BANK (U.S.A.) · QUITASUEÑO BANK (U.S.A.) · ROSALINO BANK · PEDRO BANK · PROVIDENCIA (Col.) · SAN ANDRES (Col.) · CAYOS DE ALBURQUERQUE (Col.) · BANCO GUARDIAN

VENEZUELA · CARAIBES

GREATER ANTILLES

Yucatan Channel

Inset: TRINIDAD TOBAGO

1 : 1 200 000

VENEZUELA · Gulf of Paria · TRINIDAD TOBAGO · Port of Spain · San Fernando · La Brea · Point Fortin · Sangre Grande · Princes Town · Rio Claro

ST. GEORGE · ST. ANDREW · ST. DAVID · CARONI · NARIVA · MAYARO · VICTORIA · ST. PATRICK · Gulf of Paria · Dragon's Mouths · Galera Pt. · Manzanilla Pt. · Mayaro Bay · Guayaguayare · Icacos Pt. · Cedros Bay

CHACACHACARE · MONOS · GASPAR GRANDE · PATOS

0   10   20 Km

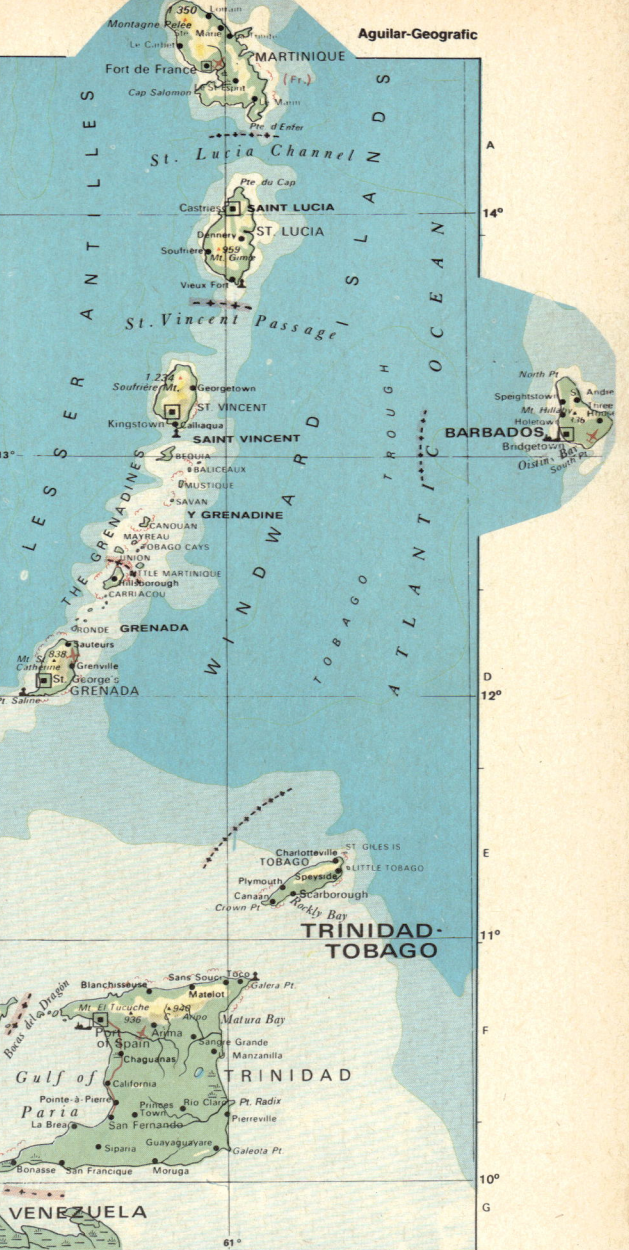
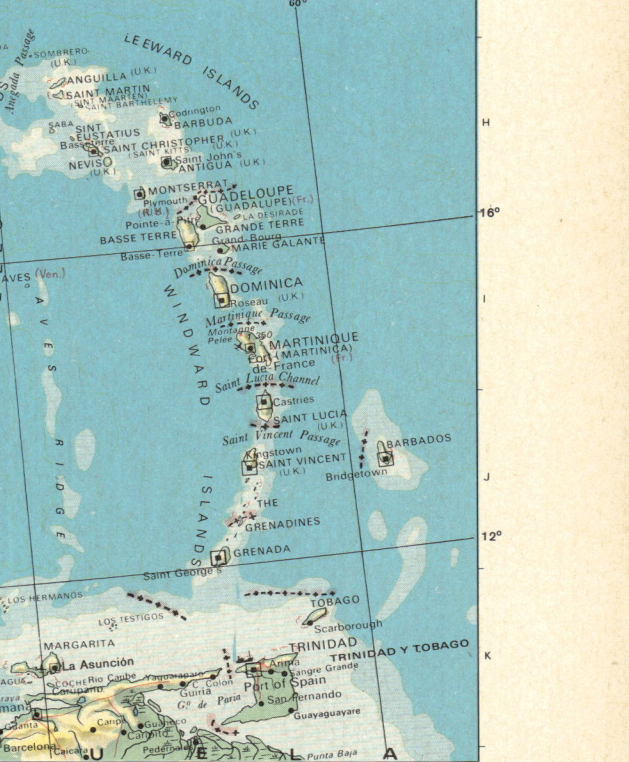

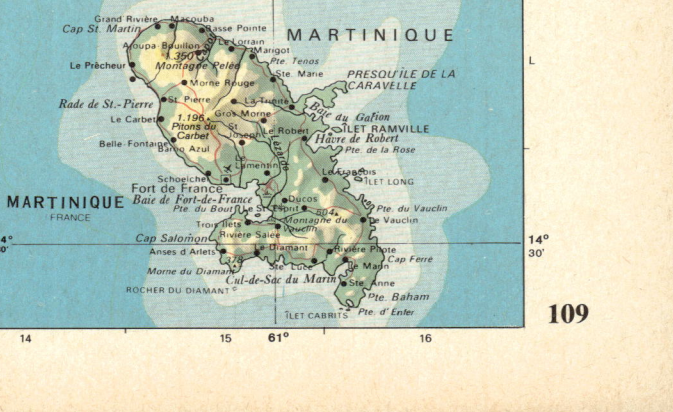

1:24,000,000 1" = 380 mi
1 cm = 240 km

ATLANTIC OCEAN

NORTH ATLANTIC

CAPE VERDE BASIN

C. VERDE

RIDGE

Fortaleza

Natal
Recife
Maceió
SALVADOR

BRAZIL

Belém
Brasília
Goiânia
MATO GROSSO
SERRA GERAL DE GOIÁS

São Luís
Teresina

Mouths of the Amazon

FRENCH GUIANA
SURINAME
GUYANA
Georgetown
New Amsterdam
Paramaribo

Manaus

VENEZUELA
CARACAS
Ciudad Bolívar
Maracaibo
Valencia
Barquisimeto

TRINIDAD - TOBAGO
BARBADOS
GRENADA
GRENADINES
St. George's

LESSER ANTILLES
MARTINIQUE (U.K.)
DOMINICA (U.K.)
GUADELOUPE
LEEWARDS
St. KITTS - ANTIGUA

VIRGEN IS.
San Juan

DOMINICAN REP.
HAITI
Santo Domingo
Port-au-Prince
HISPANIOLA

PERU

BOLIVIA
Cochabamba
CORDILLERA

COLOMBIA
BOGOTÁ
Medellín
Cali
Barranquilla
Cartagena
Santa Marta

ECUADOR
Quito
Guayaquil

Chiclayo
Trujillo
Chimbote
Callao

CORDILLERA ORIENTAL
CORDILLERA CENTRAL

PANAMA
Panamá

NICARAGUA
COSTA RICA
Managua
San José

HONDURAS
Tegucigalpa
EL SALVADOR
GUATEMALA
San Salvador

CARIBBEAN SEA

GREATER ANTILLES
CUBA
LA HABANA
Santiago de Cuba
JAMAICA
Kingston

BAHAMA
GRAN BAHAMA
ANDROS
ELEUTHERA
NASSAU

UNITED STATES
Jacksonville
Tampa
MIAMI
Charleston
Savannah

Gulf of Mexico

YUCATAN
Mérida
Campeche

MEXICO

PACIFIC OCEAN

GALAPAGOS IS.
ARCH. DE COLÓN

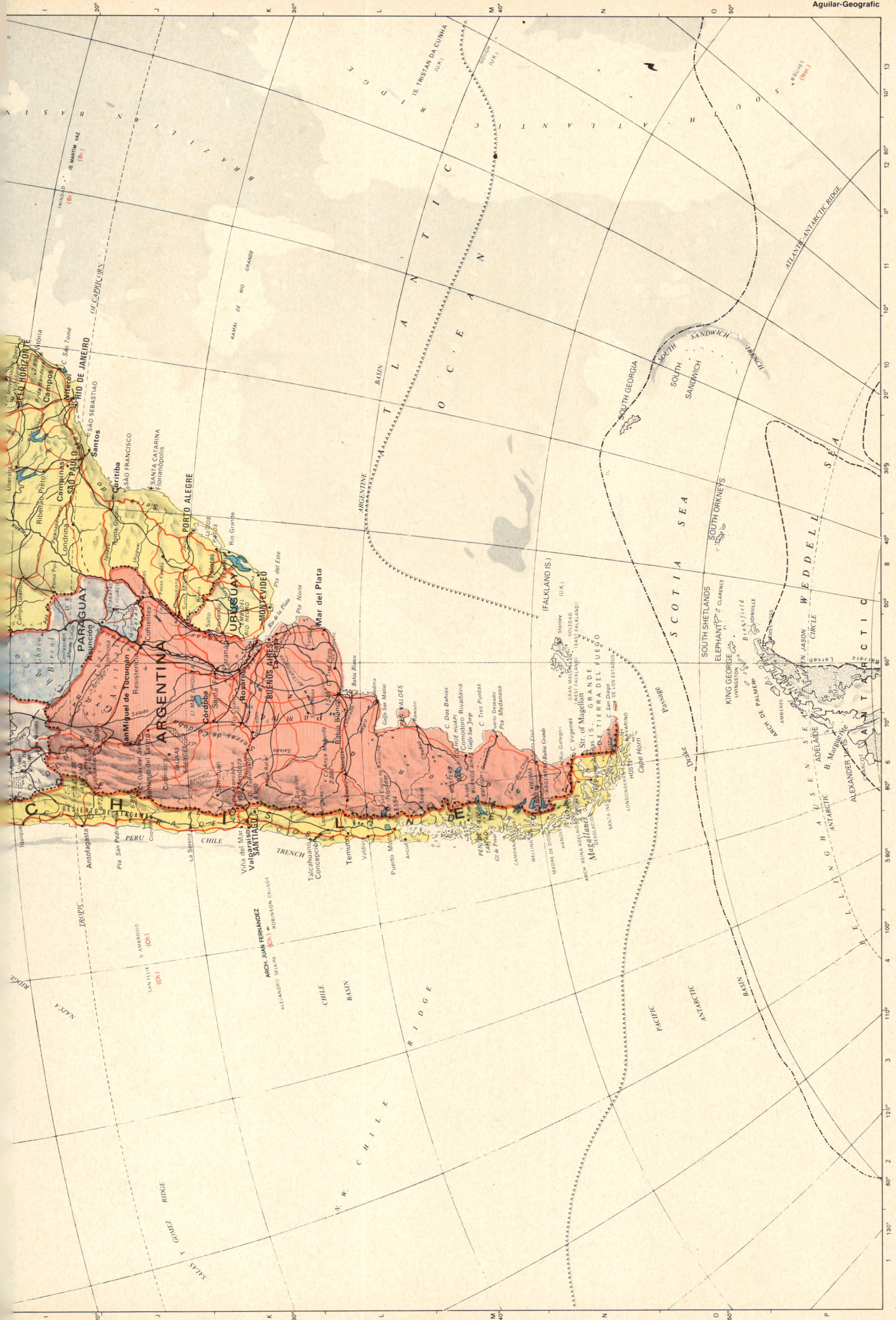

ATLANTIC OCEAN

SOUTH ATLANTIC RIDGE

ATLANTIC-ANTARCTIC RIDGE

BRAZILIAN BASIN

IS. TRISTAN DA CUNHA (U.K.)

TRINDAD (Br.) IS. MARTIM VAZ (Br.)

BR. DE CAPRICORN

BELO HORIZONTE
Campos
RIO DE JANEIRO
SÃO SEBASTIÃO
Santos
SÃO PAULO
Campinas
Ribeirão Preto
Londrina
Curitiba
SÃO FRANCISCO
SANTA CATARINA
Florianópolis
PORTO ALEGRE
Rio Grande

PARAGUAY
Asunción
URUGUAY
MONTEVIDEO
Mar del Plata
Pta. del Este

ARGENTINA
San Miguel de Tucumán
Resistencia
Corrientes
Santa Fe
Paraná
Rosario
Córdoba
BUENOS AIRES
La Plata
Bahía Blanca

CHILE
Antofagasta
PERU
Pta. San Pedro
La Serena
Viña del Mar
Valparaíso
SANTIAGO
Talcahuano
Concepción
Temuco
Valdivia
Puerto Montt
Ancud

TROPIC OF CAPRICORN

ARCH. JUAN FERNÁNDEZ (Ch.)
S. AMBROSIO (Ch.)
SAN FELIX (Ch.)
ALEJANDRO SELKIRK
ROBINSON CRUSOE

CHILE BASIN

CHILE RIDGE

N.W. CHILE RIDGE

NAZCA RIDGE

SALAS Y GOMEZ RIDGE

ARGENTINE BASIN

Golfo San Matías
Golfo San Jorge
C. Dos Bahías
Comodoro Rivadavia
C. Tres Puntas
Pta. Medanosa
Puerto Deseado
VALDÉS
NAHUEL HUAPI
Rawson

TIERRA DEL FUEGO
Str. of Magellan
Magallanes
C. Vírgenes
Río Grande
CAPE HORN
Cape Horn

FALKLAND (IS.)
GRAN MALVINAS (WEST FALKLAND) SOLEDAD (EAST FALKLAND)
Stanley

SOUTH GEORGIA

SOUTH SANDWICH
SOUTH SANDWICH TRENCH

SCOTIA SEA

SOUTH ORKNEYS

SOUTH SHETLANDS
ELEPHANT
CLARENCE
KING GEORGE
LIVINGSTON

WEDDELL SEA

ANTARCTIC CIRCLE

PALMER
ADELAIDE
ALEXANDER I IS.
ANTARCTIC

BELLINGSHAUSEN SEA

DRAKE PASSAGE

PACIFIC ANTARCTIC BASIN

DESIERTO DE ATACAMA

CARIBBEAN

HOYA DE COLOMBIA

SAN ANDRES Y PROVIDENCIA (Col.)
CAYOS DE RONCADOR (USA)
PROVIDENCIA
SAN ANDRES
CAYOS DEL E.S.E.
CAYOS DE ALBURQUERQUE

ARUBA (Hol.)
Oranjestad
LOS MONJES
PENINSULA DE LA GUAJIRA
Golfo de Venezuela
PEN. DE PARAGUANA

Sta. Marta
Riohacha
Barranquilla
Maracaibo
Cabimas
Cartagena
Valledupar
Ciudad Ojeda
Lagunillas
Barquisimeto
TRUJILLO
LAGO DE MARACAIBO
MÉRIDA
Mérida
Barinas

PANAMA
CANAL ZONE (U.S.A.)
Colón
Panama
ARCH. DE LAS PERLAS
PEN. DE AZUERO
COIBA

Golfo del Darién
Montería
BOLÍVAR
CÓRDOBA
Sincelejo
S. José de Cúcuta
TÁCHIRA
S. Cristóbal
Bucaramanga
SANTANDER

MEDELLÍN
ANTIOQUIA
CHOCÓ

CORDILLERA OCCIDENTAL
CORDILLERA CENTRAL
CORDILLERA ORIENTAL

Manizales
Pereira
Armenia
Ibagué
BOGOTÁ
CUNDINAMARCA
BOYACÁ
CASANARE
Yopal

COLOMBIA

VICHADA
META
Villavicencio
Pto. López

Buenaventura
Cali
Palmira
Tuluá
VALLE DEL CAUCA
Neiva
HUILA
TOLIMA

PACIFIC OCEAN

ARCHIPIÉLAGO DE COLÓN
GALAPAGOS IS.
DARWIN
WOLF
ISABELA
SANTA CRUZ
SAN CRISTÓBAL
ESPAÑOLA
FERNANDINA
MALPELO

Tumaco
NARIÑO
S. Juan de Pasto
PUTUMAYO
CAQUETÁ
Florencia

GUAINÍA
VAUPÉS
Mesa de Yambi
Mesa de Iguaje
Mesa de Sicayari

ESMERALDAS
IMBABURA
CARCHI
Quito
PICHINCHA
COTOPAXI
ECUADOR
NAPO
PASTAZA
MORONA
SANTIAGO

AMAZONAS

Guayaquil
GUAYAS
MANABÍ
LOS RÍOS
BOLÍVAR
CHIMBORAZO
TUNGURAHUA
CANAR
AZUAY
Cuenca
ORO
LOJA
ZAMORA
Golfo de Guayaquil
PUNA

CORDILLERA

PERÚ
LORETO
Iquitos
Marañón

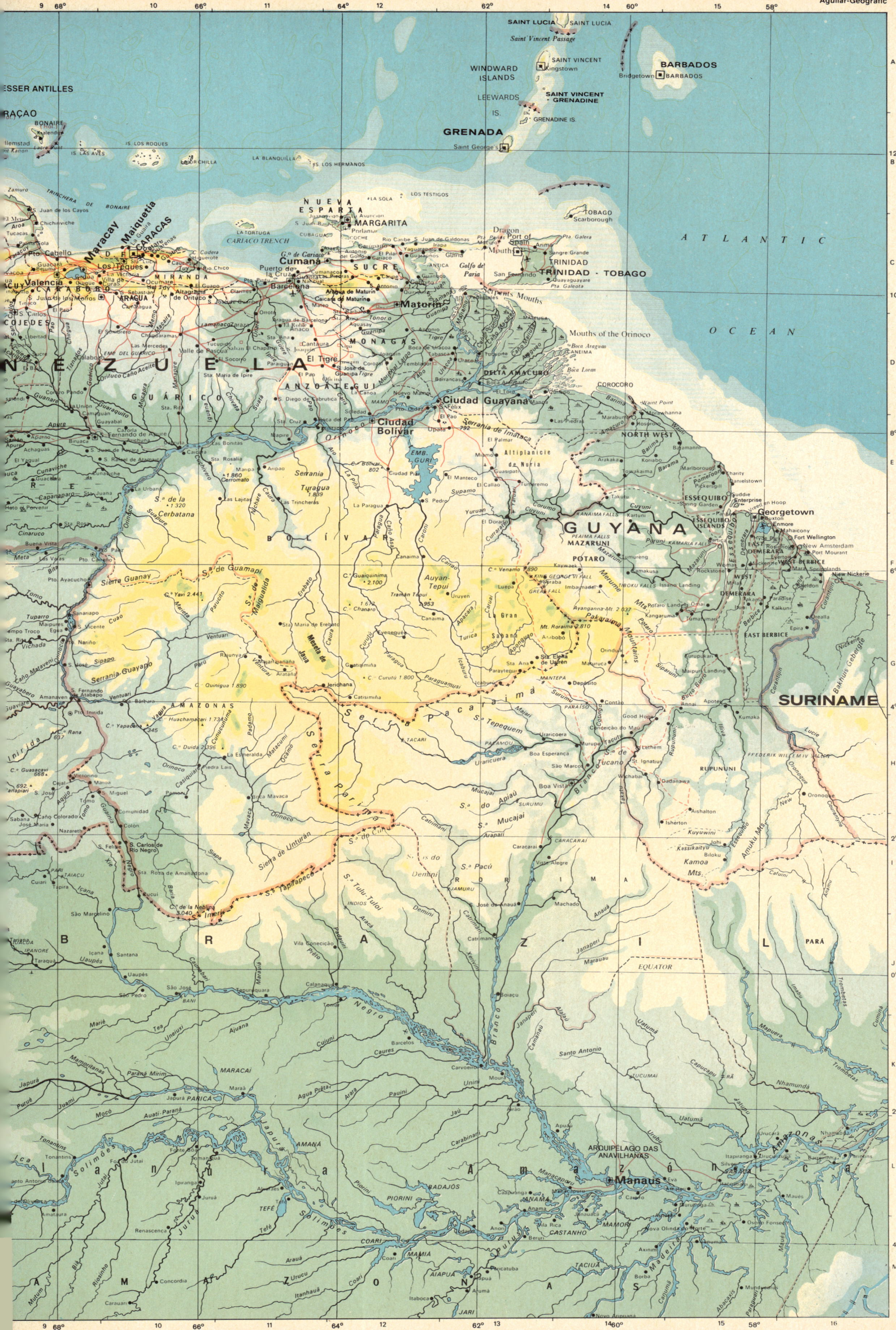

SAINT LUCIA · SAINT LUCIA
Saint Vincent Passage
WINDWARD ISLANDS
SAINT VINCENT · Kingstown
LEEWARDS IS.
SAINT VINCENT - GRENADINE
Grenadine Is.
GRENADA
Saint George's
BARBADOS · Bridgetown · BARBADOS

LESSER ANTILLES
CURAÇAO
BONAIRE
Willemstad
IS. LOS ROQUES
IS. LAS AVES
LA ORCHILLA
LA BLANQUILLA
IS. LOS HERMANOS
LA TORTUGA

TOBAGO
Scarborough

ATLANTIC

OCEAN

NUEVA ESPARTA
MARGARITA
COCHE
CUBAGUA
LA SOLA
LOS TESTIGOS

Maracay · Maiquetía · CARACAS
Los Teques
Cumaná
SUCRE
Barcelona
MONAGAS
Maturín
ANZOATEGUI
El Tigre

Dragon Mouth
Port of Spain
TRINIDAD
TRINIDAD - TOBAGO
San Fernando
Golfo de Paria
Serpents Mouths

V E N E Z U E L A
GUÁRICO
COJEDES
CARABOBO
ARAGUA
MIRANDA

Mouths of the Orinoco
DELTA AMACURO
COROCORO

NORTH WEST

Ciudad Guayana
Ciudad Bolívar
Serranía de Imataca
Altiplanicie d.z. Nuria
EMB. L. GURI

ESSEQUIBO
Georgetown
DEMERARA
WEST BERBICE
New Amsterdam
EAST BERBICE

BOLÍVAR
Serranía Tyragua
Sierra Guanay
Serranía Guayapo
Serra de Guampi

G U Y A N A
MAZARUNI
POTARO
PEAIMA FALLS
KAIETEUR FALLS
Auyán-Tepui
La Gran Sabana
Mt. Roraima 2.810
Pakaraima Mountains

SURINAME
RUPUNUNI
FREDERIK WILLEM IV MTN.

AMAZONAS
Serra Parima
Serra Tapirapecó
C.° de la Neblina 3.040
Serra de Unturán

RORAIMA
Boa Vista
Branco
S.° do Apiaú
SURUMU
Caracaraí

B R A Z I L
PARÁ

EQUATOR

Negro
Branco

ARQUIPÉLAGO DAS ANAVILHANAS
Manaus

A M A Z O N A S
Solimões
Juruá
Japurá

113

1: 6,000,000

1" = 95 mi
1 cm = 60 km

ECUADOR

COLOMB

PERU

BOLIVIA

CHILE

PARAGUAY

ARGENTINA

AMAZONAS

LORETO

ACRE

MADRE DE

Guayaquil

Cuenca

TUMBES

Piura

Iquitos

DESIERTO DE SECHURA

LAMBAYEQUE

Chiclayo

CAJAMARCA

LA LIBERTAD

Trujillo

Chimbote

ANCASH

Huánuco

Cerro de Pasco

LAGO DE JUNIN

LIMA

Callao

Huancayo

HUANCAVELICA

Cuzco

APURIMAC

AYACUCHO

Ica

Arequipa

MOQUEGU

ORURO

POTOSI

SANTA CRUZ

CHUQUISACA

TARIJA

TARAPACA

SALAR DE COIPASA

SALAR DE UYUNI

ANTOFAGASTA

BOQUERON

MILNE EDWARDS DEPTH

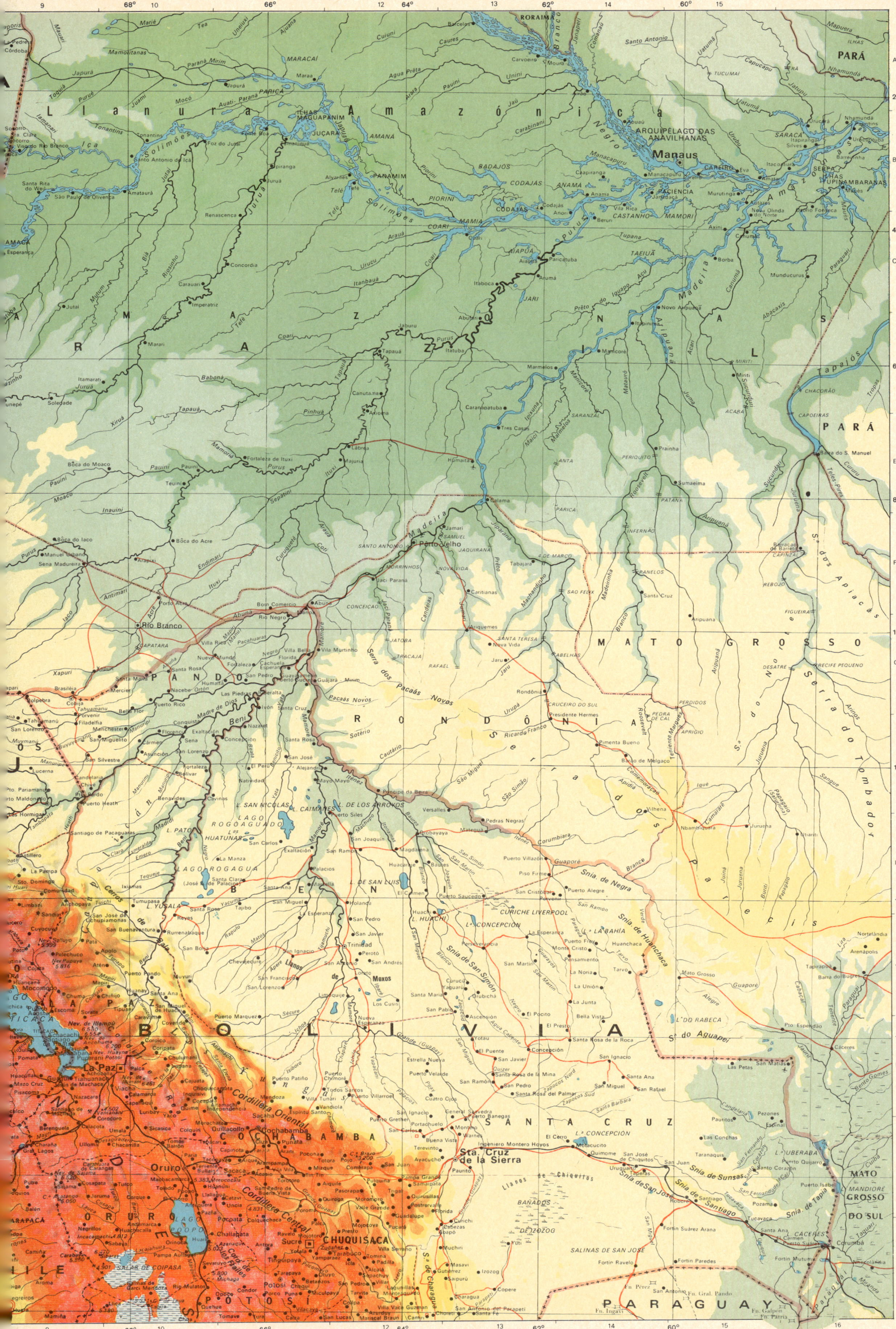

1 inch = 95 mi
1 cm = 60 km

PERU

BOLIVIA

LA PAZ
Oruro
COCHABAMBA
Sucre

SANTA CRUZ

ALTO PARAGUAY

PARAGUAY

PRESIDENTE

Iquique

Antofagasta

SALAR DE UYUNI
SALAR DE COIPASA
SALAR DE ATACAMA

JUJUY
San Salvador de Jujuy

SALTA
Salta

TUCUMÁN
SanMiguel de Tucumán

CATAMARCA
Catamarca

SANTIAGO DEL ESTERO
Santiago del Estero

CHACO

FORMOSA

Resistencia

Corrie

LA RIOJA
La Rioja

Tocopilla

SANTA FE

Caldera
Copiapó

La Serena

SAN JUAN
San Juan

CÓRDOBA
Córdoba

Santa Fé
Paraná

ENTRE RIOS

Rosario

C O R D I L L E R A   D E   L O S   A N D E S

Valparaíso
Viña del Mar
SANTIAGO

MENDOZA
Mendoza

SAN LUIS
San Luis

ARGENTINA

Gral. San Martín
Morón
Lomas de Zamora

BUENOS AIRES

LA PAMPA
Sta. Rosa

Talca
Linares

O C É A N O   P A C Í F I C O

MATO GROSSO DO SUL

BRASIL

MINAS GERAIS

Campo Grande

Uberlândia

Uberaba

Rb. Preto

São José de R. Preto

Araçatuba

Marília

Bauru

Campinas

Piracicaba

SÃO PAULO

S. Bernardo do Campo

Sorocabas

Jundiaí

Guarulhos

S. Caetano do Sul

S. Vicente

Santos

Pres. Prudente

Maringá

Londrina

Guarapuava

Ponta Grossa

Curitiba

Paranaguá

PARANÁ

SANTA CATARINA

Joinville

São Francisco

Blumenau

Itajaí

Brusque

Rio do Sul

Florianópolis

Lages

Criciúma

Laguna

RIO GRANDE DO SUL

Caxias do Sul

Passo Fundo

Santa Maria

PORTO ALEGRE

Canoas

N. Hamburgo

Pelotas

Rio Grande

CONCEPCIÓN

AMAMBAY

PARAGUAY

SAN PEDRO

CAAGUAZÚ

Villarrica

ITAPÚA

MISIONES

Posadas

URUGUAY

MONTEVIDEO

Rio de la Plata

ATLANTIC OCEAN

CAPRICORN

SÃO SEBASTIÃO

117

ATLANTIC

URUGUAY

MONTEVIDEO

BUENOS AIRES

Río de la Plata

Mar del Plata

Bahía Blanca

ARGENTINA

PAMPA

MENDOZA

Mendoza

SANTIAGO

Viña del Mar

Valparaíso

Talcahuano

Concepción

Valdivia

PEN. VALDÉS

Golfo San Matías

Golfo Nuevo

Rosario

SANTA FE

CÓRDOBA

SAN LUIS

SAN JUAN

NEUQUÉN

RÍO NEGRO

CHUBUT

Map of Southern South America (Patagonia, Tierra del Fuego and Falkland Islands)

Inset map (top right) — Antarctic region

SOUTH GEORGIA

SCOTIA SEA

SOUTH ORKNEY

CORONATION

OBSERVATORY (Arg.)

SOUTH SHETLAND

IS. FALKLAND O (U.K.) (MALVINAS)

ARGENTINA

CHILE

Río Gallegos

Magallanes

Punta Arenas

Ushuaia

TIERRA DEL FUEGO

BRITISH ANTARCTIC

WEDDELL SEA

ANTARCTIC CIRCLE

PALMER

ALEXANDER

BELLINGSHAUSEN SEA

ANTARCTICA

Ellsworth Highland

Mts. Hundred Miles

Executive Committee

1 : 30 000 000

500 Km

80°13

Main map

O C É A N O   A T L Á N T I C O

P A C I F I C   O C E A N

Falkland Islands (Malvinas)

FALKLAND ISLANDS (MALVINAS)

IS. SEBALDES

SOLEDAD

GRAN MALVINA

ROSARIO   TRINIDAD   BORBON

SAN CARLOS

GOICOECHEA

SAN RAFAEL

Bahía de la Anunciación

Pto. San Felipe

ES. DE LOS LEONES MARINOS

BEAUCHÉNE

ESTRECHO DE SAN CARLOS

BURDWOOD BANK

Mainland — Argentina

Comodoro Rivadavia

Golfo San Jorge

C. Tres Puntas

C. Blanco

Pto. Deseado

Puerto San Julián

SANTA CRUZ

GRANDE

Bahía Grande

Pampa del Desierto

Pampa del Asador

Río Santa Cruz

Río Gallegos

Río Grande

Pta. Dúngeness

BUENOS AIRES

CHUBUT

Tierra del Fuego region

GRANDE DE TIERRA DEL FUEGO

TIERRA DEL FUEGO

Pta. Arenas

Bº. San Sebastián

Río Grande

Ushuaia

PEN. DE BRUNSWICK

Punta Arenas

DAWSON

CLARENCE

SANTA INÉS

DESOLACIÓN

CAPITÁN ARACENA

HOSTE

NAVARINO

NUEVA

LENNOX

IS. WOLLASTON

Cabo de Hornos

HERMITE

Canal Beagle

STATEN ISLAND

Isla de los Estados

C. San Diego

Estrecho de Magallanes

DRAKE PASSAGE

Paso Drake

Chilean Patagonia / Archipelagos

ARCH. DE LOS CHONOS

PEN. DE TAITAO

Gº. de Penas

ARCH. GUAYANECO

PATRICIO LYNCH

ESMERALDA

WELLINGTON

MORNINGTON

CAMPANA

HANOVER

MADRE DE DIOS

DUQUE DE YORK

DIEGO DE ALMAGRO

CONTRERAS

REINA ADELAIDA

PEN. EXMOUTH

C. Pilar

Insets (bottom right)

ISLA SALA Y GÓMEZ

I. ROBINSON CRUSOE (Chile)

ALEJANDRO SELKIRK

ARCH. JUAN FERNÁNDEZ

EASTER I. (Chile)

IS. FÉLIX   IS. AMBROSIO (Chile)

GEOGRAPHICAL TERMS

BASIN: An area of land drained by a river
CABO: Cape (Spanish and Portuguese)
CANON: Canyon (Spanish)
CIMA: Mountain Peaks (Italy)
CIUDAD: City or Town (Spanish)
CONTINENTAL SHELF: A sea covered platform
CORDILLERA: Mountain range (Spanish)
FJORD: Glacial valley filled by the sea
GANGA: River (Indian)
GAWA: River (Japanese)
GEBEL: Mountain (Arabic)
GORA: Mountain (Slav.)
GUBA: Bay (Russian)
HAF: Sea (Swedish)
HO: River (Chinese)
JOCH: Pass (German)
KAHLI: Desert (Arabic)

KIANG: River (Chinese)
LAC: Lake (French)
LAGO: Lake (Spanish)
LOCH: Lake (Celtic)
MARAIS: Marsh (French)
MER: Sea (French)
MONT: Mount (French)
NAHR: River (Arabic)
OASIS: Fertile spot in a desert
OZERO: Lake (Russian)
PIZZO: Peak (Italian)
PUEBLO: Village (Spanish)
REEF: A ridge of rock or coral covered by sea
RIO: River (Portuguese, Spanish)
SELO: Village (Russian)
SIERRA: Mountain range (Spanish)
ZEE: Sea (Dutch)

GENERAL INDEX

c.,	Country		mt.,	Mount
cord.,	Mountain range		mts.,	Mounts
depr.,	Depression		mtn.,	Mountain
dist.,	District		mtns.,	Mountains
est.,	Estuary		pen.,	Peninsula
fed.,	Federal		prov.,	Province
fj.,	Fjord		rep.,	Republic
isl.,	Island		rge.,	Mountain
			volc.,	Volcano

Aachen, 42 F 3
Aaiun, 74 D 3
Ab'Uqayr, 70 I 7
Abacaxis;river, 113 M 16
Abaer, 52 B 3
Abai, 117 F 10
Abaiang;isl., 81 E 9
Abakaliki, 75 L 11
Abala, 78 C 2
Abancay, 114 I 7
Abapo, 115 M 12
Abar el Kana'is, 76 A 6
Abashiri, 62 C 6
Abashiri;river, 62 C 5
Abasolo, 106 E 8
Abau, 83 D 14
Abay;river, 77 J 10
Abbas Abad, 71 E 9
Abbeville, 103 J 6
Abbey Town, 49 G 11
Abbotsbury, 50 K 6
Abbotsford, 98 G 8
Abbottabad, 68 D 5
Abbottstown, 95 E 1
Abd-Al-Kuri;isl., 72 G 12
Abecha, 76 I 4
Abeiorral, 112 F 4
Abemama;isl., 81 E 9
Abenab, 79 J 3
Abengourou, 75 L 6
Abenra, 55 N 4
Abeokuta, 75 L 9
Aberaeron, 50 E 2
Aberargie, 49 A 11
Abercarn, 50 G 5
Abercorn, 78 F 8
Abercrombie, 98 D 3
Aberdeen, 94 C 3
Aberdovey, 50 D 3
Aberedw, 50 E 4
Aberffraw, 50 A 2
Aberford, 49 K 16
Aberfoyle, 49 A 9
Abergavenny, 50 G 5
Abergele, 49 M 11
Abernathy, 97 K 9
Abernethy, 49 A 11
Abersoch, 50 B 2
Abersychan, 50 G 5
Aberthaw, 50 H 4
Abertillery, 50 G 5
Aberystwyth, 50 D 3
Abha, 70 K 3
Abidjan, 75 M 6
Abilene, 96 H 11
Abingdon, 51 G 9
Abisko, 52 E 5
Abitau;river, 89 E 9
Abitibi;river, 90 H 5
Abiy Adi, 77 I 11
Abminga, 82 G 7
Abohar, 68 F 5
Aboisso, 75 M 7
Abomey, 75 L 9
Abong Mbang, 76 M 1
Aborlan, 65 E 10
Abou De-ia, 76 J 3
Abra Pampa, 116 D 4
Abrams, 99 G 10
Abrego, 112 D 6
Abri, 76 F 7
Abring, 66 E 13
Absecon, 95 F 8
Abu Dawm, 76 G 8
Abu Dawud el Sibakh, 77 C 13
Abu Deleiq, 76 H 8
Abu Dhabi, 76 M 16
Abu Dis, 76 F 8
Abu Gabra, 76 J 6
Abu Ghirban, 76 F 8
Abu Hamed, 76 F 8
Abu Hammad, 77 D 13
Abu Haraz, 76 F 8
Abu Higar, 77 I 9
Abu Matariq, 76 J 6
Abu Sultan, 77 E 15
Abu Suwen, 77 D 15
Abu Zabad, 76 I 7
Abu el Namrus, 77 G 12
Abufari, 115 D 13
Abujao;river, 114 F 6
Abukuma;river, 62 G 12
Abul-a, 75 K 10
Abulug, 65 A 11
Abumombazi, 78 A 5
Abuna, 115 F 11
Abuna;river, 115 F 11
Abuta, 62 E 2
Abwong, 76 J 8
Abyad, 76 H 6
Abydos, 82 F 3
Abyei, 76 J 7
Abymes, 109 M 12
Abyn-, 53 I 7
Acacias, 112 G 6
Acahay, 117 F 9
Acaill;isl., 47 K 2
Acajutla, 108 J 4
Acala, 107 L 14
Acambam, 107 J 9
Acanceh, 107 I 15
Acandi, 112 D 3
Acapetahua, 107 M 14
Acaponeta, 106 H 6
Acapulco, 107 L 9
Acaray;river, 117 E 10
Acari, 114 J 5
Acari;river, 114 J 6
Acarigua, 112 D 8
Acarigua;river, 113 D 9
Acatlan, 107 K 10
Acatzingp, 107 J 11
Acayucan, 107 K 12
Accra, 75 M 8
Accrington, 49 K 14
Acebuches, 106 D 8
Acedevo, 112 I 4
Acegua, 117 J 11
Achacachi, 115 K 9
Achadh an luir, 48 K 3
Achaguas, 113 E 9
Achalpur, 68 L 7
Acheng, 61 C 14
Achiras, 114 E 4
Achiyacu;river, 114 C 4
Achnasheen, 46 F 7
Achocalla, 115 K 10
Acipayam, 70 C 1
Aciraale, 39 M 10
Ackerman, 103 L 7
Acklington, 49 E 15
Acklins;isl., 109 F 10
Acle, 51 C 16
Acme, 103 N 4
Acobamba, 114 H 6
Acobambilla, 114 H 5
Acomayo, 114 F 5
Acomb, 49 J 16
Aconcagua;prov, 116 K 2
Aconchi, 106 C 4
Acoponeta;river, 106 H 6
Acora, 115 K 9
Acornhock, 79 L 8
Acos, 114 G 4
Acostambo, 114 H 5
Acre;river, 110 G 6
Acre;state, 114 F 7
Acteon;isl., 81 J 16
Acton, 102 D 10
Acton Turville, 50 H 7
Acton Vale, 90 K 8
Acumal, 107 I 16
Acuna, 117 I 9
Acworth, 104 H 3
Ada, 97 J 12
Adair, 98 M 8
Adairsville, 104 H 3
Adak;isl., 84 C 1
Adale, 77 M 13
Adam, 71 K 9
Adams, 98 B 2
Adams Landing, 88 F 7
Adamstown, 95 C 4
Adamville, 103 I 8
Adarama, 77 G 9
Adavale, 83 G 10
Adda;river, 38 D 4
Addagala, 77 J 12
Adderbury, 51 F 9
Addis, 103 O 4
Addis Derra, 77 J 10
Addis-Ababa, 77 J 10
Addison, 101 E 11
Adel, 98 K 5
Adelaida;isl., 33 C 6
Adelaide River, 82 B 6
Adelia Maria, 116 K 5
Adelphi, 100 L 2
Aden, 70 4
Adi Arkay, 77 I 10
Adi Remoz, 77 I 10
Adi Ugri, 77 H 10
Adi;isl., 65 C 13
Admiral, 89 L 9
Admiralty Gulf;gulf, 82 C 4
Admiralty Islands;isl., 83 A 13
Admiralty;isl., 86 G 3
Adonara;isl., 65 M 12
Adraj;oasis, 70 K 6
Adrano, 39 M 9
Adre, 76 I 4
Adrian, 99 K 14
Adriatic Sea, 69 I 1
Adrinn, 105 J 5
Adur;river, 51 J 12
Aduwa, 77 H 10
Advian, 98 D 1
Adyca;river, 57 E 12
Adzope, 75 L 6
Ae;river, 49 E 11
Aero;isl., 55 O 5
Aertaimian, 61 C 11
Aertunquke, 60 G 7
Aetna, 102 D 8
Afghanistan;c., 71 G 11
Afgoi, 77 M 13
Afmadu, 78 C 13
Afodo, 77 J 9
Afognak;isl., 86 E 1
Afram;river, 75 L 7
Afton, 96 E 4
Agades, 74 H 11
Agaire, 75 K 10
Agano;river, 62 G 11
Agar, 68 K 6
Agara, 68 H 7
Agawa, 99 C 13
Agboville, 75 L 6
Agere Hiywet, 77 J 10
Agere Maryam, 77 L 10
Aghadowey, 48 F 4
Aghalee, 48 H 5
Aginskoje, 57 I 11
Agly;river, 45 L 9
Agnibilekrou, 75 L 7
Agra;river, 37 C 10
Agrado, 112 I 4
Agri;river, 39 J 11
Agricola, 103 O 8
Agriento;prov, 39 N 8
Agrigento, 39 N 8
Agrihan;isl., 80 B 5
Agrio;river, 118 D 4
Agropoli, 39 J 9
Agua Blanca, 116 D 5
Agua Caliente de Vaca, 106 E 4
Agua Caliente;river, 115 K 13
Agua Clara, 112 G 6
Agua Colorado, 106 G 3
Agua Leon, 106 C 1
Agua Preta;river, 113 K 11
Agua Prieta, 106 B 4
Aguada Cecilio, 118 F 6
Aguada de Guerra, 118 F 5
Aguadas, 112 F 4
Aguadilla, 109 H 13
Aguan;river, 108 I 5
Aguanaval;river, 106 G 7
Aguanish, 91 H 13
Aguanus;river, 91 G 12
Aguapey;river, 117 H 9
Aguarague;cord., 114 L 4
Aguaray, 116 C 5
Aguaray Guazu;river, 117 E 9
Aguarico;river, 112 J 4
Aguas Blancas, 116 E 2
Aguas Calientes, 114 F 6
Aguas Dulces, 117 L 11
Aguas Vivas;river, 37 D 10
Aguasay, 113 D 12
Aguascalientes, 106 H 8
Aguayo, 116 J 4
Aguaytia, 114 F 5
Aguaytia;river, 114 F 5
Aguelhoc, 74 G 8
Aguila;isl., 119 M 9
Aguilar, 97 I 7
Aguilares, 116 G 4
Aguililla, 106 K 8
Agujita, 106 D 8
Agusan;river, 65 E 13
Aguzuut, 61 E 10
Ahad, 77 E 11
Aheqi, 60 E 3
Ahmadpur East, 68 G 4
Ahmed Wal, 68 F 1
Ahmedabad, 68 K 4
Ahoghill, 48 G 5
Ahome, 106 F 4
Ahoskie, 104 E 11
Ahsport, 103 I 6
Ahtari, 53 L 8
Ahuacatlan, 106 I 7
Ahuachapan, 108 J 4
Ahvenanmaa;isl., 53 O 6
Ahwar, 70 M 5
Aiapua, 113 M 13
Aiari;river, 112 I 8
Aibetsu, 62 C 4
Aidia, 83 C 13
Aigua, 117 L 10
Aihui, 61 B 13
Aija, 114 F 4
Aijal, 69 J 16
Aikawa, 62 G 10
Aiken, 105 I 6
Aileron, 82 F 7
Ailigandi, 112 D 3
Aillik, 91 D 13
Ails Craig, 99 I 16
Ailsa Craig;isl., 48 E 7
Aimogasta, 116 H 4
Ain Ben Tili, 74 E 5
Ain Dar, 70 I 6
Ain Galakka, 76 G 3
Ain el Heiz, 76 C 6
Ainabo, 77 J 13
Ainsworth, 98 L 7
Aioun el Atrouss, 74 H 4
Aipe, 112 H 5
Aiquile, 115 L 11
Air Force;isl., 87 E 11
Airabu;isl., 64 H 5
Airao, 113 K 13
Airbangis, 64 I 2
Aird Hills, 83 C 13
Airdrie, 49 C 10
Aire;river, 49 K 15
Airville, 95 E 5
Aisega, 83 B 14
Aisen;prov, 119 I 3
Aisen;river, 118 H 3
Aishalton, 113 H 15
Aishinik, 86 F 3
Aitape, 83 A 12
Aitkin, 98 D 5
Aitsu, 63 M 1
Aix-les-Bains, 45 H 12
Aiyansh, 88 G 2
Aizenay, 44 F 5
Aizuwakamatsu, 62 G 12
Ajaccio, 45 M 15
Ajaccio Golfe d;gulf, 45 M 15
Ajaju;river, 112 I 6
Ajalpan, 107 K 11
Ajan, 57 G 14
Ajana, 82 H 2
Ajax Mountain, 96 C 3
Ajdabiya, 76 B 3
Ajdovscina, 38 C 8
Aji;isl., 62 F 13
Ajigasawa, 62 C 12
Ajman, 70 I 8
Ajmer, 68 I 6
Ajon;isl., 57 C 14
Ajoupa-Bouillon, 109 L 15
Ajtos, 40 G 11
Ajuana;river, 113 K 11
Ajuchitlan del Progreso, 107 K 9
Akabira -, 62 C 3
Akan, 62 D 6
Akankohan, 62 D 6
Akasha, 76 F 7
Akayu, 62 G 12
Akcay;river, 41 N 13
Akchar;des., 74 G 3
Akdjoudt, 74 G 3
Akedamuhe;river, 60 H 6
Akelamo, 65 H 14
Aken, 103 M 1
Akernes, 55 J 3
Akesai, 60 F 7
Akesu, 60 E 3
Akesuhe;river, 60 E 3
Aketi, 78 B 5
Akhisar, 41 L 12
Akiavik, 86 D 4
Akimisk;isl., 87 J 12
Akimiski;isl., 90 F 5
Akin, 102 F 7
Akinum, 83 B 14
Akita, 62 D 12
Akjak, 86 C 1
Akkastugan, 52 E 5
Akkeshi, 62 D 6
Aklera, 68 J 6
Akobo, 76 K 8
Akobo;river, 77 K 9
Akola, 68 M 7
Akonolinga, 75 M 13
Akordat, 77 H 10
Akot, 68 M 7
Akpatok;isl., 91 A 10
Akron, 96 G 8
Aksum, 77 H 10
Aktogay, 56 J 7
Akulurak, 86 C 1
Akumyri, 52 A 3
Akure, 75 L 10
Akuse, 75 L 8
Ala Shan;des., 58 H 9
Alabama;river, 103 N 8
Alabama;state, 93 I 11
Alabat;isl., 65 C 11
Alachua, 105 M 5
Alagan, 60 E 5
Alah;river, 65 F 12
Alaharma, 53 K 7
Alahuixtlan;river, 107 K 9
Alajarvi, 53 K 8
Alaknanda;river, 68 F 8
Alalau;river, 113 J 14
Alamagan;isl., 80 B 5
Alameda, 96 E 3
Alamitos, 106 D 7
Alamo, 94 J 7
Alamogordo, 97 L 6
Alamor, 112 M 2
Alamos, 106 E 4
Alamos de Pena, 106 C 5
Alamos;river, 106 D 8
Alamosa, 97 I 7
Alandsbro, 53 L 4
Alanson, 99 F 13
Alaorg, 55 L 5
Alapaha, 105 K 4
Alapaha;river, 105 L 5
Alaquines, 107 H 9
Alashanshamo;des., 60 F 8
Alaska Range;cord., 86 D 1
Alaska;gulf, 84 E 3
Alaska;pen., 84 D 2
Alaskan Peninsula;pen., 86 D 1
Alausi, 112 L 2
Alava;prov, 37 B 9
Alavus, 53 L 8
Alazeja;river, 57 C 13
Alba, 102 F 1
Alba Posse, 117 G 10
Alba Iulia, 40 C 8
Albacete;prov, 37 H 9
Alban, 112 G 5
Albania;c., 41 I 4
Albany, 82 J 3
Albany;river, 90 G 4
Albardon, 116 J 3
Albemarle, 104 F 8
Alberdi, 116 F 8
Alberene, 104 C 10
Alberga;river, 82 G 7
Albert Lea, 98 H 5
Albert Nile;river, 78 B 9
Alberti, 116 L 7
Alberton, 91 J 12
Albertville, 104 H 1
Albia, 98 L 6
Albightsville, 95 A 5
Albion, 98 K 1
Albreda, 88 J 5
Albuquerque, 97 J 6
Alburtis, 95 B 5
Alca, 114 J 7
Alcala, 65 A 11
Alcamo, 39 M 7
Alcester, 50 E 8
Alco, 103 N 2
Alcoa, 104 F 4
Alconbury Hill, 51 D 12
Alcorn College, 103 M 4
Alcorta, 116 K 7
Aldabra;isl., 73 J 8
Aldama, 106 D 6
Aldamas Los-, 107 F 10
Aldan, 57 G 12
Aldan;river, 57 F 12
Aldbourne, 50 H 8
Aldbrough, 49 H 15
Aldeburgh, 51 E 16
Alden, 100 G 7
Alder Creek, 101 F 10
Alderley Edge, 49 M 14
Aldermaston, 51 H 10
Alderney;isl., 47 P 9
Aldershot, 51 I 10
Alderson, 104 C 7
Alderson, 88 L 8
Aldham, 47 K 9
Aldora, 105 I 3
Aldrich, 98 E 4
Aldsworth, 50 G 8
Ale;river, 49 D 12
Aledo, 98 L 8
Aleg, 74 H 3
Alegre;river, 115 J 15
Alegrete, 117 I 10
Alejander;isl., 111 P 6
Alejandra, 116 I 7
Alejandro Roca, 116 K 5
Alejandro Selkirk;isl., 111 K 4
Alejo Ledesma, 116 K 6
Alejuela, 108 L 6
Aleksandrovsk-Sakhalinskiy, 57 H 15
Aleksandrow Kujawski, 43 D 12
Aleksinac, 40 G 6
Aleman, 114 B 5
Aleria, 45 L 16
Alert, 87 A 10
Alert Bay, 88 J 2
Alerta, 114 G 7
Alesund, 54 E 3
Aleutian Range;cord., 86 E 1
Aleutians;isl., 84 B 1
Alexander, 96 B 8
Alexander Arhipelago, 86 G 2
Alexander;archp., 84 E 3
Alexandra Fiord, 87 B 10
Alexandria, 40 F 9
Alexandria, 115 H 11
Alexandria Bay, 101 E 9
Alexandria;prov, 38 D 3
Alexandroupolis, 41 J 10
Alexis, 98 L 8
Alexis Creek, 86 J 3
Alexis;river, 91 F 14
Alfatar, 41 F 11
Alfo Park, 104 H 2
Alfold, 51 J 11
Alford, 51 A 13
Alfred, 100 H 7
Alfreton, 51 A 9
Algard, 55 J 2
Algarrobal, 116 H 2
Algarrobo, 116 G 1
Algarrobo del Aguila, 116 4
Algena, 77 G 10
Alger, 99 H 14
Algeria, 34 M 4
Algodon;river, 114 B 7
Algodones, 106 A 1
Algoma, 99 G 11
Algonquin Provincial Park, 90 K 5
Algood, 104 E 2
Algorta, 117 J 9
Ali Sabieh, 77 J-12
Aliakmon;river, 41 K 6
Alibey;isl., 41 L 11
Alibori;river, 75 J 9
Alicante;prov, 37 H 11
Alice Arm, 88 G 2
Alice Springs, 82 F 7
Alice;river, 83 B 12
Aliceville, 103 L 8
Alicudi;isl., 39 L 9
Aliganj, 68 H 8
Aligarh, 68 H 7
Aline, 61 B 13
Alima;river, 78 C 2
Alindao, 76 L 4
Alingsas, 55 K 7
Alipur, 68 G 3
Alipura, 68 J 8
Aliquippa, 100 J 5
Alisos;river, 106 B 3
Alix, 88 J 7
Allagash, 101 A 15
Allagash;river, 101 A 15
Allakaket, 86 C 2
Allamo, 99 M 11
Allan Water, 90 G 1
Allan;river, 49 A 10
Allard, 87 J 15
Allardt, 104 E 3
Allegan, 99 J 12
Allegheny;river, 100 J 6
Allemands, 103 P 5
Allen, 65 C 12
Allen;river, 49 G 14
Allendale, 102 E 8
Allendale Town, 49 G 14
Allende, 106 E 6
Allenheads, 49 G 14
Allenhurst, 105 O 7
Allentown, 95 B 5
Allenville, 99 E 13
Alliance, 96 F 8
Alligator;river, 104 F 13
Allison, 98 J 6
Allison Harbour, 88 J 2
Alloa, 49 B 10
Allonby, 49 G 11
Alloway, 48 E 8
Alma, 96 G 12
Almer, 50 K 7
Almeria;prov, 37 J 9
Almhult, 55 M 8
Almimu Kolpos-;gulf, 41 P 2
Almirantazgo;isl., 80 F 5
Almirante, 108 L 7
Almirante Montt;gulf, 119 M 3
Almirante Tamandare, 117 F 13
Almiros, 41 L 7
Almond, 99 H 9
Almond;river, 49 C 11
Almora, 68 G 8
Almyra, 103 J 4
Aln;river, 49 D 15
Alnham, 49 E 14
Alnmouth, 49 E 15
Alnwick, 49 D 15
Aloja La-, 116 F 5
Alonnisos;isl., 41 L 8
Alonsa, 89 K 13
Alonso;river, 117 E 12
Alpachiri, 118 C 7
Alpena, 99 F 14
Alpine, 97 M 8
Alresford, 51 J 10
Alrewas, 50 C 8
Alright;isl., 91 J 13
Alsen, 98 A 1
Alston, 49 G 13
Alsuta, 62 D 3
Alta, 52 C 8
Alta Gracia, 116 J 5
Alta Vista, 118 D 8
Altaelva;river, 52 C 8
Altagracia, 112 C 7
Altagracia de Orituco, 113 C 10
Altai, 60 D 7
Altai;cord., 56 I 7
Altamachi, 115 K 11
Altamachi;river, 115 K 10
Altamaha;river, 105 K 6
Altamira, 107 H 10
Altamirano, 107 L 14
Altamont, 98 G 2
Altamura, 39 I 11
Altamura;isl., 106 F 4
Altaquer, 112 I 3
Altar, 106 B 3
Altar;des., 106 B 2
Altar;river, 106 B 3
Altata, 106 G 5
Altavista, 104 D 9
Altenburg, 102 F 6
Altenkirchen, 42 A 8
Altha, 105 L 2
Altheimer, 103 J 4
Alto Baudo, 112 F 3
Alto Chicapa, 78 G 4
Alto Molocue, 79 I 11
Alto Pelado, 116 K 4
Alto Pencoso, 116 K 4
Alto Purus;river, 114 G 8
Alto Rio Serguerr, 118 H 4
Alto Yurua;river, 114 F 7
Alton, 51 I 10
Altona, 42 C 6
Altoona, 98 G 7
Altrincham, 49 M 14
Altura, 98 H 7
Aluk, 76 K 6
Alula, 77 I 15
Alumine, 118 E 3
Alumine;river, 118 E 3
Alunhe;river, 69 H 13
Alva, 49 B 10
Alvarado, 98 B 3
Alvaraes, 115 B 12
Alvdal, 54 F 6
Alvdalen, 54 G 8
Alvear, 117 H 9
Alveston, 50 H 6
Alvin, 99 E 10
Alvsborg;land., 55 K 7
Alvsbyn, 52 H 7
Alwar, 68 H 7
Alwinton, 49 E 14
Am Dam, 76 I 4
Am Timan, 76 J 4
Amachonga;cord., 114 D 3
Amacuro;river, 113 E 14
Amacuzac;river, 107 K 10
Amadi, 76 L 7
Amagi, 63 L 1
Amahai, 65 K 14
Amakusa-Nada;sea, 63 N 1
Amakusa-Shotb;isl., 63 N 1
Amal, 55 J 7
Amalfi, 112 E 5
Amaluza, 112 M 2
Amambay;prov, 117 D 9
Amami;isl., 59 I 13
Amana;river, 113 C 13
Amanaven, 113 G 9
Amanaveni;river, 112 H 8
Amantea, 39 K 10
Amara Abu Sin, 77 H 9
Amaraca;isl., 115 C 9
Amaranth, 89 K 13
Amarillo, 97 J 9
Amarkantak, 69 L 9
Amarwara, 68 L 8
Amatan ., 107 L 14
Amatari, 115 B 15
Amataura, 115 B 10
Amatepec, 107 K 9
Amazone;river, 110 F 8
Amb, 68 D 5
Ambah, 68 I 8
Ambala, 68 F 7
Ambalema, 112 G 5
Ambam, 75 N 13
Ambanja, 79 L 12
Ambar, 114 G 4
Ambarchik, 57 C 14
Ambares-etagrave, 44 I 6
Ambato, 112 K 2
Ambatondrazaka, 79 M 12
Ambelau;isl., 65 K 13
Amberg, 99 F 10
Ambergate, 51 A 9
Amberley, 99 H 16
Ambesh, 68 B 5
Ambidedi, 75 I 3
Ambilobe, 79 L 13
Amble, 49 E 15
Ambler, 95 D 6
Ambleside, 49 I 12
Ambo, 114 G 5
Ambohidratrimo, 79 N 12
Ambohimahasoa, 79 O 12
Ambon, 65 K 14
Ambon;isl., 65 K 14
Amborompotsy, 79 O 12
Ambositra, 79 N 12
Ambovombe, 79 P 12
Amboy, 98 H 5
Ambridge, 100 J 5
Ambriz, 78 F 1
Ambrizete, 78 F 1
Ambrose, 105 K 5
Ambuar, 65 B 13
Ameca, 106 I 7
Ameghino, 116 L 6
Amelia, 103 P 5
Amelia;isl., 105 L 7
Amenia, 101 H 11
America, 103 K 1
American Falls, 94 F 8
American Fork, 96 G 3
American Samoa, 83 F 15
Americus, 105 K 3
Amersham, 51 G 11
Amery, 98 F 6
Amery;barr., 33 K 3
Amesbury, 50 I 8
Amet, 68 J 5
Amfipolis, 41 J 8
Amga, 57 F 13
Amga;river, 58 D 11
Amguid, 74 E 10
Amgun;river, 57 H 14
Amhara;reg, 77 I 10
Amherst, 91 K 12
Amherst;isl., 91 J 13
Amherstberg, 99 J 15
Amherstdale, 104 C 6
Aminga, 116 H 4
Aminuis, 79 L 4
Amirantes;isl., 73 J 13
Amite;river, 103 O 5

Beira,

Beirut, 70 E 3

Bradenton Beach,

Bradfield, 49 M 15
Bradfield Combust, 51 E 14
Bradford, 49 K 15
Bradley, 103 L 2
Bradleyville, 102 G 3
Bradwell, 51 G 14
Bragado, 116 L 7
Braganza, 105 L 5
Braggs, 103 M 10
Brahmani;river, 69 M 12
Brahmaur, 66 G 13
Braintre, 51 F 14
Braithwaite, 103 P 6
Bramley, 49 M 16
Bramon;isl., 53 L 4
Brampton, 49 F 13
Bramwell, 104 D 7
Brancaster, 51 B 14
Branchdale, 95 B 3
Branco;river, 115 I 14
Brandenburg, 102 E 10
Brandon, 49 G 15
Brandsby, 49 J 16
Branford, 105 M 5
Braniewo, 43 A 13
Bransk, 43 C 16
Branson, 102 G 2
Branston, 51 A 11
Brantford, 90 M 5
Brantley, 105 K 1
Brasher Falls, 101 D 10
Brasilia, 115 G 9
Brass, 75 M 10
Bratsk, 57 H 10
Brattleboro, 101 G 12
Brava;isl., 75 P 1
Bravo del Norte;river, 107 D 9
Braxton, 103 M 6
Bray;isl., 87 E 11
Braymer, 102 D 2
Brazean, 86 J 5
Brazil, 102 D 9
Brazo de Loba;river, 112 D 5
Brazzaville, 78 D 2
Brda;river, 43 B 12
Bre, 47 L 6
Brea Pozo, 116 H 5
Breamish;river, 49 D 14
Breaux Bridge, 103 O 4
Breckenridge Hill, 102 C 2
Brecknock;pen., 119 N 4
Breezewood, 100 K 7
Breford, 103 I 4
Bregenz, 42 K 5
Brekken, 53 K 1
Brekkhas, 54 G 2
Bremangerlandet;isl., 54 F 1
Bremen, 100 K 3
Brentford, 51 H 11
Brentwood, 51 G 13
Brenzett, 51 J 14
Brese, 102 E 6
Bressanone, 38 B 6
Bressay;isl., 46 C 9
Bressuire, 44 F 7
Brest, 44 D 3
Bretangne;pen., 34 H 4
Breton Woods, 95 D 9
Breueh;isl., 64 G 1
Brevard, 104 G 5
Brewerton, 101 G 9
Brewster, 105 P 6
Brewton, 103 N 9
Bria, 76 K 4
Briar Creek, 95 A 3
Bribie;isl., 83 G 12
Brickaville, 79 N 13
Brickeys, 102 F 6
Bride, 49 I 9
Bridge of Allan, 49 B 10
Bridgeboro, 105 K 4
Bridgehampton, 95 A 13
Bridgend, 48 C 5
Bridgeport, 104 G 2
Bridgeton, 104 G 12
Bridgetown, 91 L 2
Bridgeville, 101 M 10
Bridgewater, 91 L 12
Bridlington, 47 K 10
Bridport, 101 F 11
Bridqeton, 95 F 6
Bridsboro, 95 C 4
Brielle, 95 D 9
Brieronest, 89 L 10
Brighouse, 49 L 15
Brighstone, 51 K 9
Brighton, 51 K 12
Brightshade, 104 D 4
Brightwell, 51 F 15
Brignogan, 44 D 3
Brikama, 75 J 1
Brindisi, 39 I 12
Brindisi;prov, 39 I 12
Brinkley, 103 J 4
Brinklow, 51 D 9
Brinkmann, 116 I 6
Brinson, 105 L 3
Brintoodarna, 53 N 2
Brion;isl., 91 I 13
Broad;river, 104 H 5
Broadback;river, 90 G 6
Broaddus, 103 N 1
Broadford, 104 D 6
Broadstairs, 51 H 16
Broadview, 89 K 11
Brock;isl., 86 B 7
Brockenhurst, 51 K 9
Brockport, 100 G 7
Brockton, 101 H 14
Brockville, 90 L 7
Brockway, 100 I 6
Bruce Rock, 82 I 3
Brumath, 45 C 13
Brundidge, 105 K 1
Brunei, 64
Brunswick, 101 E 14
Brunswick;pen., 119 N 4
Brus Laguna, 108 I 6
Brusque, 117 G 14
Brussels, 42 F 1
Bryant, 103 J 3
Bryne, 55 J 2
Bryson City, 104 F 4
Brzeg, 43 F 12
Brzesko, 43 G 14
Brzozow, 43 H 15
Bu'ayrat el Hsun El-, 76 A 2
Buatan, 64 I 3
Buba, 75 J 2
Bucaramanga, 112 E 6
Buchaman, 100 M 2
Buchanan, 75 L 4
Buchans, 91 H 15
Buchardo, 116 L 6
Bucharest, 77
Buchholf, 42 C 6
Buck Mountain, 95 A 4
Buck Run, 95 E 4
Buck;isl., 109 B 9
Buckden, 49 I 14
Buckfield, 101 E 14
Buckhannon, 100 L 5
Buckingham, 90 K 7
Bucksport, 101 D 15
Budak, 68 I 2
Budapest, 43 K 13
Budhir, 52 B 4
Buding, 64 J 5
Budjala, 78 B 4
Buea, 75 M 12
Buen Pasto, 118 H 4
Buena, 95 F 7
Buena Vista, 115 L 12
Buena Vista, 113 F 9
Buena Vista, 104 C 9
Buenaventura, 112 G 3
Buenavista, 107 G 10
Buenavista, 114 F 3
Buenavista de Cuellar, 107 K 10
Bueno;river, 118 E 2
Buenos Aires, 112 H 4
Buerjin, 60 C 5
Buesaco, 112 I 3
Bufareh, 65 B 15
Buffalo, 102 F 3
Buffalo Narrows, 89 H 9
Buffalo River, 86 H 6
Buffalo;river, 102 H 3
Buford, 104 H 3
Bug;river, 43 D 15
Buga, 112 G 4
Buglandar, 66 D 11
Bugsuk;isl., 65 E 9
Buguluna, 56 F 4
Buhut, 77 B 12
Buin, 83 C 16
Buiochar, 48 J 3
Buionach, 48 M 1
Buitama, 112 F 6
Buji, 83 C 12
Bujumbura, 78 D 8
Buka;isl., 36 F 6
Bukavu, 78 D 7
Bukene, 78 D 9
Bukittinggi, 64 I 2
Bukoba, 78 C 9
Bulan, 65 C 12
Bulandshahr, 68 H 7
Buldan, 41 M 13
Buldibuyo, 114 E 4
Buler, 68 F 5
Bulgan, 60 D 6
Bulgaria, 60 C 8
Bulgroo, 83 G 10
Bulhar, 77 J 12
Bull;isl., 105 J 9
Bulloo;river, 83 H 10
Bulls Gap, 104 E 5
Bulnes, 118 C 3
Bulo Burti, 77 M 13
Bulongji, 60 F 7
Bulukumba, 65 L 11
Bumba, 78 B 5
Bumbuli, 78 D 4
Bumthang, 69 H 15
Bun Brosnai, 48 L 2
Buna, 78 B 12
Bunbury, 82 J 2
Bunceton, 102 E 3
Bundaberg, 83 G 12
Bundey;river, 82 E 7
Bundi, 68 J 6
Bundooma, 82 F 7
Bunessan, 48 A 5
Bunga;river, 75 K 12
Bungay, 51 D 16
Bungoma, 78 C 10
Bunji, 66 C 10
Bunju;isl., 65 H 10
Bunker Group;isl., 83 F 12
Bunker Hill, 86 B 1
Bunkie, 103 O 3
Bunkris, 53 M 1
Bunnell, 105 N 7
Bunny, 51 B 10
Buntingford, 51 F 12
Bur Acaba, 77 M 12
Bur Gavo, 78 C 13
Bur Said, 77 A 15
Bur Tawliq, 77 G 16
Bura, 78 C 12
Buram, 76 J 5
Burao, 77 J 13
Buras, 103 P 6
Burathum, 69 H 11
Burdekin;river, 83 E 11
Bure, 77 K 9
Bure;river, 51 C 16
Burea, 54 B 12
Bureya;river, 57 H 14
Burgaw, 104 G 10
Burgeo, 91 I 15
Burgess Hill, 51 J 12
Burgh le Marsh, 51 A 13
Burghfield, 51 H 10
Burgos, 107 F 10
Burhaniye, 41 L 11
Burias;isl., 65 C 12
Burin, 91 I 16
Burin Peninsula;pen., 91 J 16
Buriti;river, 115 I 15
Burka, 77 L 11
Burke;river, 83 F 9
Burketown, 83 D 9
Burkeville, 103 N 2
Burley, 49 K 15
Burlington, 90 M 5
Burma, 59 K 9
Burnett;river, 83 G 12
Burney, 102 D 11
Burnham on Crouch, 51 G 14
Burnie, 83 L 10
Burnley, 49 K 14
Burnopfield, 49 G 15
Burnsall, 49 J 14
Burnside, 104 D 3
Burntisland, 49 B 11
Burntwood;river, 89 H 12
Burra, 75 K 12
Burrel, 41 I 4
Burro-Burro;river, 113 G 15
Burrsville, 95 H 4
Burrundie, 82 B 6
Burruyacu, 116 F 5
Burrwood, 103 P 7
Bursa, 41 K 13
Burscough, 49 L 12
Burton, 49 J 13
Burton Latimer, 51 D 11
Burton in Lonsdale, 49 J 13
Burtus, 77 F 11
Buru;isl., 65 K 13
Burundi, 73 J 9
Bury, 49 L 14
Bury Saint Edmunds, 51 E 14
Busango, 78 F 6
Busar, 69 J 11
Bush;river, 48 F 4
Bushey, 51 G 12
Bushman Pits, 79 J 6
Bushmills, 48 E 4
Businga, 78 B 4
Busira;river, 78 C 4
Busko Zdroj, 43 G 14
Bussa, 75 K 10
Busselton, 82 J 2
Bustamante, 107 E 9
Busu-Djanoa, 78 B 4
Busuanga, 65 D 10
Busuanga;isl., 65 D 10
Busunu, 75 K 7
Buta, 78 B 6
Buta Ranquil, 118 C 4
Bute;isl., 48 C 7
Butler, 100 J 5
Buttermere, 49 H 11
Button Islands;isl., 90 A 11
Buttville, 95 A 6
Butuan, 65 E 13
Butung;isl., 65 K 12
Butwal, 69 H 10
Buxton, 49 M 16
Buyuyomano;river, 115 H 9
Byas;river, 68 E 6
Bychawa, 43 F 16
Bydgoszcz, 43 C 12
Byfield, 51 E 9
Bygdiz, 54 G 4
Bygland, 55 J 3
Byglandsfjand, 55 J 3
Byhalia, 103 J 6
Bykle, 55 I 3
Bynginlet, 90 K 5
Byram, 103 M 5
Byrdstown, 104 E 2
Byrness, 49 E 13
Byron, 105 J 4
Byron;isl., 119 J 2
Byronville, 105 J 4
Byske alv;river, 54 A 11
Bythorn, 51 D 11
Bythorn, 51 D 11
Byton, 43 G 13
Byton, 43 G 13
Caacupe, 117 E 9
Caaguazu, 117 F 10
Caaguazu;prov, 117 E 10
Caapiranga, 115 B 14
Caapucu, 117 F 9
Caazapa, 117 F 9
Caazapa;prov, 117 F 9
Cabacal;river, 115 J 16
Cabalian, 65 D 13
Caballococha, 114 C 8
Cabana, 114 F 4
Cabanaconde, 114 J 7
Cabanatuan, 65 B 11
Cabano, 91 J 10
Cabano, 87 K 14
Cabarruyan;isl., 65 B 10
Cabezas, 115 M 12
Cabildo, 116 K 2
Cabimas, 112 C 7
Cabinda, 78 E 1
Cable, 98 E 7
Cabo Blanco, 119 J 6
Cabo Pantoja, 114 A 5
Cabo Raso, 118 H 6
Cabool, 102 G 4
Cabot, 103 I 4
Cabri, 89 K 9
Cabruta, 113 E 10
Cabure, 112 B 8
Cabuyaro, 112 G 6
Cacador, 117 G 13
Cacahoatan, 107 M 14
Cacahuatepec, 107 L 10
Cacak, 40 F 4
Cacapava do Sul, 117 I 11
Cacapon;river, 100 L 7
Cacaro;river, 113 F 11
Cacequi, 117 I 10
Caceres, 112 E 5
Cachari, 116 M 8
Cache;river, 102 G 7
Cachi, 116 F 4
Cachira, 112 E 6
Cachoeira do Sul, 117 J 11
Cachuela Esperanza, 115 G 11
Cacolo, 78 G 4
Caconda, 78 H 2
Cadbury, 50 K 4
Cadeje, 106 E 2
Cadereyta, 107 F 9
Cadereyta de Montes, 107 I 9
Cadillac, 89 L 9
Cadiz, 65 D 12
Cadomin, 88 J 6
Cadott, 98 G 7
Caduyari;river, 112 I 8
Caen, 44 C 7
Caerleon, 50 G 5
Caernafon, 50 A 2
Caerphilly, 50 H 5
Caersws, 50 D 4
Cafre;river, 112 H 5
Cafuini;river, 113 I 15
Cagaan-Ovoo, 61 D 9
Cagaan-Uur, 60 B 8
Cagayan Islands;isl., 65 E 11
Cagayan Sulu;isl., 65 G 10
Cagayan de Oro, 65 E 12
Cagayan;river, 65 A 11
Cagda, 57 G 13
Cagliari;prov, 39 K 2
Cagua., 113 C 9
Caguas, 109 H 13
Cahaba;river, 103 L 9
Cahuapanas, 114 C 4
Cahuide, 114 B 4
Cahuinari;river, 112 K 7
Cahuinaries;river, 112 K 7
Cai;river, 117 H 12
Caibarien, 108 F 8
Caicara, 113 C 12
Caicara;river, 112 E 8
Caicedonia, 112 G 4
Caicos Islands;isl., 109 F 11
Caicos;isl., 85 K 11
Caihue, 118 F 3
Cailloma, 114 J 7
Caina, 114 G 5
Caioa, 114 D 8
Cairbre, 48 M 3
Cairlinn, 48 J 5
Cairn Ryan, 48 F 8
Cairns, 83 D 11
Cairo, 101 H 11
Cairo El-, 112 G 4
Caisle-an na Finne, 48 G 2
Caislean Shiurdain, 48 M 3
Caislean an Bharraigh, 47 K 3
Caister-on-Sea, 51 C 16
Caiundo, 79 I 3
Caiza, 115 M 11
Cajabamba, 114 E 4
Cajacay, 114 G 4
Cajamarca, 114 E 3
Cajaro;river, 112 E 8
Cajatambo, 114 G 4
Cajon El-, 116 G 4
Cajon;river, 116 G 4
Cajones;river, 107 K 11
Cajuata, 115 K 10
Calabar, 75 M 11
Calabozo, 113 D 10
Calabozo;ensda., 112 B 7
Calacota, 115 L 9
Caladh na Sionainne, 48 M 1
Calafate, 119 L 3
Calagua;isl., 65 C 12
Calais, 44 A 8
Calama, 115 E 13
Calamar, 112 C 5
Calamarca, 115 K 10
Calamian Group;isl., 65 D 10
Calamian;isl., 80 D 1
Calanaque, 113 J 12
Calanscio;des., 72 E 7
Calapan, 65 C 11
Calarasi, 41 F 11
Calarca, 112 G 4
Calasetta, 39 K 2
Calauag, 65 C 11
Calayan;isl., 65 A 11
Calbayog, 65 D 12
Calbuco, 118 F 2
Calca, 114 I 7
Calcasieu;river, 103 O 2
Calceta, 112 K 2
Calchaqui, 116 I 7
Caldbeck, 49 G 12
Calder;river, 49 L 15
Caldera, 116 G 1
Caldew;river, 49 G 12
Caldicot, 50 H 6
Caldiran, 70 C 6
Caldwell, 94 E 6
Caldy;isl., 50 G 1
Caledon, 48 I 3
Caledonia, 91 L 12
Calentura, 106 B 1
Calera, 103 L 10
Calera Victor Rosales, 106 H 8
Caleta Clarencia, 119 M 4
Caleta Coloso, 116 E 1
Caleta Josefina, 119 N 5
Caleta Olivia, 119 I 5
Calexico, 94 M 7
Calf of Man;isl., 48 J 8
Calfax, 98 D 2
Calgary, 88 K 7
Calhoun, 103 M 3
Calhoun City, 103 K 7
Calhoun Falls, 104 H 5
Cali, 112 H 4
Calico Rock, 102 H 4
Caliente, 94 I 8
Califon, 95 B 7
California;gulf, 106 1
California;state, 92 F 2
Calilegua, 116 E 5
Calion, 103 L 3
Calkini, 107 I 15
Call, 103 O 1
Callahan, 105 M 6
Callander, 49 A 9
Callao, 104 C 12
Callaway, 98 D 3
Callianua, 109 C 15
Callicoon, 101 I 10
Callington, 50 L 2
Calliope, 83 F 12
Callis, 77 K 14
Callison Ranch, 88 E 1
Callon;river, 48 4
Calmar, 88 J 7
Calmelli, 106 D 2
Calnali, 107 I 10
Calne, 50 H 8
Caloosahatchee;river, 105 N 10
Calotmul, 107 I 16
Calstock, 90 H 4
Caltagirone, 39 N 9
Caltanissetta, 39 N 8
Caltanissetta;prov, 39 N 8
Calthwaite, 49 G 12
Caluango, 78 F 4
Calucinga, 78 G 2
Calumet, 99 D 10
Calvary, 105 L 3
Calvert, 103 N 8
Calvert City, 102 G 8
Calvert;river, 82 D 8
Calvillo, 106 I 8
Calzada, 114 D 4
Cam;river, 51 E 13
Camabatela, 78 F 2
Camacho, 106 G 8
Camaguan, 113 D 9
Camaguey, 109 G 9
Camaguey;archp., 108 F 8
Camaguey;prov, 108 G 8
Camaleon, 106 F 8
Camana, 114 K 7
CamanaVu;river, 113 K 13
Camaqua;river, 117 I 11
Camaquia, 117 I 12
Camarare;river, 115 I 15
Camargo, 97 I 10
Camarones, 118 H 6
Camas, 94 D 3
Camatagua, 113 C 10
Cambay, 68 L 4
Cambellpore, 71 G 14
Camberville, 89 J 12
Camblaya;river, 114 L 3
Cambona, 78 G 10
Camboriu, 117 G 14
Camborne, 47 P 6
Cambridge, 96 G 10
Cambridge, 101 M 9
Cambridge Bay, 86 E 8
Cambridge City, 102 C 11
Cambridge Springs, 100 I 5
Cambridpe, 101 H 14
Camdem, 101 F 9
Camden, 101 E 15
Camel;river, 50 L 1
Camelford, 50 K 1
Cameron, 97 J 3
Cameron Falls, 99 A 10
Cameron;isl., 86 C 8
Cameroon;c., 72 H 5
Camiguin;isl., 65 A 11
Camilla, 105 L 3
Camina, 116 B 2
Camiri, 115 M 12
Camisea, 114 H 7
Camisea;river, 114 H 7
Camlaren, 88 C 8
Camline Beach, 104 H 11
Cammack, 103 J 3
Cammore, 88 K 6
Camooweal, 82 E 8
Camorure;river, 112 C 8
Camp Creek, 104 D 7
Camp Crook, 96 D 8
Camp Hill, 105 J 2
Camp Point, 98 M 7
Camp Shelby, 103 N 7
Camp Springs, 95 H 1
Camp Verde, 97 J 3
Campamento, 112 G 1
Campana, 116 L 8
Campana;isl., 119 J 2
Campbele, 98 E 3
Campbell River, 88 K 2
Campbellford, 90 L 6
Campbellpore, 68 D 4
Campbells Creek, 91 I 14
Campbellsport, 99 I 10
Campbellsville, 104 D 2
Campbellton, 91 J 11
Campbeltown, 48 D 6
Campeche, 107 J 15
Campeche;state, 107 J 15
Campeon, 104 C 4
Campo, 75 N 12
Campo Boscan, 112 C 7
Campo Domino, 114 H 7
Campo Esperanza, 116 C 8
Campo Gallo, 116 F 6
Campo Largo, 117 F 13
Campo Los Andes, 116 L 3
Campo Menonitas, 106 C 5
Campo Nacional, 106 A 1
Campo Troco, 113 G 9
Campo del Cielo, 116 G 5
Campoalegre, 112 H 5
Campobasso;prov, 38 H 9
Campobello, 104 G 5
Campos Novos, 117 G 12
Campozano, 112 K 1
Campti, 103 M 2
Campuya, 114 A 5
Campuya;river, 114 A 5
Camuya;river, 112 6
Canaan, 91 K 12
Canada Harbour, 91 G 15
Canada Honda, 116 K 4
Canada Oruro, 116 C 5
Canada de Gomez, 116 K 6
Canada;c., 86
Canadaigua, 100 G 8
Canadian, 97 J 10
Canadian;river, 92 H 7
Canadys, 105 I 7
Canaelan, 106 G 6
Canagua, 112 E 7
Canagua;river, 112 D 8
Canaiejas, 116 L 4
Canaima, 113 F 12
Canainiktok;river, 91 E 12
Canajoharie, 101 G 10
Canal Flats, 88 L 6
Canals, 116 K 5
Cananari;river, 112 J 7
Cananea, 106 B 4
Cananeia, 117 F 14
Cananguchal, 112 I 4
Canar, 112 L 2
Canar;prov, 112 L 2
Canarreos;archp., 108 G 7
Canas, 114 M 3
Canastota, 101 G 9
Canberra, 83 J 10
Canbrook, 86 K 5
Canchas, 116 F 2
Cancun;isl., 107 I 15
Candala, 77 I 14
Candarave, 114 K 8
Candeias;river, 115 F 12
Candela, 107 E 9
Candelaria, 107 K 15
Candelaria;river, 107 E 9
Candido de Abreu, 117 E 12
Candle, 86 B 1
Cando, 89 J 9
Candon, 65 A 11
Candor, 101 H 9
Cane Valley, 104 D 2
Caneima;isl., 113 D 14
Canela, 117 H 13
Canelas, 106 F 5
Canelones, 117 L 9
Canelones;prov, 117 L 9
Canelos, 112 K 3
Canete, 118 D 2
Canete;river, 114 I 5
Caney Fork;river, 104 F 2
Caneyville, 104 D 1
Canfield, 100 J 4
Cangallo, 114 I 6
Cangamba, 78 H 4
Cangandala, 78 G 2
Cangombe, 78 H 4
Cangucu, 117 J 11
Cangumbe, 78 G 4
Canguro;isl., 80 L 4
Caniapiscau;river, 91 C 9
Canicado, 79 L 9
Canicatti, 39 N 8
Canisteo, 100 H 7
Canisteo;river, 100 H 8
Canistota, 98 H 2
Canitas, 106 G 8
Canle;river, 113 F 16
Cannelton, 102 F 10
Canning, 91 K 12
Cannock, 50 C 8
Cannon Falls, 98 G 6
Cannon;river, 98 G 5
Cannonball;river, 96 C 9
Cannonsburg, 103 N 4
Cano Aceite;river, 113 D 10
Cano Asa;river, 113 F 12
Cano Cocumita;river, 113 D 13
Cano Colorado, 113 H 9
Cano Igues;river, 112 D 8
Cano Macareo;river, 113 D 13
Cano Manamo;river, 113 D 13
Cano Mariusa;river, 113 D 13
Cano Mataveni;river, 113 G 9
Cano;isl., 108 L 6
Canoa, 112 J 1
Canoas, 117 I 12
Canoas;river, 117 G 13
Canoinhas, 117 F 13
Canol, 88 A 4
Canon City, 96 H 7
Canonbie, 49 F 12
Canonsburg, 100 K 5
Canoochee;river, 105 K 6
Canora, 89 K 11
Canosa di Puglia, 39 I 10
Canouan;isl., 109 C 15
Canover, 99 E 9
Cansahcab, 107 I 16
Canso, 91 K 13
Canta, 114 H 4
Cantabric;cord., 36 B 5
Cantaro, 108 K 2
Cantaura, 113 D 11
Canterbury, 51 I 15
Cantilan, 65 E 13
Cantin;prov, 118 D 2
Canton, 98 I 2
Cantonnement, 103 O 9
Canuan;river, 112 J 5
Canuma, 115 C 15
Canuma;river, 115 C 15
Canutama, 115 D 12
Canutillo, 97 L 6
Canvey, 51 H 14
Canvey;isl., 51 H 14
Canyon, 97 J 9
Canyon City, 94 E 5
Canzar, 78 F 4
Caonach, 48 L 1
Caoueza, 112 G 5
Cap-Chat, 87 K 15
Cap-aux-Meules, 91 J 13
Capachica, 115 J 9
Capaia, 78 F 4
Capanaparo;river, 113 E 9
Capanema, 117 F 11

Cherwell;river, 51 F 9
Chesapeake, 104 D 12
Chesapeake City, 95 F 4
Chesham, 51 G 11
Cheshunt, 51 G 12
Chesilhurst, 95 E 7
Chesley, 90 L 4
Chesnee, 104 G 6
Chester, 50 A 6
Chester le Street, 49 G 15
Chester;river, 95 G 4
Chesterfield, 49 M 16
Chesterfield Inlet, 89 B 14
Chesterfield;isl., 80 H 7
Chesterton, 99 K 11
Chestertown, 101 F 11
Chesterville, 90 L 7
Chestnut, 103 M 2
Cheswold, 95 G 5
Cheta;river, 57 E 9
Chetek, 98 F 7
Chetenham, 47 M 9
Cheticamp, 91 J 13
Chetopa, 97 I 12
Chetumal, 107 J 16
Chevak, 86 C 1
Chevejecure, 115 J 11
Chevy Chase, 95 H 1
Chewelah, 94 B 6
Cheyenne, 96 F 7
Cheyenne Wells, 96 H 8
Cheyenne;river, 96 E 8
Chezacut, 88 J 3
Chhabra, 68 J 7
Chhast, 66 B 8
Chhatarpur, 68 J 8
Chhota Andaman;isl., 67 N 10
Chhota Nicobar;isl., 67 O 12
Chhoti Sadri, 68 J 5
Chiapa de Corzo, 107 L 13
Chiapas;state, 107 L 14
Chiapilla, 107 L 14
Chiautla de Tapia, 107 K 10
Chibabava, 79 K 9
Chibougamau, 90 H 8
Chibuto, 79 L 9
Chicago, 99 K 11
Chicago Center, 98 F 6
Chicago Hights, 99 K 11
Chical-Co, 116 M 4
Chicama, 114 E 3
Chicama;river, 114 E 3
Chicamocha;river, 112 E 6
Chicapa;river, 78 F 4
Chicas;river, 114 J 7
Chichagof;isl., 86 G 2
Chichas;cord., 114 K 2
Chichester, 51 K 10
Chichiriviche, 113 C 9
Chickahominy;river, 104 D 11
Chickamauga, 104 G 2
Chickasaw, 103 O 8
Chickasawhay;river, 103 N 8
Chicken, 86 E 3
Chickmagalur, 67 N 4
Chiclayo, 114 D 2
Chico;river, 78 H 6
Chicoana, 116 E 4
Chicomo, 79 L 9
Chicomuselo, 107 M 14
Chicontepec de Tejada, 107 I 10
Chicot, 103 L 4
Chicoutimi, 91 I 9
Chidester, 103 K 2
Chiefland, 105 N 5
Chieti;prov, 38 H 8
Chigorodo, 112 E 4
Chiguana, 114 L 1
Chiguaza, 112 K 3
Chigwell, 51 G 13
Chihfeng, 61 E 12
Chihuahua, 106 D 6
Chihuahua;state, 106 D 6
Chihuahuilla, 106 C 3
Chikalda, 68 L 7
Chikaskia;river, 97 I 11
Chikwawa, 79 I 10
Chilapa, 107 K 10
Chilas, 68 C 5
Chilca, 114 H 4
Chilca;cord., 114 J 7
Chilcotin;river, 88 J 3
Childers, 83 G 12
Childs, 105 M 11
Chile, 111 I 6
Chile Chico, 119 3
Chile;c., 111 I 6
Chilecito, 116 H 3
Chiles, 66 C 9
Chilete, 114 E 3
Chilham, 51 I 15
Chilhowie, 104 E 6
Chili;river, 114 K 8
Chill.An-, 48 M 4
Chilla, 112 L 2
Chillan, 118 C 3
Chillar, 118 C 9
Chillicothe, 99 L 9
Chilliculco, 115 K 9
Chillingham, 49 D 14
Chilliwak, 88 L 3
Chiloe;isl., 118 G 2
Chiloe;prov, 118 G 3
Chilok, 57 I 11
Chilonga, 78 G 8
Chilton, 99 H 10
Chiluaje, 78 F 5
Chim, 112 F 7
Chima, 67 L 13
Chimaltenango, 108 I 3
Chimaltitan, 106 I 7
Chiman, 112 D 2
Chimbero, 116 G 2
Chimbo;river, 112 K 2
Chimborazo;prov, 112 L 3
Chimbote, 114 F 3
Chimbu, 83 B 13
Chimichagua, 112 D 5
Chimore;river, 115 K 11
Chimpay, 118 D 6
China, 60 F 3
China Grove, 104 F 8
China;c., 60 F 3
Chinacota, 112 E 6
Chinandega, 108 J 5
Chinca, 112 J 2
Chincha Alta, 114 I 5
Chincha;isl., 114 I 5
Chinchaza;river, 88 F 5
Chinchilla, 83 G 12
Chinchina, 112 G 4
Chinchipe;river, 112 M 2
Chindwin;river, 67 I 10
Chingalpo, 114 F 4
Chinganaza, 114 C 4
Chingoles, 116 I 2
Chingoroi, 78 H 1
Chinguetti, 74 G 3
Chinijo, 115 J 9
Chiniot, 68 E 5
Chinipas, 106 E 4
Chinipas;river, 106 E 4
Chinko;river, 76 L 5
Chinon, 44 F 7
Chinook, 94 D 3
Chinquapin, 104 G 11
Chinu, 112 D 4
Chipalcingo, 107 L 10
Chipao, 114 I 6
Chipewyan Lake, 88 G 7
Chipley, 105 J 3
Chipman, 91 K 11
Chipola;river, 105 L 2
Chippenham, 50 H 7
Chippewa Falls, 98 G 7
Chippewa;river, 98 G 7
Chipping, 49 K 13
Chipping Camden, 50 F 8
Chipping Norton, 51 F 9
Chipping Ongar, 51 G 13
Chipping Sodbury, 50 H 7
Chiquian, 114 G 4
Chiquila, 107 H 16
Chiquimula, 108 I 4
Chiquinquira, 112 F 6
Chira;river, 114 C 2
Chiramba, 79 I 10
Chiran, 63 O 1
Chirang, 69 H 14
Chirawa, 68 H 6
Chirdon Burn;river, 49 F 13
Chireno, 103 N 1
Chirfa, 74 G 13
Chirgua;river, 113 D 9
Chiricahua National Monument, 97 L 4
Chiriguana, 112 D 6
Chirikof;isl., 84 D 2
Chirinos, 114 C 3
Chirinos;river, 114 C 3
Chiriqui;gulf, 108 M 7
Chirnside, 49 C 14
Chiromo, 79 I 10
Chisapan, 69 H 11
Chisholm, 88 H 7
Chishtian Mandi, 68 G 5
Chita, 114 K 2
Chitek, 89 I 9
Chitembo, 78 H 3
Chitina, 86 E 2
Chitorgarh, 68 J 5
Chitose, 62 D 3
Chitral, 66 B 6
Chitral;river, 66 C 6
Chitre, 112 E 1
Chitud;isl., 40 E 13
Chiuchiu, 116 D 2
Chiumbe;river, 78 F 4
Chiume, 78 H 4
Chivacoa, 113 C 9
Chivata;river, 113 D 11
Chivay, 114 J 8
Chive, 115 H 9
Chivilcoy, 116 L 7
Chiwefwe, 78 H 8
Chloride, 97 J 2
Choapa;river, 116 J 1
Choapan, 107 L 12
Chocaya, 114 L 2
Chochola, 107 I 15
Chociwel, 43 C 10
Choconta, 112 G 6
Chocowinity, 104 F 11
Chocta Whatchee;river, 105 M 1
Choctaw', 103 M 8
Chodecz, 43 D 13
Chodziez, 43 C 11
Choele Choel, 118 E 6
Choique, 118 D 8
Choiseul;isl., 80 F 7
Chojna, 43 D 9
Chojnice, 43 B 12
Chojnow, 43 F 10
Chokio, 98 F 3
Cholet, 44 F 6
Cholila, 118 G 3
Chollerford, 49 F 14
Chollerton, 49 F 14
Cholo, 79 I 10
Cholula, 107 J 10
Choluteca, 108 J 5
Chomu, 68 H 6
Chomutov, 42 G 8
Chonchi, 118 G 2
Chone, 112 K 2
Chongos Alto, 114 H 5
Chongos Bajo, 114 H 5
Chongoyape, 114 D 3
Chontabamba, 114 G 5
Chontali, 114 D 3
Chontayacu;river, 114 F 4
Chop Gate, 49 I 16
Chopin;river, 117 F 12
Choptank;river, 95 H 4
Choquecamata, 115 K 11
Choreti, 115 M 12
Chorley, 49 L 13
Chorog, 56 K 5
Choroszcz, 43 C 16
Chorrillos, 114 H 4
Chorzele, 43 C 14
Chos Malal, 118 C 4
Chosica, 114 H 4
Choszczno, 43 C 10
Chota, 68 L 5
Chota Udaipur, 68 L 5
Chotan, 68 I 3
Choteau, 96 B 4
Chotila, 68 K 3
Chovd, 60 C 6
Chovd gol;river, 60 D 6
Chowan;river, 104 E 12
Choya, 116 H 5
Christchurch, 50 K 8
Christiana, 95 D 4
Christiana;river, 95 F 4
Christiansburg, 104 D 8
Christianshab, 87 C 13
Christiansted, 109 B 9
Christina;river, 88 H 8
Christmans, 95 A 5
Christmas Creek, 82 D 5
Christmas Island, 91 K 13
Christmas;isl., 59 P 11
Christopher, 102 F 7
Chuadanga, 69 J 12
Chubut;prov, 118 G 4
Chubut;river, 118 G 6
Chuchungas;river, 114 D 3
Chucuma, 116 J 4
Chucunaque;river, 112 D 3
Chudleigh, 50 L 4
Chuhuichupa, 106 C 4
Chukchi Sea;sea, 86 A 2
Chukchi;sea, 57 A 14
Chukotat;river, 90 A 7
Chula, 104 D 10
Chuld, 61 D 9
Chulmleigh, 50 J 3
Chulo, 116 G 2
Chulucanas, 114 C 2
Chulumani, 115 K 10
Chulym;river, 58 F 7
Chuma, 115 J 9
Chumaerhe;river, 60 G 6
Chumar, 68 E 8
Chumatang, 68 D 8
Chumbicha, 116 H 4
Chumpi, 114 J 6
Chumuckla, 103 O 9
Chuna Huasi, 116 I 5
Chunchi, 112 L 2
Chunchula, 103 O 8
Chunian, 68 F 5
Chuntang, 69 H 13
Chuquibamba, 114 J 7
Chuquibambilla, 114 I 7
Chuquicamata, 116 D 2
Church Hill, 95 G 4
Church Point, 103 O 3
Church Stretton, 50 D 6
Churchill, 50 I 6
Churchill;river, 89 F 13
Churchs Ferry, 98 B 1
Churchville, 95 F 3
Churin, 114 G 4
Churqui, 116 F 5
Churu, 68 H 6
Churubusco, 99 L 13
Churuguara, 112 C 8
Churumuco, 106 K 8
Chuschi, 114 I 6
Chute-aux-Outardes, 91 10
Chuvica, 114 L 2
Chuviscar;river, 106 D 6
Cianaga Prieta, 106 E 5
Cicero, 101 G 9
Cide ', 70 A 3
Ciechanow, 43 C 14
Ciego de Avila, 108 F 8
Cienaga, 112 C 5
Cienaga de Oro, 112 D 4
Cienfuegos, 108 F 8
Cieplice Slaskie, 43 F 10
Cifra, 105 N 6
Cihanbeyli, 70 C 3
Cihuatlan, 106 J 6
Ciinele;river, 40 F 9
Cill Airme, 47 M 3
Cill Anna, 48 M 3
Cill Bearaigh, 48 L 4
Cill Bhairrfhinn;cast., 48 H 1
Cill Bheagain, 48 M 2
Cill Bhride, 48 M 4
Cill Chainnigh, 47 L 5
Cill Chainnigh. cond., 47 M 5
Cill Chaoi, 47 L 3
Cill Dara, 47 L 5
Cill Dealga, 48 L 3
Cill Dheaglain, 48 L 4
Cill Droichid, 48 M 4
Cill Inion Leinin, 48 M 5
Cill Maodhog, 48 M 3
Cill Mhantain, 47 L 6
Cill Tuama, 48 L 1
Cimarron, 96 H 10
Cimarron;river, 97 I 11
Cimiearra;river, 112 E 5
Cimpeni, 40 C 7
Cimpia Turzii, 40 C 8
Cimpulung Moldovenesc, 40 B 9
Cinaruco;river, 113 F 9
Cincinnati, 100 L 1
Cincinnatus, 101 H 9
Cinco Chanares, 118 F 6
Cinco Saltos, 118 D 5
Cinderford, 50 G 6
Cine;river, 41 N 12
Cintalapa de Figueroa, 107 L 13
Cintra, 116 J 6
Cintra;gulf, 74 F 2
Cionn Atha Gad, 48 M 3
Cipolletti, 118 D 5
Circle, 96 B 7
Circleville, 100 K 2
Circulo Polar Artico, 57 D 13
Ciro, 39 K 11
Cirpan, 40 H 9
Cirtojani, 40 E 9
Cisnadie, 40 D 8
Cisneros, 112 F 5
Cisnes;river, 118 H 3
Cita, 57 I 11
Citronelle, 103 N 8
City Point, 98 H 8
Ciucea, 40 B 7
Ciudad Allende, 107 D 9
Ciudad Bolivar, 113 E 12
Ciudad Bolivia, 112 D 7
Ciudad Camargo, 107 F 10
Ciudad Cuauhtemoc, 107 M 14
Ciudad Delicias, 106 D 6
Ciudad Gracias, 108 I 6
Ciudad Guayana, 113 D 13
Ciudad Guerrero, 106 D 5
Ciudad Hidalgo, 107 M 14
Ciudad Ixtepec, 107 L 12
Ciudad Juarez, 106 B 6
Ciudad Lerdo, 106 F 7
Ciudad Madero, 107 H 10
Ciudad Mante, 107 H 10
Ciudad Manuel Doblado, 106 I 8
Ciudad Melchor Musquiz, 106 D 8
Ciudad Mier, 107 E 10
Ciudad Miguel Aleman, 107 E 10
Ciudad Netzahualcoyotl, 107 J 10
Ciudad Obregon, 106 E 4
Ciudad Ocampo, 107 H 10
Ciudad Ojeda, 112 C 7
Ciudad Piar, 113 E 12
Ciudad Victoria, 107 G 10
Ciudad de Nutrias, 112 D 8
Ciudad de Valles, 107 H 10
Ciudad del Carmen, 107 K 14
Ciudad del Maiz, 107 H 9
Clachan, 48 C 6
Clacton on Sea, 51 G 15
Cladich, 48 A 7
Claiborne, 95 H 3
Clain;river, 44 F 7
Clamency, 45 E 10
Clanabogan, 48 H 2
Clanton, 103 L 10
Clanwilliam, 79 O 4
Claonadh, 48 M 4
Clar Chlainne Mhuiris, 47 K 3
Clara, 103 N 7
Clara;river, 83 D 9
Claraz, 118 D 10
Clare, 99 H 13
Claremont, 98 E 1
Claremont Islands;isl., 83 C 10
Claremore, 97 I 12
Clarence, 98 K 8
Clarence Town, 109 F 10
Clarence;isl., 119 N 4
Clarendon, 97 J 9
Clarenville, 91 I 16
Claresholm, 88 L 7
Clarinda, 98 L 4
Clarines, 113 C 11
Clarington, 100 K 4
Clarion, 100 I 6
Clarion;isl., 106 K 1
Clarion;river, 100 I 6
Clark, 98 F 2
Clark Fork;river, 94 C 8
Clarkdale, 97 J 3
Clarke City, 91 H 10
Clarke;isl., 83 L 10
Clarkesville, 104 G 4
Clarkfield, 98 G 3
Clarks, 98 K 1
Clarks Harbour, 91 M 11
Clarks Summit, 101 I 9
Clarksburg, 95 D 8
Clarksdale, 103 J 5
Clarkston, 94 D 6
Clarksville, 97 J 13
Clarkton, 102 H 6
Claro;river, 115 I 10
Claudy, 48 G 3
Claveria, 65 A 11
Clawton, 50 K 2
Claxton, 88 H 1
Claxton Bay, 108 L 2
Clay, 100 M 4
Clay Center, 98 L 1
Clay City, 102 E 8
Clay Cross, 51 A 9
Claydon, 51 E 15
Claymont, 95 E 5
Clayoquot, 88 L 2
Claypool, 99 L 12
Clayton, 51 J 12
Clayton le Moors, 49 K 14
Cle Elum, 94 C 4
Clear Hills, 88 G 6
Clear Lake, 98 F 6
Clear Prairie, 88 G 5
Clearbrook, 98 C 4
Clearfield, 96 F 3
Clearfield Creek, 100 J 6
Clearwater, 98 J 1
Clearwater;river, 89 G 9
Cleator, 49 H 11
Cleire;isl., 47 N 3
Clemente;isl., 119 I 2
Clementon, 95 E 6
Clementson, 98 B 5
Clendenin, 100 M 4
Cleobury Mortimer, 50 D 6
Cleobury North, 50 D 6
Cleona, 95 C 3
Clermont, 83 F 11
Clevedon, 50 H 6
Cleveland, 97 I 12
Cleveland Heights, 100 I 4
Cleveland Tontine, 49 I 16
Clevelandia, 117 F 12
Cleveley, 49 K 12
Clever, 102 G 2
Clewiston, 105 N 11
Cley, 51 B 15
Cliara;isl., 47 K 2
Clifton, 49 H 13
Clifton Forge, 104 C 8
Climax, 89 L 9
Climton, 101 H 13
Clinch;river, 104 F 3
Clinchco, 104 D 6
Clinton, 88 K 4
Clinton, 102 G 7
Clintonville, 99 G 10
Clintwood, 104 D 5
Clio, 105 K 2
Clipperton;isl., 85 M 5
Clitheroe, 49 K 13
Cliza, 115 L 11
Clochan. An-, 48 M 1
Clochar, 48 K 1
Clogh Mills, 48 F 5
Cloghy, 48 I 7
Cloich Oir, 48 H 1
Cloirtheach, 48 M 1
Cloncurry, 83 E 9
Cloose, 88 L 2
Clophill, 51 F 11
Cloquet, 98 D 6
Clorinda, 117 F 9
Closeburn, 49 E 10
Closter, 95 A 9
Cloters, 102 F 9
Clough, 48 I 6
Clover, 104 E 10
Cloverdale, 94 I 3
Cloverport, 104 C 1
Clovis, 97 K 8
Cloyfin, 48 F 4
Cluain Aodha, 48 M 4
Cluain Bolg, 48 M 3
Cluain Dolcain, 48 M 4
Cluain Eois, 48 J 3
Cluain Fearta, 48 M 1
Cluain Meala, 47 M 4
Cluain Mhic Nois, 48 M 1
Cluden;river, 49 F 10
Clul, 40 B 8
Clun, 50 D 5
Clun;river, 50 D 5
Clwyd;river, 50 A 5
Clyattville, 105 L 4
Clyde, 88 I 7
Clyde;river, 48 B 8
Clydebank, 49 C 9
Clyman, 99 I 10
Clymer, 100 J 5
Clynnog-fawr, 50 B 2
Clyo, 105 J 7
Clyst Saint Mary, 50 K 4
Coaebridge, 49 C 10
Coagh, 48 H 4
Coahuayana;river, 106 K 7
Coahuayutla, 106 K 8
Coal City, 99 L 10
Coal Hill, 103 I 2
Coal River, 88 D 3
Coal;river, 88 D 3
Coalcoman, 106 K 7
Coalcoman;river, 106 K 7
Coaldale, 95 A 4
Coalisland, 48 H 4
Coalville, 51 C 9
Coari, 115 C 13
Coari;river, 115 C 12
Coasa, 115 I 9
Coast Mountains;cord., 88 E 1
Coast Ranges;cord., 94 C 3
Coastal Trench, 88 J 1
Coatepec, 107 J 11
Coatepec Harinas, 107 K 9
Coatesville, 95 D 4
Coatham, 49 H 16
Coaticook, 91 L 9
Coatzacoalcos, 107 K 12
Coatzacoalcos;river, 107 K 12
Coauila;state, 106 E 8
Cobalt, 90 J 5
Coban, 108 I 4
Cobb;isl., 104 D 13
Cobbtown, 105 J 6
Cobden, 102 F 7
Cobham, 104 C 10
Cobham;river, 89 I 14
Cobija, 115 G 9
Cobo, 118 D 10
Coboconk, 90 L 5
Cobourg, 90 L 6
Cobquecura, 118 C 2
Cobre, 94 G 8
Cobres, 116 E 4
Cobua, 78 G 10
Coc Hill, 90 L 6
Coca, 112 J 4
Coca;river, 112 J 3
Cocachacra, 114 K 7
Cocapata, 115 K 10
Cochabamba, 115 L 11
Cochamarce, 114 G 4
Cochapeti, 114 G 4
Cochas, 114 G 4
Cochayuc, 114 H 7
Coche;isl., 113 C 12
Cochons;isl., 109 M 12
Cochran, 105 J 4
Cochrane, 98 H 7
Cochrane, 90 I 5
Cochrane;river, 89 F 11
Cochranville, 95 E 4
Cockburn Island, 99 E 14
Cockburnspath, 49 B 13
Cockenzie, 49 B 12
Cockerham, 49 J 12
Cockermouth, 49 H 11
Cockeysville, 95 F 2
Cocle del Norte, 112 D 1
Coco;isl., 67 M 11
Coco;river, 108 J 6
Cocoa, 105 O 7
Cocoa Beach, 105 O 8
Cocodrie, 103 P 5
Cocolamus, 95 B 1
Cocomorachic, 106 D 5
Cocorit, 106 D 4
Cocos;isl., 85 N 9
Cod;isl., 91 C 12
Codajas, 115 B 13
Codajas;isl., 115 B 13
Codazzi, 112 C 6
Coderre, 89 L 10
Codihue, 118 D 4
Codpa, 116 A 1
Codrington, 109 C 12
Codroy, 91 I 14
Cody, 96 D 5
Coeburn, 104 D 5
Coel River, 86 G 4
Coelemu, 118 C 2
Coello, 112 G 5
Coen, 83 C 10
Coetivy;isl., 73 J 13
Coeur d Alene, 96 B 2
Coevorden, 42 D 3
Coffee, 105 K 5
Coffee Springs, 105 L 1
Coffeyville, 97 I 12
Coforo, 112 B 7
Cogdell, 105 L 5
Coggeshall, 51 F 14
Cogollo;river, 112 C 6
Cogon;river, 75 J 2
Cohansey;river, 95 F 6
Coharie Creek;river, 104 G 10
Cohengua;river, 114 G 6
Cohocton, 100 H 8
Cohocton;river, 100 H 8
Cohoes, 101 G 11
Cohutta, 104 G 3
Coiba;isl., 108 M 7
Coig;river, 119 L 5
Coihaique, 119 I 3
Coihaique Alto, 119 I 3
Coill Droma, 48 F 2
Coill an Chollaigh, 48 K 3
Coin, 98 L 4
Cojata, 115 J 9
Cojbalsan, 61 C 11
Cojedes, 113 C 9
Cojedes;river, 113 D 9
Cojedes;state, 113 D 9
Cojutepeque, 108 J 4
Cokato, 98 F 5
Cokeville, 96 F 4
Cokurdach, 57 D 13
Colbrir, 116 M 2
Colby, 96 G 9
Colca;river, 114 J 8
Colcabamba, 114 H 6
Colchester, 51 F 14
Cold Ashton, 50 H 7
Cold Lake, 88 I 8
Cold Norton, 51 G 14
Cold Spring, 98 F 5
Coldspring, 97 M 13
Coldstream, 49 D 13
Coldwater, 97 I 10
Coldwater, 99 K 13
Coldwater;river, 103 J 6
Coldwell, 97 I 11
Cole Camp, 102 E 12
Colebrook, 100 I 4
Coleman, 88 L 7
Coleman;river, 83 C 9
Colemin, 104 E 12
Coleraine, 48 F 4
Colesburg, 98 J 7
Coleshill, 50 D 8
Colfax, 103 N 3
Colfax, 94 C 6
Colima, 106 J 7
Colima;state, 106 J 7
Colina, 116 K 2
Colinet, 91 I 16
Colipa, 115 M 12
Coll;isl., 46 G 5
Collaguasi, 116 C 2
Collann, 48 K 4
College Grove, 104 F 1
College Park, 105 I 3
Collegeville, 95 D 5
Collie, 82 J 2
Collier City, 105 O 10
Collingbourne Kingston, 50 I 8
Collingdale, 95 E 6
Collingswood, 95 E 6
Collingwood, 83 F 9
Collins, 90 G 2
Collinston, 103 L 4
Collinsville, 83 E 11
Collinwood, 103 I 8
Collipulli, 118 D 3
Collon Cura;river, 118 E 4
Colmar, 45 D 13
Colmesneil, 103 N 1
Colmonell, 48 F 8
Colne, 49 K 14
Colnett, 106 B 1
Cologne, 95 F 7
Colombia, 107 E 9
Colombia;c., 112 G 6
Colon, 117 D 9
Colon;archp., 112 F 1
Colona, 82 I 7
Colonel Hill, 109 F 10
Colonia Carlos Pellegrini, 117 H 9
Colonia Caseros, 116 J 8
Colonia Dora, 116 H 6
Colonia Elisa, 116 G 8
Colonia Gonzalez Ortega, 106 G 7
Colonia Josefa, 118 E 6
Colonia Tata Cua, 116 H 8
Colonia Urdaniz, 116 L 4
Colonia Veinticinco de Mayo, 118 D 5

Colonia Yucatan, 107 I 16
Colonia del Sacramento, 117 L 9
Colonia; prov, 117 K 9
Colonsay; isl., 48 B 5
Coloradas Las-, 118 E 4
Colorado Desert, 97 K 1
Colorado Springs, 96 H 7
Colorado; river, 97 K 2
Colorado; state, 92 G 6
Colotlan, 106 H 7
Colotlan; river, 106 H 7
Colouechaca, 115 L 11
Colouemarca, 114 I 7
Colouiri, 115 L 10
Colouitt, 105 L 3
Colsterworth, 51 C 11
Colstrip, 96 C 6
Coltauco, 116 L 2
Colton, 98 H 2
Colts Neck, 95 C 9
Columbia, 98 E 1
Columbia City, 99 L 13
Columbia Mountains; cord., 88 I 4
Columbia; river, 92 D 3
Columbus, 96 C 5
Columbus Grove, 100 J 1
Columbus Junction, 98 L 7
Colville, 94 B 6
Colville; river, 86 B 3
Colwan, 98 H 2
Colwyn Bay, 49 M 10
Colyford, 50 K 5
Comala, 106 J 7
Comalapa, 107 M 14
Comalcaico, 107 K 13
Comales, 107 F 10
Comallo, 118 F 4
Comandante Fontana, 116 E 8
Comandante Leal, 116 I 4
Comandante Luis Piedrabuena, 119 K 5
Comandante Nicanor Otamendi, 118 D 10
Comandante Portillo, 119 N 6
Comandante Salas, 116 L 3
Comanesti, 40 C 10
Comayagua, 108 J 5
Combahee; river, 105 J 7
Combarbala, 116 J 1
Comber, 48 H 6
Comedero, 106 G 5
Comemoracao; river, 115 H 14
Comer, 104 H 5
Comet; river, 83 F 11
Comfort, 104 G 11
Comfrey, 98 H 4
Cominto, 103 K 4
Comiso, 39 N 9
Comitan, 107 L 14
Comite; river, 103 O 5
Commentry, 45 G 9
Commerce, 104 H 4
Commerce, 102 G 1
Como, 103 J 6
Como; prov, 38 C 4
Comodoro Rivadavia, 119 I 5
Comoe; river, 75 L 6
Comondu, 106 E 2
Comores; isl., 30 H 1
Comoros; c., 79 K 11
Compass, 95 D 4
Compeer, 88 K 8
Compostela, 106 I 6
Comrie, 49 A 10
Comsel, 103 M 9
Comstok, 101 F 11
Comunidad, 113 H 10
Cona; river, 57 G 10
Conakry, 75 K 2
Conambo, 112 K 4
Conambo; river, 112 K 4
Conasauga; river, 104 G 3
Concaran, 116 K 5
Concarneau, 44 E 4
Concepcion, 112 J 4
Concepcion de la Norma, 106 G 8
Concepcion de la Sierra, 117 G 10
Concepcion del Bramador, 106 J 6
Concepcion del Oro, 106 G 8
Concepcion del Uruguay, 116 J 8
Concepcion; prov, 117 D 9
Concepcion; river, 106 B 2
Conception Junction, 98 M 4
Conchali, 116 K 2
Conchi, 116 D 2
Conchillas, 116 K 8
Conchos; river, 106 C 6
Concord, 101 G 13
Concordia, 96 G 11
Concordia, 117 G 12
Concordia sulla, 116 J 8
Condamine, 83 G 12
Condar, 112 K 7
Conde, 98 F 1
Conde-sur-Noireau, 44 C 7
Condersport, 100 I 7
Condon, 94 D 4
Condor, 115 M 11
Condor; cord., 114 C 3
Condoto, 112 G 4
Conecuh; river, 105 K 1
Conejera; isl., 37 G 13
Conejos, 97 I 6
Conemaugh; river, 100 J 6
Conesa, 116 K 7
Conestoga, 95 D 3
Coneto, 106 F 6
Conflict Group; isl., 83 D 15
Confluence, 100 K 6
Confuso; river, 116 E 8
Congaree; river, 104 H 7
Congleton, 50 A 7
Congresbury, 50 H 6
Conisbrough, 49 L 16
Coniston, 90 K 5
Conkal, 107 I 15
Conklin, 88 H 8
Conlig, 48 H 6
Conmhaigh, 48 G 2
Connahs Quay, 50 A 5
Connaught, 99 A 16
Conneaut, 100 H 5
Conneaut Lake, 100 I 5
Conneautville, 100 I 5
Connecticut; river, 101 I 12
Connecticut; state, 93 E 14
Connell, 94 C 5
Connellsville, 100 K 5
Connersville, 102 C 11
Cono Niyeu, 118 F 5
Cononaco, 112 K 4
Cononaco; river, 112 K 4
Conorochite; river, 113 H 10
Conover, 104 F 7
Conquista, 115 H 10
Conrad, 96 B 4
Conrath, 98 F 8
Conshohocken, 95 D 6
Constancia La-, 106 G 7
Constanta, 40 F 13
Constanta; prov, 40 F 12
Constantine, 99 K 13
Contact, 94 G 7
Contamana, 114 E 5
Continental, 99 L 14
Contoy; isl., 107 H 16
Contreras; isl., 119 M 2
Contulmo, 118 D 2
Contumaza, 114 E 3
Convencion, 112 D 6
Convent, 103 P 5
Converse, 99 M 12
Conway, 49 M 10
Conway Castle; cast., 49 M 10
Conway Springs, 97 I 11
Conway; river, 50 A 3
Conwyl Elfed, 50 F 2
Conyers, 105 I 3
Conyngham, 95 A 4
Coober Redy, 82 H 7
Cook, 82 H 6
Cook Islands; isl., 83 E 16
Cookeville, 104 E 2
Cooks, 99 E 12
Cooks Hammock, 105 M 4
Cooks Harbour, 91 F 15
Cookstown, 48 H 4
Cooktown, 83 C 10
Coolawanyah, 82 F 2
Coolidge, 105 L 4
Coolpardie, 82 I 4
Coolville, 100 L 3
Coon Rapids, 98 F 6
Cooper; river, 83 G 9
Cooperstown, 98 C 2
Coopersville, 99 I 12
Cooroy, 83 G 12
Coos Bay, 94 F 2
Coosawhatchie; river, 105 J 7
Copala, 107 L 10
Copalyacu; river, 114 B 5
Cope, 96 G 8
Copeland, 105 O 11
Copeland; isl., 48 H 6
Copenhagen, 55 N 7
Copetonas, 118 D 9
Copiap-o, 116 G 2
Copiapo; river, 116 G 1
Coplay, 95 B 5
Coporaoue, 114 J 8
Coppell, 90 H 4
Copper Center, 86 E 2
Copper Cliff, 99 D 16
Copper Harbor, 99 C 10
Copper Mountain, 88 L 4
Copper; river, 86 E 2
Copperfield, 83 F 11
Coppermine, 86 E 6
Coppermine; river, 88 A 7
Copplestone, 50 K 3
Coquet; river, 49 E 14
Coquimbo; prov, 116 J 1
Coquinbana, 116 H 1
Corabia, 40 F 9
Coracora, 114 J 6
Coral Gables, 105 O 12
Coral Sea; sea, 83 D 13
Coranzuli, 116 D 3
Coraopolis, 100 J 5
Corato, 39 I 11
Corazon El-, 112 K 2
Corbin, 104 D 4
Corby, 51 B 11
Corcaigh, 47 M 3
Corcovado; gulf, 118 G 2
Cordele, 105 K 4
Cordesville, 105 I 8
Cordilleras; prov, 117 E 9
Cordillo Downs, 83 G 9
Cordoba, 106 E 7
Cordova, 86 E 2
Corfe Castle, 50 L 7
Corfu; isl., 34 L 8
Corhampton Waltham, 51 J 10
Corigliano Calabro, 39 K 11
Corinna, 83 L 10
Corinth, 100 M 1
Coripata, 115 K 10
Corisco; isl., 75 N 11
Corleone, 39 M 8
Corlield, 83 E 10
Corlu, 41 I 12
Cornelia, 104 H 4
Cornelius, 104 F 7
Cornell, 98 F 7
Corner Brook, 91 H 14
Cornhill., 49 C 14
Corning, 98 M 3
Corninq, 98 L 4
Cornwall, 90 L 7
Cornwall Peninsula; pen., 47 O 7
Cornwall; isl., 87 B 9
Cornwallis; isl., 87 C 9
Coro, 112 B 8
Corocoro; isl., 113 D 14
Coroico, 115 K 10
Coroico; riv., 115 J 10
Corona, 97 K 7
Coronado; isl., 106 E 3
Coronation, 88 J 8
Coronation Gulf; gulf, 86 E 7
Coronation; gulf, 84 D 6
Coronation; isl., 33 C 8
Coronda, 116 J 7
Coronel, 107 H 9
Coronel Baigorria, 116 K 5
Coronel Bogado, 117 G 10
Coronel Cornejo, 116 D 5
Coronel Dorrego, 118 D 8
Coronel Eugenio Garay, 117 F 10
Coronel Fraga, 116 J 6
Coronel Francisco Sosa, 118 E 7
Coronel Moldes, 116 F 4
Coronel Oviedo, 117 F 9
Coronel Portillo, 114 B 4
Coronel Pringles, 118 D 8
Coronel Segovia, 116 L 4
Coronel Suarez, 118 C 8
Coronel Vidal, 118 D 10
Corongo, 114 F 4
Coroue, 115 L 10
Corovoda, 41 J 4
Corozal, 112 D 5
Corozo Pando, 113 D 9
Corpus, 117 G 10
Corr an Mha, 48 J 3
Corr na Fola, 48 M 1
Corral, 118 E 2
Corral de Bustos, 116 K 5
Correctionville, 98 J 3
Corrie Common, 49 F 11
Corrientes, 116 G 8
Corrientes; prov, 116 H 8
Corrientes; river, 114 B 5
Corrigin, 82 I 3
Corry, 100 H 6
Corse; isl., 34 K 6
Corsica; isl., 45 L 14
Corston, 50 H 6
Cortadera, 116 K 2
Cortaderas, 116 M 3
Corte, 45 L 13
Corte Alto, 118 F 2
Cortez, 97 I 5
Cortina d'Ampezzo, 38 B 6
Cortland, 100 I 5
Corumba, 115 M 16
Corumbiara; river, 115 I 13
Corumo; river, 113 E 13
Corunna, 100 G 1
Corvallis, 94 E 3
Corwallis; isl., 84 C 7
Corwen, 50 B 4
Coryai, 108 L 3
Corzuela, 116 G 7
Cosala, 106 G 5
Cosamaloapan, 107 K 12
Cosapata, 115 L 9
Cosenza, 39 K 11
Cosenza; prov, 39 K 10
Coshocton, 100 K 3
Cosmoledo; isl., 73 K 12
Cosmos, 98 G 4
Cosne-sur-Loire, 45 E 9
Cosnipata; river, 114 I 8
Cosnipate, 114 I 8
Cosouin, 116 J 5
Costa Rica, 106 D 3
Costa Rica; c., 108 L 6
Costa de la Luz; costa, 36 J 5
Costa; cord., 116 F 1
Costwold Hills; mts, 47 N 9
Coxxs Cove, 91 H 14
Cotabambas, 114 I 7
Cotabato, 65 F 12
Cotacocha, 112 J 4
Cotagaita, 114 L 3
Cotahuasi, 114 J 7
Cotahuasi; river, 114 J 7
Cotajes; river, 115 K 10
Cotapaita; river, 114 L 2
Cotentin; pen., 34 H 4
Cothi; river, 50 F 3
Coti; river, 115 F 11
Cotingo; river, 113 H 14
Cotmeana; river, 40 E 9
Cotopaxi; prov, 112 K 2
Cottage Grove, 94 E 3
Cottageville, 105 J 8
Cotter, 102 H 3
Cottingham, 51 D 10
Cotton Plant, 103 I 4
Cotton Valley, 103 L 2
Cottondale, 105 L 2
Cottonport, 103 N 3
Cottonton, 105 J 2
Cottonwood, 96 H 12
Cottonwood; river, 98 G 4
Cotuhe; river, 112 L 8
Cotui, 109 H 12
Coulee Dam, 94 B 5
Coulman; isl., 33 G 2
Coulonge; river, 90 K 6
Coulterville, 102 E 6
Council, 96 D 1
Council Bluffs, 98 L 3
Council Grove, 96 H 12
Coupeville, 94 B 3
Courantyne; river, 113 H 15
Courtenay, 88 K 2
Courtland, 103 J 9
Courtney, 104 F 7
Courtright, 99 I 16
Coushatta, 103 M 2
Coutances, 44 C 6
Couva, 108 L 2
Couva; river, 108 L 2
Covadonga; isl., 119 K 2
Cove City, 104 G 11
Covenas, 112 D 4
Covendo, 115 J 10
Coventry, 51 D 9
Covesville, 104 C 9
Covington, 99 D 10
Cowall; pen., 48 B 7
Cowan, 104 G 1
Cowbit, 51 C 12
Cowbridge, 50 H 4
Cowden, 102 D 7
Cowdenbeath, 49 B 11
Cowes, 51 K 9
Coweta, 97 I 12
Cowfold, 51 I 12
Cowpens, 104 G 6
Cox; river, 82 C 7
Coxs Bazar, 69 L 16
Coxsackie, 101 H 11
Coy Aike, 119 L 5
Coyame, 106 C 6
Coyucade Benitez, 107 L 9
Coyuquilla, 106 L 8
Cozad, 96 F 10
Cozumel, 107 I 16
Cozumel; isl., 107 I 16
Crab Orchard, 104 D 3
Craig, 105 P 11
Craig Healing Springs, 104 D 8
Craighouse, 48 C 6
Craigmyle, 88 K 7
Craigsville, 100 M 4
Craii, 49 A 13
Craik, 89 K 10
Craiova, 40 F 8
Craley, 95 D 2
Cranberry, 100 I 5
Cranberry Portage, 89 H 12
Cranbrook, 51 J 14
Cranbury, 95 C 8
Crandall, 103 M 8
Crandon, 99 F 10
Crane, 94 F 5
Cranston, 101 H 13
Crarae, 48 7
Crary, 98 B 1
Crasna, 40 C 12
Crater Lake National Park, 94 F 3
Crathorne, 49 H 16
Craven Arms, 50 D 6
Cravo Norte, 112 F 9
Cravo Sur; river, 112 G 7
Crawford, 49 D 10
Crawfordsburn, 48 H 6
Crawfordsville, 99 M 12
Crawley, 51 I 12
Creal Springs, 102 F 7
Creciente; isl., 106 G 2
Crediton, 50 K 4
Cree River, 89 G 9
Cree; river, 49 F 9
Creetown, 49 G 9
Creighton, 89 H 11
Cremona, 88 K 7
Cremona; prov, 38 D 4
Crenshaw, 103 J 5
Crescent, 97 I 11
Crescent Beach, 104 H 10
Crescent City, 94 G 2
Cresco, 98 I 7
Crespo, 116 J 7
Cressona, 95 B 3
Creston, 88 L 6
Crestview, 103 O 10
Creswell, 104 F 12
Crete, 98 E 2
Creuse; river, 44 F 7
Crianlarich, 48 A 8
Criccieth, 49 B 2
Criciuma, 117 H 13
Crickhowell, 50 F 5
Cricklade, 50 G 8
Cridersville, 99 L 14
Crieff, 49 A 10
Cripple, 86 C 1
Crisfield, 101 M 9
Crisnejas; river, 114 E 3
Cristobal, 112 D 2
Crisul Aib; river, 40 C 6
Crisul Negru; river, 40 C 6
Crivitz, 99 F 10
Croa; river, 41 I 6
Crocker, 102 F 3
Crockernwell, 50 K 3
Crocketford, 49 F 10
Crocodile; river, 79 L 7
Crocus Hill, 109 B 11
Crofton, 98 I 2
Croglin, 49 G 13
Croisic, 44 F 5
Croksville, 100 K 3
Cromarty, 90 B 1
Cromer, 51 B 15
Cromwell, 98 D 6
Crook, 49 G 15
Crooked; isl., 109 F 10
Crooked; river, 94 E 4
Crooks, 99 B 9
Crookston, 98 C 3
Crooksville, 99 M 16
Croom, 105 O 5
Cros Domhnaigh, 48 J 2
Crosa Caoil, 48 K 3
Crosby, 49 G 11
Crosby Ravensworth, 49 H 13
Crosbyton, 97 K 9
Cross Gates, 50 E 4
Cross Hands, 50 G 3
Cross Lake, 89 H 13
Cross Timbers, 102 F 2
Cross; isl., 101 D 16
Cross; river, 75 L 11
Crosse La-, 99 L 11
Crossett, 103 L 4
Crossgar, 48 I 6
Crosshill, 48 E 8
Crossmichael, 49 F 10
Crossnore, 104 F 6
Crossville, 104 F 2
Crotone, 39 K 12
Crouch; river, 51 G 14
Crow Lake, 89 L 15
Crow Wing; river, 98 D 4
Crowborough, 51 J 13
Crowder, 103 K 6
Crowland, 51 C 12
Crowley, 103 O 3
Crown Point, 99 K 11
Crownsville, 95 H 2
Crowthorne, 51 H 10
Croxton Keyrial, 51 B 10
Croydon, 51 H 12
Crozet, 104 C 9
Crozon, 44 D 3
Crucero, 115 J 9
Cruces, 106 D 7
Crudgington, 50 C 6
Crudwell, 50 G 7
Cruger, 103 L 6
Cruillas, 107 G 10
Crumpton, 95 G 4
Cruz Alta, 116 K 6
Cruz Bay, 109 B 9
Cruz De la-; river, 112 K 6
Cruz Grande, 107 L 10
Cruz del Eje, 116 I 5
Cruzeiro do Sul, 114 E 7
Cruzen; isl., 33 E 2
Crystal Beach, 95 F 4
Crystal City, 89 L 13
Crystal Falls, 99 E 10
Crystal Lake, 99 J 10
Crystal Springs, 103 J 2
Csongrad, 43 L 14
Csongrad; komit., 43 L 14
Csurgo, 43 M 12
Cu, 56 J 5
Cu; river, 56 J 5
Cuaca; river, 110 E 4
Cuadro Nacional, 116 L 3
Cuajinicuilapa, 107 L 10
Cualac, 107 K 10
Cuanavale; river, 78 H 4
Cuangar, 79 I 3
Cuango, 78 E 2
Cuango; river, 78 F 3
Cuanza; river, 78 F 2
Cuao; river, 113 G 10
Cuaro, 117 I 9
Cuarto; river, 116 K 6
Cuatro Cianagas, 106 E 8
Cuatro Ojos, 115 K 12
Cuauhtamoc, 106 J 7
Cuautepec, 107 L 10
Cuautitlan, 107 J 10
Cuautla Morelos, 107 K 10
Cuba, 98 M 1
Cuba City, 98 J 8
Cuba; c., 108 F 7
Cubagua; isl., 113 C 12
Cubango; river, 79 J 4
Cucao, 118 G 2
Cuchi, 78 H 3
Cuchi; river, 78 H 3
Cuchillo-Co, 118 D 7
Cuchivero; river, 113 E 10
Cuckfield, 51 J 12
Cuckney, 49 M 16
Cucui, 113 I 10
Cucurpe, 106 C 3
Cucurrupi, 112 G 3
Cudahy, 99 I 11
Cudworth, 89 J 10
Cue, 82 H 3
Cuemani; river, 112 J 6
Cuenca, 112 L 2
Cuenca; prov, 37 F 9
Cuencame de Ceniceros, 106 F 7
Cuernavaca, 107 K 10
Cuevo, 114 L 4
Cuiari, 113 I 9
Cuiba; river, 115 L 16
Cuicas, 112 C 8
Cuicatlan, 107 K 11
Cuil Dabhcha, 48 E 3
Cuil Mhuine, 47 J 4
Cuilapa, 108 J 3
Cuilo, 78 F 4
Cuio, 78 H 1
Cuito Cuanavae, 79 I 4
Cuito; river, 79 I 4
Cuiuni; river, 113 K 11
Cujar; river, 114 G 7
Cul du Graty, 116 G 7
Culbertson, 96 B 7
Culebra; isl., 109 H 14
Culebras, 114 G 3
Culgaith, 49 H 13
Culiacan, 106 F 5
Culiacan Rosales, 106 F 5
Culiacan; river, 106 F 5
Culion; isl., 65 D 10
Cullaville, 48 J 4
Cullen, 103 L 2
Cullman, 103 J 10
Cullompton, 50 K 4
Cullybackey, 48 G 5
Culo; river, 78 F 4
Culpeper, 100 M 7
Culpina, 114 L 3
Culross, 49 B 10
Culter, 49 D 11
Culuunchoroot, 61 B 11
Culver, 99 L 12
Cuma, 78 H 2
Cumana, 113 C 11
Cumanacoa, 113 C 12
Cumaria, 114 G 6
Cumberland, 88 K 2
Cumberland City, 102 H 8
Cumberland Islands; isl., 83 E 11
Cumberland Peninsula; pen., 87 E 12
Cumberland; isl., 105 L 7
Cumberland; river, 93 H 10
Cumbernauld, 49 B 10
Cumbre, 106 C 5
Cumikan, 57 G 14
Cumina; river, 113 K 16
Cumming, 104 H 3
Cummington, 101 H 12
Cumnock, 49 D 9
Cumpan; river, 107 K 14
Cumpas, 106 C 4
Cumrew, 49 G 13
Cumuripa, 106 D 4
Cumuto, 108 L 3
Cuna; river, 57 G 9
Cunare, 112 I 6
Cunaviche, 113 E 10
Cunaviche; river, 113 E 9
Cunchiyacu, 112 K 4
Cunco, 118 D 3
Cuncumen, 116 J 2
Cuncunul, 107 I 16
Cunduacan, 107 K 13
Cunene; river, 79 I
Cuneo; prov, 38 E 2
Cunhinga; river, 78 G 2
Cunucunuma; river, 113 H 11
Cunupia, 108 L 2
Cuona, 60 J 6
Cupar, 49 A 12
Cupica; gulf, 112 F 3
Cuprija, 40 F 6
Curacao; isl., 112 A 9
Curacautin, 118 D 3
Curaco; river, 118 D 6
Curahuasi, 114 I 7
Curanilahue, 118 D 2
Curanipe, 116 M 1
Curanja; river, 114 G 7
Curaray; river, 112 K 4
Curaru, 116 M 6
Cure; river, 112 K 7
Curepipe, 79 O 10
Curepto, 116 M 1
Curiapo, 113 D 13
Curichi, 115 M 13
Curico, 116 M 2
Curico; prov, 116 M 1
Curipaya; river, 112 I 5
Curiplaya, 112 J 5
Curitiba, 117 F 13
Curitibanus, 117 G 13
Curiuja; river, 114 G 7
Curlew, 94 B 5
Curran, 100 D 2
Currarate; river, 113 F 13
Current; river, 102 F 4
Currie, 94 H 7
Currituck, 104 E 13
Curtea de Arges, 40 E 9
Curtis, 103 K 2
Curtis; isl., 83 H 15
Curuca; river, 114 C 7
Curucu, 115 J 13
Curuguaty, 117 E 10
Curuquete; river, 115 F 11
Cururu; river, 115 E 16
Curutu rio, 113 G 12
Curuzu Cuatia, 116 H 8
Curwensville, 100 J 7
Cusaram, 106 E 5
Cushabatay; river, 114 E 5
Cushendall, 48 F 5
Cushendun, 48 F 5
Cushman, 102 H 4
Cusiana; river, 112 G 6
Cusihuiriachic, 106 D 5
Cusino, 99 E 12
Cusis Los-, 115 J 12
Cusseta, 105 J 3
Custer, 96 E 8
Cut Bank, 94 B 8
Cut Knife, 89 J 9
Cutchogue, 95 A 13
Cutervo, 114 D 3
Cuthbert, 105 K 3
Cutler Ridge, 105 O 12
Cutucu; cord., 112 L 3
Cuturi; river, 113 G 12
Cuvo; river, 78 G 1
Cuyabeno, 112 J 4
Cuyahoga Falls, 100 I 4
Cuyahoga; river, 100 I 4
Cuyamaca; river, 94 K 5
Cuyo; isl., 65 D 11
Cuyocuyo, 115 J 9
Cuyuni; river, 113 E 14
Cuzco, 114 I 7
Cynthiana, 100 M 1
Cypress, 103 N 2
Cyprus, 70 D 2
Czar, 88 J 8
Czarnkow, 43 D 11
Czechoslovakia, 43 H 9
Czersk, 43 B 12
Czestochowa, 43 F 13
Czopa, 43 C 10
Czplinek, 43 C 11
D'Arcy, 88 K 3
Dabajuro, 112 C 7
Dabakala, 75 K 6
Daban, 61 D 12
Dabancheng, 60 D 5
Dabas, 43 K 14
Dabat, 77 I 10
Dabaton, 75 J 4
Dabe, 108 M 2
Dabeiba, 112 E 4
Dable, 43 E 13
Dabola, 75 K 3
Dabou, 75 M 6
Daboya, 75 K 7
Dabrowa Biarostocka, 43 B 16
Dabrowa Tarnowska, 43 15
Dacca, 67 I 9
Dachainan, 60 F 7
Dachia, 114 L 8
Dadal, 61 C 10
Dadanawa, 113 H 14
Dade City, 105 O 6
Dadeldhura, 69 G 9

Drini;river, 40 H 4
Drinkwater, 89 L 10
Drivstua, 54 E 5
Drobak, 55 I 5
Droichead Atha, 48 K 5
Droim Bile, 48 J 4
Droim Lis, 48 K 1
Droim ar Snamh, 48 J 1
Droing, 48 K 2
Droitwich, 50 E 7
Dromad, 48 K 1
Dromore, 48 H 2
Drones, 48 F 5
Dronfield, 49 M 16
Drosei, 39 J 4
Drosh, 66 C 6
Drovanaja, 56 D 8
Druid, 50 B 4
Drumburgh, 49 G 12
Drumheller, 88 K 7
Drummond, 94 C 8
Drummond Island, 100 C 2
Drummond;isl., 99 E 14
Drummondville, 90 K 8
Drummone, 48 H 8
Drumore, 95 E 3
Drumquin, 48 H 2
Drums, 95 A 4
Drumskinny, 48 H 2
Druzba, 56 J 7
Druzina, 57 D 13
Drva;river, 38 C 12
Drvar, 40 E 1
Drweca;river, 43 C 13
Dry Fork;river, 100 L 5
Dry Prong, 103 N 3
Dry Tortugas;isl., 105 P 9
Dry;river, 82 C 6
Dryden, 89 K 15
Dryfe;n-o, 49 F 11
Drygalsky;isl., 33 K 4
Drymen, 49 B 9
Drynoo, 102 F 3
Drysdale;river, 82 C 5
Drywood, 88 L 7
Dschang, 75 M 12
Du Pont, 105 L 5
Duaca, 112 C 8
Duaringa, 83 F 11
Dubach, 103 L 3
Dubawni;river, 86 H 7
Dubawnt;river, 89 D 10
Dubh;river, 48 K 3
Dublin, 103 K 5
Dubois, 96 D 3
Dubreka, 75 K 3
Dubrovnik, 40 H 2
Dubuque, 98 J 8
Duc of Braganca, 78 F 2
Duchesne, 96 G 4
Duchess, 83 E 9
Duck Bay, 89 J 12
Duck Hill, 103 K 6
Duck Lake, 89 J 10
Duck;river, 102 H 8
Ducktown, 104 G 3
Ducos, 109 M 15
Duda;river, 112 H 5
Duddington, 51 C 11
Duddon;river, 49 I 11
Dudignac, 116 M 7
Dudinka, 57 E 9
Dudley, 50 D 7
Dudo, 77 J 15
Dudwa, 69 H 9
Dudypta;no, 57 E 9
Duekoue, 75 L 5
Duff;isl., 80 G 8
Duffield, 51 B 9
Dufresne Lake, 91 G 11
Dugdemona Bayou;river, 103 M 3
Duhau, 116 M 6
Duk Fadiak, 76 K 8
Dukhmeis, 77 B 11
Duki, 68 F 2
Dula, 78 A 4
Dulac, 103 P 5
Dulan, 60 G 7
Dulansi, 60 G 8
Dulce;gulf, 108 M 6
Dulce;river, 116 I 6
Dullabchara, 69 J 16
Dulovo, 41 F 11
Duluth, 98 D 7
Dulverton, 50 J 4
Dumaguete, 65 E 12
Dumai, 64 H 3
Dumaran;isl., 65 E 10
Dumaring, 65 H 10
Dumas, 97 J 9
Dumbarton, 49 B 9
Dumbridge, 50 L 4
Dumfries, 49 F 11
Dumoga-Ketjil, 65 I 12
Dumont, 98 E 3
Dun Buinne, 48 M 4
Dun Dealgan, 48 J 4
Dun Droma, 48 M 5
Dun Fionnachaidh, 48 E 1
Dun Garbhan, 47 M 4
Dun Laoghaire, 48 M 5
Dun Leire, 48 K 1
Dun Seachlainn, 48 L 4
Dun an Ri, 48 K 4
Dun-sur-Auron, 45 F 9
Duna;river, 43 L 13
Dunafo-Idvar, 43 L 13
Dunaj;river, 43 J 12
Dunajec;river, 43 H 14
Dunarea Veche;river, 40 E 12
Dunarii;delta, 40 E 13
Dunav;river, 40 E 4
Dunbar, 49 B 13
Dunblane, 49 A 10
Duncan, 88 L 3
Duncannon, 95 C 1
Duncansville, 100 J 7
Dunchurch, 51 D 9
Duncombe, 49 K 13
Dundalk, 95 G 2
Dundas, 87 B 11
Dundas;isl., 88 G 1
Dundee, 46 H 8
Dundee;isl., 33 C 7
Dundit, 77 D 12
Dundonald, 48 H 6
Dundrennan, 49 G 10
Dundrum, 48 I 6
Dundurn, 89 K 10
Duneaton;river, 49 D 10
Dunedin, 105 P 5
Dunfermline, 49 B 11
Dungannon, 48 H 4
Dungarpur, 68 K 5
Dungiven, 48 F 3
Dungu, 78 A 7
Dunham, 51 A 10
Dunhua, 61 D 14
Dunhuang, 60 F 7
Dunkerque, 45 A 9
Dunkerton, 98 J 7
Dunkirk, 99 M 13
Dunkwa, 75 L 7
Dunlap, 98 K 3
Dunmara, 82 D 7
Dunmurry, 48 H 5
Dunn, 104 G 10
Dunnamanagh, 48 G 2
Dunnan Gall, 48 G 1
Dunnellon, 105 N 5
Dunning, 96 F 9
Dunnville, 87 M 13
Dunoon, 48 B 8
Dunqunab, 77 E 10
Dunragit, 48 G 8
Dunrankin, 99 A 14
Dunraon, 69 J 11
Duns, 49 C 13
Dunseith, 96 A 9
Dunstable, 51 F 11
Dunster, 50 I 4
Dunwich, 51 E 16
Duobukulche;river, 61 B 13
Duokecheng, 60 H 3
Duolun, 61 E 12
Duomaer, 60 G 3
Duomula, 60 G 3
Duparquet, 90 I 5
Dupo, 102 E 6
Dupree, 96 D 9
Duque de Braganca, 78 F 2
Duquette, 98 E 6
Dur, 66 A 7
Durand, 96 D 13
Durango, 97 I 5
Durango;state, 106 G 6
Durant, 98 L 2
Durants Neck, 104 E 12
Durazno, 117 K 9
Durbuy, 42 G 2
Durfuyad, 103 I 9
Durg, 69 M 9
Durham, 49 G 15
Duri, 64 H 3
Durisdeer, 49 E 10
Durlas, 47 L 4
Durness, 46 E 7
Dursley, 50 G 7
Durston, 50 J 5
Duru, 78 A 7
Dushore, 101 I 9
Dusi, 66 C 9
Dutton, 99 16
Duvno, 40 F 1
Dwarka, 68 K 2
Dyan, 48 I 3
Dyckesville, 99 G 11
Dyer Bay, 99 F 16
Dyersburg, 102 H 7
Dyersvile, 98 J 7
Dyje;river, 43 I 11
Dymchurch, 51 J 15
Dysart, 49 B 12
Dzag, 60 C 8
Dzamun-uud, 61 D 11
Dzargalant, 60 C 8
Dzavchan, 60 C 7
Dzavchan gol;river, 60 C 6
Dzibalchen, 107 J 15
Dzidzantun, 107 I 16
Dzierzoniow, 43 F 11
Dzinst, 60 D 8
Dzitas, 107 I 16
Dzitbalche, 107 I 15
Dziwnow, 43 B 9
Dzusaly, 56 I 4
Dzuungov, 60 B 7
Dzuunmod, 61 C 9
Eadan Doire, 48 M 3
Eads, 96 H 8
Eagle, 86 E 3
Eagle Grove, 96 E 13
Eagle Lake, 97 M 12
Eagle Mills, 103 K 3
Eagle Pass, 97 N 10
Eagle River, 99 C 10
Eagle Rock, 104 D 8
Eagle;river, 91 F 14
EagleBend, 98 E 4
Eagletown, 103 J 1
Eagleville, 98 M 5
Ealing, 51 H 11
Eamont;river, 49 H 12
Eanainh-no, 48 J 3
Earby, 49 K 14
Earith, 51 D 12
Earle, 103 I 5
Earlington, 102 G 9
Earlston, 49 D 12
Earlton, 90 J 5
Eartville, 98 J 7
Earsdon, 49 F 15
Easdale, 48 A 6
Easington, 49 G 16
Easingwold, 49 J 16
Easley, 104 G 5
East Aurora, 100 G 7
East Bangor, 95 A 6
East Barnet, 51 G 12
East Bend, 104 E 8
East Brady, 100 J 6
East Branch Delaware;
river, 101 H 10
East Branch;river, 101 C 15
East Brent, 50 I 5
East Brewton, 103 N 9
East Broughton, 91 K 9
East Grand Forks, 96 B 11
East Pen;isl., 89 F 16
East Tawas, 100 E 2
EastAngus, 91 K 9
Eastbourne, 47 O 11
Eastend, 86 L 6
Eastland, 97 L 11
Eastmain, 87 K 12
Eastmain;river, 84 G 9
Easton, 101 J 10
Eastport, 100 D 1
Eaton Rapids, 100 G 1
Eberbach, 42 H 5
Ebersberg, 42 J 7
Eberswalde, 43 D 9
Ebingen, 42 J 5
Ebro;delta, 37 E 12
Echarate, 114 H 7
Echo Bay, 99 D 14
Eckernf-orde, 42 A 6
Eclectic, 105 J 1
Econfina;river, 105 M 4
Economy, 91 K 12
Ecru, 103 J 7
Ecuandureo, 106 J 8
Ed, 77 H 11
Edah, 82 H 2
Edam, 89 J 9
Eddies Cove, 91 G 15
Eddys, 99 I 16
Eddyville, 98 L 6
Ede, 42 D 2
Edea, 75 M 12
Edefors, 52 G 6
Eden, 96 F 5
Eden Valley, 98 F 4
Eden;river, 49 A 12
Edenton, 104 E 12
Ederny, 48 H 2
Edesville, 95 G 3
Edgar, 98 M 1
Edgar Springs, 102 F 4
Edgard, 103 P 5
Edgartown, 101 I 14
Edge;isl., 58 B 5
Edgedeog Rothes Fjord, 87 C 15
Edgefield, 104 H 6
Edgegrove, 95 E 1
Edgeley, 98 D 1
Edgemont, 96 E 8
Edgerly, 103 O 2
Edgerton, 98 H 3
Edgewater, 105 O 7
Edgewood, 88 L 5
Edh Dhahi-bat, 74 C 12
Edhessa, 41 J 6
Edinburg, 102 D 7
Edinburgh, 49 B 11
Edirne, 41 I 11
Edison, 105 K 3
Edisto Island, 105 J 8
Edisto;river, 105 I 8
Edith, 105 L 5
Edlingham, 49 E 15
Edmond, 97 J 11
Edmonds, 94 C 3
Edmonton, 88 I 7
Edmundston, 91 J 10
Ednam, 49 D 13Edr
Edson, 88 I 6
Edstone, 50 E 8
Eduardo Castex, 116 M 5
Edward, 104 F 12
Edward River, 83 C 9
Edward VII;pen., 33 F 3
Edwards, 101 E 10
Edwards;river, 98 L 8
Edwardsburg, 99 K 12
Edwardsville, 102 E 6
Eek, 86 C 1
Eel;river, 94 H 3
Eergunahe;no, 61 B 12
Effingham, 102 D 7
Effort, 95 A 5
Egadi;isl., 34 L 6
Eganville, 90 K 6
Egegik, 86 D 1
Egeland, 98 A 1
Egg Harbar, 99 G 11
Egg Harbor City, 95 F 7
Eggers;isl., 87 E 16
Eggleston, 49 H 14
Egham, 51 H 11
Egilsstadhir, 52 B 4
Eglingham, 49 D 14
Eglinton;isl., 86 C 6
Egremont, 49 H 11
Egua, 113 G 9
Egypt, 105 J 6
Ehen;river, 49 H 11
Ehren, 105 O 5
Ehrhardt, 105 I 7
Ei, 63 O 1
Eibiswald, 43 L 10
Eide, 54 G 3
Eidfjord, 54 H 3
Eidsvoll, 54 H 6
Eigersund, 55 J 2
Eigg;isl., 46 G 6
Eighty Mile Beach, 82 E 3
Eikefjord, 54 F 2
Eil, 77 K 14
Einasleigh, 83 D 10
Einasleigh;river, 83 D 10
Eiseb;river, 79 K 4
Eitzen, 98 I 7
Eiutla de Crespo, 107 L 11
Ejeda, 79 P 11
Ejima-Shoto;isl., 63 K 6
Ejk, 43 15
Ejole, 54 A 1
Ejura, 75 7
Ekalaka, 96 C 7
Ekenas, 53 O 8
Ekhinadhes Nisoi;isl.,
41 M 5
Eklo, 95 E 2
Eksjo, 55 K 9
Ektrask, 53 I 6
Ekwan;river, 90 F 4
El Salvador;c., 85 M 7
Elasson, 41 K 6
Elba, 102 H 3
Elbasan, 41 J 4
Elbe;river, 42 B 5
Elberfel, 102 E 9
Elberton, 104 H 5
Elblag, 43 B 13
Elbow, 89 K 10
Elbow Lake, 98 E 3
Eldersburg, 95 F 1
Eldon, 98 L 6
Eldora, 98 J 6
Eldorado, 89 E 9
Eldoret, 78 C 10
Eldred, 102 D 6
Electric Mills, 103 L 8
Elek, 43 L 15
Elena, 40 G 10
Elephant;isl., 33 C 7
Eleuthera;isl., 109 E 9
Eleven Point;river, 102 G 5
Elfin Cove, 86 G 2
Elga, 54 F 6
Elgin, 46 F 8
Elhovo, 40 H 11
Eliand, 49 L 15
Elias Pina, 109 H 11
Eliaville, 105 J 3
Elida, 97 K 8
Elie, 49 A 12
Elila;river, 78 D 6
Elisabeth, 100 L 4
Elizabeth, 102 E 10
Elizabeth City, 104 E 12
Elizabeth Islands;isl., 101 I 14
Elizabethtown, 100 L 8
Elizabethville, 95 B 2
Elk Fork;river, 102 D 4
Elk Garden, 104 D 6
Elk Lake, 90 J 5
Elk Mills, 95 F 4
Elk Mound, 98 G 7
Elk Neck, 95 F 4
Elk Point, 88 I 8
Elk Rapids, 99 G 13
Elk River, 96 B 2
Elk Valley, 104 E 3
Elk;river, 100 M 4
Elkader, 98 J 7
Elkhart, 99 M 9
Elkhart, 97 I 9
Elkhart, 99 K 12
Elkhorn, 89 L 12
Elkhorn City, 104 D 5
Elkhorn;river, 98 K 2
Elkin, 104 E 7
Elkins, 100 L 5
Elkland, 100 H 8
Elkmont, 103 J 9
Elko, 88 L 6
Elkton, 98 G 3
Elkville, 104 F 7
Ellamar, 86 E 2
Ellastone, 50 B 8
Elle;river, 44 D 4
Ellef Ringnes;isl., 86 B 8
Ellenburg, 101 D 11
Ellendale, 98 E 1
Ellenebad, 68 G 6
Ellensburg, 94 C 4
Ellenton, 105 I 6
Ellenville, 101 I 10
Ellerbe, 104 G 8
Ellerbee, 105 M 6
Ellesmere, 50 B 6
Ellesmere Port, 49 M 12
Ellesmere;isl., 87 B 10
Ellettsville, 102 D 10
Ellice Islands;isl., 83 E 14
Ellice;isl., 81 F 9
Ellice;river, 89 A 10
Ellicott City, 95 G 2
Ellicottville, 100 H 6
Ellijay, 104 G 3
Ellington, 49 E 15
Ellinwood, 96 H 10
Elliot, 104 H 8
Elliot Key;isl., 105 O 12
Elliot Lake, 90 J 4
Ellis, 96 H 10
Ellison Bay, 99 F 11
Elliston, 104 D 8
Ellisville, 103 N 7
Elloree, 105 I 7
Ellsinore, 102 G 5
Ellsworth, 96 H 11
Ellwood City, 100 J 5
Elm, 95 D 3
Elm City, 104 F 10
Elma, 88 I 6
Elmer, 95 F 6
Elmhurst, 99 K 10
Elmira, 91 J 13
Elmodel, 105 K 3
Elmora, 100 J 6
Elmore, 105 J 1
Elmsta, 54 H 11
Elmwood, 99 L 9
Elnora, 102 E 9
Elobey;isl., 75 N 12
Elora, 104 G 1
Elorn;river, 44 D 3
Elorza, 112 E 8
Elota, 106 G 5
Elota;river, 106 G 5
Eloy, 97 L 3
Elqui;river, 116 I 2
Elrod, 98 F 2
Elrose, 89 K 9
Elroy, 98 H 8
Elsa, 88 A 1
Elsas, 99 A 14
Elsberry, 102 D 5
Elsdon, 49 E 14
Elsey, 82 C 7
Elsmere, 95 E 5
Elston, 102 E 3
Elton, 103 O 3
Elvan;river, 49 E 10
Elveden, 51 D 14
Elverum, 54 G 6
Elwick, 49 G 16
Elwood, 99 M 13
Elwood, 95 F 7
Elworthy, 50 I 4
Ely, 98 C 7
Ely, 51 D 13
Elyaburg, 95 A 2
Elyanfoot, 49 D 10
Elyria, 100 I 3
Elysian Fields, 103 M 1
Embar, 91 G 11
Embarcacion, 116 D 5
Embarras Portage, 88 F 8
Embarrass;river, 102 E 8
Embira;river, 114 E 8
Embleton, 49 D 15
Embreeville, 104 F 5
Embu, 78 C 11
Emde, 102 C 4
Emerado, 98 B 2
Emerald, 83 F 11
Emerald;isl., 86 B 7
Emero;river, 115 I 10
Emerson, 98 A 2
Emery, 96 H 4
Emigsville, 95 D 2
Emilio V. Bunge, 116 L 6
Emiliano Zapata, 107 K 14
Emilio Mitre, 116 M 4
Emily, 98 D 5
Emin, 60 C 4
Eminabad, 68 E 5
Eminence, 102 E 11
Emirau;isl., 83 A 14
Emlenton, 100 I 7
Emmaboda, 55 M 9
Emmaus, 95 B 5
Emmaus Junction, 95 B 5
Emmet, 99 I 15
Emmet, 83 F 10
Emmet., 100 G 3
Emmetsburg, 98 I 4
Emmett, 94 E 6
Emmitsburg, 100 L 8
Emmorton, 95 F 3
Emo, 98 A 5
Empalme, 107 F 10
Empalme, 106 D 3
Empedrado, 116 G 8
Empexa, 114 L 1
Empingham, 51 C 1
Empire, 99 G 12
Emporia, 96 H 12
Emporium, 100 I 7
Empress, 88 K 8
Emsworth, 51 K 10
Emu, 83 F 12
Enambu, 112 I 8
Enbetsu, 62 B 3
Encampment, 96 F 6
Encanada, 114 E 3
Encantada de, 106 C 6
Encantado, 117 H 12
Encarnacion, 117 G 10
Encarnacion de Diaz, 106 I 8
Enchi, 75 L 7
Encinillas, 106 C 6
Encon, 116 K 3
Encontrados, 112 D 6
Encruzilhada do Sul, 117 I 12
Endau, 64 H 4
Endaxo, 88 H 3
Ende, 65 M 11
Endeavor, 99 H 9
Enderbury;isl., 81 F 12
Enderby, 88 K 5
Enderlin, 98 D 2
Endicott, 101 H 9
Endimari;river, 115 F 10
Endrick;river, 49 B 9
Ene, 96 H 12
Ene;river, 114 H 6
Energia, 118 D 9
Enfaula, 97 J 12
Enfield, 51 G 12
Engaru, 62 C 5
Engelhard, 104 F 13
Enggano;isl., 64 L 3
Engineer, 88 D 1
England, 103 J 4
Englee, 91 G 15
Englehart, 90 J 5
Englevale, 98 D 2
Englewood, 88 K 2
English, 102 E 10
English River, 98 A 8
Enid, 97 I 11
Enkoping, 55 I 10
Enmons, 98 I 5
Enna, 39 M 9
Enna;prov, 39 M 9
Ennadai, 86 H 8
Ennaidi, 89 D 11
Enningdal, 55 J 6
Enns;river, 43 J 9
Eno, 53 K 12
Enontekio, 52 D 8
Enoree;river, 104 G 6
Enosburg Falls, 101 D 12
Ensenada de los Muertos, 106 G 4
Ensign, 99 E 11
Ensley, 103 O 9
Entebbe, 78 C 9
Enterprise, 94 D 6
Entre Rios, 78 H 11
Entre Rios;prov, 116 J 8
Entreistle, 88 I 7
Enu;isl., 65 D 14
Enugu, 75 L 11
Envigado, 112 F 4
Envira, 114 E 8
Enyelle, 78 B 3
Enying, 43 L 13
Eochaill, 47 M 4
Eola, 103 O 3
Eolia, 102 D 5
Epanomi, 41 K 7
Epene, 78 B 3
Epes, 103 L 8
Ephraim, 96 G 4
Ephrata, 94 C 5
Epinal, 45 D 12
Epiphany, 98 H 2
Epoufette, 99 E 13
Epping, 51 G 13
Epps, 103 L 4
Epsom, 51 I 11
Epu Pel, 118 C 7
Epukiro, 79 K 4
Epulu;river, 78 B 8
Epuyen-, 118 G 3
Epworth, 98 J 8
Equateur;prov, 78 B 3
Equatorial Guinea, 75 N 11
Era;river, 76 L 7
Eranga, 78 C 3
Erath, 103 P 3
Erawadi Myitwanya;delta, 67 L 11
Erawadi;river, 67 I 11
Erd, 43 K 13
Erdene, 60 D 7
Erdenecagaan, 61 D 11
Erdenemandal, 60 C 8
Ere, 114 A 7
Erfjord, 55 I 2
Ergani, 70 C 4
Ergene;river, 41 J 11
Ergig;river, 76 J 2
Erickson, 89 K 12
Ericsburg, 98 B 6
Eridu, 105 M 4
Erie, 98 K 8
Eriean, 99 J 16
Erigavo, 77 J 14
Erikousa;isl., 41 K 3
Eriksdale, 89 K 13
Erilla, 118 D 3
Erin, 102 H 9
Erkowit, 77 F 10
Erlanget, 100 L 1
Erlanghe, 43 E 6
Erldunda, 82 F 7
Erlenbach, 42 L 3
Erlistoun, 82 H 4
Erme;river, 50 L 3
Ermil Post, 76 I 6
Ermine Street, 51 C 11
Ernabella Mission, 82 G 7
Ernee, 44 D 6
Eromanga, 83 G 10
Eros, 103 M 3
Erraid;isl., 48 A 5
Errego, 79 I 11
Errol, 49 A 11
Erseka, 41 K 5
Ersi, 43 K 13
Erskine, 98 C 3
Erstein, 45 D 13
Ertai, 60 D 6
Erval, 117 J 11
Erwin, 104 E 5
Erwood, 50 F 4
Erzgebirge;cord., 42 G 8
Esashi, 62 B 4
Esbjerg, 55 N 3
Escalante, 96 H 3
Escalante Desert, 96 H 2
Escalante;river, 96 H 4
Escalon, 106 E 7
Escambia;river, 103 O 9
Escanaba, 99 F 11
Escanaba;river, 99 E 11
Escandon, 107 H 10
Escarcega, 107 J 15
Escatawpa;river, 103 O 8
Escaut;river, 45 B 10
Escuinapa, 106 H 6
Escuintla, 107 M 14
Escute, 114 D 2
Eseka, 75 M 12
Esfahan, 70 F 8
Esher, 51 H 11
Esira, 79 P 12
Esk North;river, 49 C 11
Esk South;river, 49 C 12
Esk;river, 49 E 12
Esker, 91 F 11
Eskifjordhur, 52 B 5
Eskilsarter, 55 J 7
Eskilstuna, 55 I 9
Eskimo Point, 89 D 13
Esmeralda;river, 115 I 10
Esmeraldas, 112 I 2
Esmeraldas;prov, 112 J 2
Espalion, 45 I 9
Espanola;isl., 112 H 2
Espanola, 90 K 4
Espartillar, 118 C 8
Esperanca, 115 C 9
Esperance, 82 J 4
Esperanza, 114 F 8
Esperanza;isl., 119 L 3
Espinal, 115 L 16
Espinazo, 107 E 9
Espinillo, 116 E 8
Espinosa, 106 B 1
Espiritu Santo, 115 K 11
Espiritu Santo;isl., 106 G 3
Espita, 107 I 16
Espoo, 53 O 9
Espy, 95 A 3
Esquel, 118 G 3
Esquimah, 88 L 3
Esquina, 116 I 8
Esquiu, 116 H 5
Essequibo;river, 113 F 15
Essex, 99 J 15
Essex Junction, 101 E 12
Esslingen am Neckar, 42 I 5
Est Glacier Park, 94 B 8

Estancia,97 J 6
Estanislao del Campo, 116 E 7
Estcourt,79 N 8
Esteio,117 I 12
Esteli,108 J 5
Estell Manor,95 G 7
Estelline,98 G 2
Esterhazy,89 K 12
Estevan,89 L 11
Estevan Point,88 K 2
Estherville,98 I 4
Estill,105 J 7
Esto,105 L 2
Eston,89 K 9
Estrella Nueva,115 K 12
Eszahi-Kozephegyseg; range,43 J 13
Etah,68 H 8
Etal,49 D 14
Etawah,67 I 4
Etchojoa,106 E 4
Ethel,103 L 7
Ethelbert,89 K 12
Ethiopia,77 J 10
Ethridge,103 I 9
Etla,107 L 11
Eton,51 H 11
Etosha Pan;lake,79 J 3
Etowah;river,104 H 2
Ettington,50 E 8
Ettn-ck Bridge End, 49 D 12
Ettrick,98 H 7
Ettrick;river,49 D 12
Etwall,51 B 9
Euchanio,49 E 10
Eucla,82 I 6
Euclid,98 B 3
Eudom,102 D 1
Eudora,103 L 4
Eufaula,105 K 2
Eugmo;isl.,53 J 7
Eulonia,105 K 7
Eumka,87 A 9
Eunice,97 L 8
Eura,53 N 7
Eurajoki,53 N 7
Eureka,94 H 7
Eureka River,88 G 5
Eureka Springs,102 H 2
Europe,103 K 7
Eutaw,103 L 8
Eutawville,105 I 8
Eva,102 H 8
Evadale,103 O 1
Evan;river,49 E 11
Evans,103 N 2
Evanston,96 F 4
Evansville,98 E 3
Evart,99 H 13
Evarts,104 E 5
Eveaham,50 F 8
Eveleth,98 C 6
Evenequen,113 G 12
Evening Shade,102 H 4
Evenlode;river,51 G 9
Evensk,57 D 15
Evercreech,50 I 6
Everett,100 K 7
Everett City,105 K 6
Everglades,105 O 11
Evergreen,103 N 9
Everton,102 D 11
Evinayong,75 N 12
Evington,104 D 9
Evje,55 J 3
Evron,44 D 7
Evros;river,41 J 10
Evrotas;river,41 O 6
Ewen,99 E 9
Ewes;river,49 E 12
Ewing,98 J 1
Ewo,78 C 2
Exaltacion,115 H 10
Excelsior Springs,102 D 2
Exe;river,50 K 4
Exeland,98 F 7
Exeter,50 K 4
Exira,98 K 4
Exminster,50 K 4
Exmore,104 D 13
Exmouth,50 L 4
Exmouth Gulf,82 F 1
Exmouth Gulf;gulf,82 F 1
Exmouth;pen.,119 K 2
Experiment,105 I 3
Exshaw,88 K 6
Exstew,88 H 2
Exter,102 G 2
Exton,50 J 4
Eye,51 15
Eyemouth,49 C 14
Eynsham,51 G 9
Eyrabakki,52 C 2
Eyre,82 I 5
Ezab el Basarta,77 A 14
Fabala,75 K 4
Fabens,97 L 6
Fabersham,51 I 14
Faborg,55 N 5
Facatativa,112 G 5
Facundo,118 H 4
Fada,76 G 4
Faden,91 E 11
Fafa,75 I 8
Fafan;river,72 H 11
Fafen;river,77 K 12
Fagaras,40 D 9
Fagelsjo,54 F 8
Fagernes,54 G 5
Fagersta,54 H 9
Fahraj,71 H 10
Faiche,48 L 1
Failon,94 H 5
Faiport,100 G 8
Fair Bluff-,104 H 9
Fair Haven,101 F 11
Fair Oaks,99 L 11
Fair Play,102 F 2
Fair Port,104 C 12
Fair;isl.,46 D 9
Fairbanks,86 D 2
Fairborn,99 M 14
Fairburn,105 I 3
Fairbury,98 M 2
Fairchild,98 G 8
Fairfax,98 G 4
Fairfield,94 F 7
Fairford,50 G 8
Fairhaven,101 I 14
Fairhope,103 O 8
Fairland,102 G 1
Fairlee,101 F 12
Fairmont,100 L 5
Fairmount,99 M 13
Fairplay,96 H 6
Fairport,99 F 11
Fairview,83 C 10
Fais;isl.,80 D 4
Faison,104 G 10
Faith,96 D 8
Faithorn,99 F 11
Faizabad,67 J 5
Fajardo,109 H 13
Fajelsjo,53 M 2
Fajou;isl.,109 L 12
Fakaofo;isl.,81 G 11
Fakaofu;isl.,83 E 15
Fakenham,51 B 14
Fakfak,65 B 13
Falaise,44 C 7
Falam,67 I 10
Falckner,118 E 5
Falcon;state,112 C 8
Faleme;river,75 I 3
Falher,88 H 6
Falis Church,95 H 1
Falkenberg,55 L 7
Falkirk,49 B 10
Falkland,49 A 11
Falkner,103 J 7
Falkoping.,55 K 7
Falkville,103 J 9
Fall River,101 I 14
Falls City,98 M 3
Falls Creek,100 I 6
Falmer,51 J 12
Falmouth,47 P 6
Falster;isl.,55 O 6
Falstone,49 E 13
Falticeni,40 B 10
Falun,53 N 3
Famatina,116 H 3
Fanara,77 E 15
Fancy Farm,102 G 7
Fandriana,79 N 12
Fangatau;isl.,81 I 16
Fangdoushan;mount, 35 172 E 7
Fangzheng,61 C 14
Fank,68 E 3
Fannett,103 P 1
Fannin,103 M 6
Fanning;isl.,81 E 13
Fano;isl.,55 N 3
Fanpak,76 J 8
Fanskur,77 A 13
Faqus,77 C 14
Far Hills,95 B 7
Faradje,78 A 8
Farafangana,79 O 12
Farafra Wahat el-;oasis, 76 C 6
Farah,71 G 11
Farah;river,71 G 11
Faralion de Pajaros;isl., 80 B 5
Faranah,75 K 4
Farberg,52 H 4
Farcuhar;isl.,73 J 12
Fareham,51 K 10
Farewell,86 D 1
Fargo,98 D 3
Fargu,105 L 5
Faribault,98 H 6
Faridabad,66 H 5
Faridkot,68 F 6
Farila,54 F 9
Farilhoes;isl.,36 G 2
Farim,75 J 2
Farina,102 E 7
Faringdon,51 G 9
Faringehavn,87 E 14
Farkwa,78 E 10
Farmensville,102 D 6
Farmer City,99 M 10
Farmer;isl.,90 B 5
Farmerville,103 L 3
Farmingdale,95 D 9
Farmington,88 G 5
Farmville,104 D 10
Farnborough,51 I 10
Farndon,50 A 6
Farne Islands;isl., 49 D 15
Farner,104 G 3
Farnham,50 J 7
Farningham,51 H 13
Farnworth,49 L 13
Faro;river,76 K 1
Farrars;river,83 F 9
Farrel;isl.,119 M 2
Farrell,100 I 5
Farrukhabad,68 H 8
Farsala,41 L 6
Farsi',71 F 11
Farsis,77 D 12
Farsund,55 K 3
Farthingnoe,51 F 9
Farwell',97 K 8
Fasano,39 I 12
Fasher. El,76 I 5
Fatahabad,68 G 6
Fatehgarh,68 H 8
Fatehpur,67 I 5
Fathain,48 F 2
Fattjaure,54 B 8
Fatu Hiva;isl.,81 H 16
Fatural,65 D 13
Faughan;river,49 F 3
Faulkton,96 D 10
Fauske,52 F 4
Faustino M. Parera, 116 K 8
Favara,39 N 8
Fawcett,88 I 7
Fawley,51 K 9
Fawn Grove,95 E 3
Fawn;river,89 G 16
Faxafloi;gulf,52 B 1
Faxalven;river,54 D 9
Faxon,102 H 8
Faya,76 G 3
Fayette,98 J 7
Fayetteville,102 H 1
Fayid,77 E 15
Fazilka,68 F 5
Federacion,116 I 8
Feeny,48 G 3
Feijo,114 E 8
Feinne;river,48 K 4
Feiring,54 H 6
Felbridge,51 I 12
Felch,99 E 10
Feldbach,43 L 11
Felderkirchen,43 L 9
Feldkirch,42 K 5
Felipe Carrillo Puerto, 107 J 15
Felix U. Gomez,106 C 6
Felixstowe,51 F 15
Fell,52 A 3
Fellsmere,105 P 7
Felton,49 E 14
Fence,99 F 10
Fengzhen,61 F 11
Fenhe;river,61 G 11
Feni,69 K 15
Feni Islands;isl.,83 B 16
Fennimore,98 I 8
Fenny Stratford,51 F 11
Fenton,99 I 14
Fenwick,49 D 9
Fenyang,61 G 11
Ferdinand,102 E 9
Ferdows,71 F 10
Ferfer,77 L 13
Fergus Falls,98 E 3
Fergusson;isl.,83 C 15
Ferkessedougou,75 K 6
Ferlach,43 L 9
Fermi,41 J 10
Fernandez,116 G 5
Fernandina Beach,105 L 7
Fernandina;isl.,112 G 1
Ferndale,95 B 6
Fernhurst,51 J 11
Fernie,86 K 5
Fernwood,103 N 5
Ferozepore,68 F 6
Ferrenafe,114 D 2
Ferrera;prov,38 D 6
Ferreyra,116 J 5
Ferriday,103 N 4
Ferrol;pen.,114 F 3
Ferrum,104 E 8
Ferryhill,49 G 15
Fertile,98 C 3
Feshi,78 E 5
Fessenden,96 B 10
Festus,102 E 6
Fetesti,40 E 12
Fetlar;isl.,46 B 10
Feuilles;river,90 C 8
Ffestiniog,50 B 3
Ffostrasol,50 E 2
Fhia;river,48 G 1
Fiambala,116 G 3
Fianarantsoa,79 O 12
Fidra;isl.,49 B 12
Field,88 K 6
Fieldale,104 E 8
Fielding,89 J 9
Fier,41 J 3
Fifield,98 E 8
Fiuerte Olimpo,117 C 9
Fiji Sea;sea,83 G 14
Fiji;Coln.,81 H 10
Fiji;sea,81 J 9
Filadelfia,115 H 9
Filer,96 E 2
Filiasi,40 E 8
Filicudi;isl.,39 L 9
Filingure,75 I 9
Filipstad,53 O 2
Fillmore,100 H 7
Fillmore,89 L 11
Filton,50 H 6
Fimi;river,78 D 3
Finaghy,48 H 5
Findlay,99 L 15
Fine,101 E 10
Finecastle,104 D 8
Finger,103 I 7
Finke,82 G 7
Finke;river,82 G 7
Finksburg,95 F 1
Finland;c.,98 C 7
Finland;c.,98 C 7
Finlay Forks,88 G 3
Finlayno,88 H 4
Finley,98 C 2
Finmark,99 B 9
Finne;river,48 G 2
Finnegan,88 K 8
Finschhafen,83 B 14
Finse,54 G 3
Finsterwalde,43 E 9
Fintona,48 H 3
Fiodh Aluine,48 M 3
Fiodhari Atha-,48 K 2
Fionach,48 J 1
Fionnphort,48 A 5
Fire Islands;isl.,95 B 11
Fire River,99 A 14
Fire;isl.,95 B 11
Firenze;prov,38 F 5
Firmat,116 K 7
Firozabad,68 H 8
Firozpur Jhirka,68 H 7
First;river,104 F 6
Firth,98 L 2
Fischerton,83 D 10
Fisher,103 I 5
Fisher;isl.,101 I 13
Fishguard,47 M 7
Fishguard,50 F 1
Fishhook,102 C 5
Fiskardho,41 M 4
Fiskenasset,87 E 14
Fitchburg,101 G 13
Fitz Roy,119 I 5
Fitzcarrald,114 H 8
Fitzgerald,105 K 4
Fitzroy Crossing,82 D 4
Fitzroy;river,82 D 4
Fitzwilliam;isl.,90 K 4
Five Lanes,50 K 1
Five Points,105 I 2
Fivemiletown,48 I 2
Fizi,78 D 8
Fjallesen;mount,52 F 6
Flager Beach,105 N 7
Flagler,96 G 8
Flagstaff,97 J 3
Flaherty;isl.,90 D 6
Flak Fork;river,103 M 1
Flakstadoy;isl.,52 E 3
Flambeau;river,98 F 8
Flamenco,116 G 1
Flamenco;isl.,118 E 8
Flamingo,105 P 11
Flanagin Town,108 L 2
Flanders,98 A 7
Flannan Islands;isl., 46 E 5
Flat,86 C 1
Flat Bay,91 I 14
Flat River,102 F 5
Flat Rock,102 D 10
Flat;river,100 F 1
Flatey;isl.,52 A 3
Flateyri,52 A 1
Flathead Range,94 B 8
Flathead;river,94 B 7
Flatholm;isl.,50 H 5
Flatrock;river,102 D 10
Flax Bourton,50 H 6
Fleemason;isl.,103 P 7
Fleet,51 I 10
Fleet;river,49 G 9
Fleetwood,95 C 4
Flekkefjord,55 J 2
Flekkeroy;isl.,55 K 3
Fleming,104 D 5
Flemingsburg,100 M 1
Flemington,95 C 7
Flensburg,42 A 6
Flers,44 D 7
Flesberg,54 H 5
Fletcher,99 M 14
Fletcher;isl.,33 D 4
Fleur de Lys,91 G 15
Flimby,49 H 11
Flin Flon,89 H 11
Flinders;isl.,83 L 1
Flinders;river,83 D 9
Flint,49 M 12
Flint;isl.,81 H 14
Flint;river,99 I 15
Flippin,102 H 3
Flisa,54 H 6
Floasen,53 L 1
Flockton,49 L 15
Flomaton,103 N 9
Floodwood,98 D 6
Floore,51 E 10
Flor de Punga,114 D 6
Flora,54 F 2
Flora,99 M 12
Floral City,105 O 5
Florala,105 L 1
Floraville,83 D 9
Floreana,112 H 1
Florencia,115 H 10
Flores,108 H 4
Flores Las-,116 M 8
Flores Magon,107 L 14
Flores;isl.,80 G 1
Flores;sea,80 G 1
Florianopolis,117 G 14
Florida,108 G 8
Florida City,105 O 12
Florida Keys;isl., 105 P 10
Florida Peninsula;pen., 93 J 12
Florida;pen.,85 K 9
Florida;prov,117 K 9
Florida;state,93 J 12
Floridana Beach,105 P 8
Florido;river,106 E 6
Florien,103 N 2
Florin,95 D 2
Florina,41 J 5
Floris,98 L 6
Flowers Cove,91 G 14
Flowery Branch,104 H 4
Floyd,104 E 8
Floyd;river,98 I 3
Floydada,97 K 9
Flumendosa;river,39 K 3
Flushing,95 B 9
Fnjoska;river,52 B 3
Foam Lake,89 K 11
Foca,40 G 3
Focsani,40 D 11
Fogelsville,95 B 5
Foggaret ez Zoura,74 D 9
Foggia;prov,38 H 9
Foggo,75 J 12
Fogo,91 H 16
Fogo;isl.,91 H 16
Fohnsdorf,43 K 10
Folegandros;isl.,41 O 9
Foley,87 E 11
Foley;isl.,87 E 11
Foleyet,90 I 4
Folkestone,47 N 12
Folkingham,51 B 11
Folkston,105 L 6
Folkstone,104 G 11
Follette La-,104 E 4
Foloyet,99 B 15
Folsom,103 O 6
Fonda,98 J 4
Fonddu Lac,99 H 10
Fonseca,112 C 6
Fonseca;gulf,108 J 4
Font;river,49 E 14
Fontana,104 F 4
Fontas,88 F 4
Fontas;river,88 F 5
Fonte Boa,113 L 10
Fonualei;isl.,83 F 15
Fonualei;isl.,81 H 11
Fonyod,43 L 12
Forbach,45 C 13
Forcados,75 M 10
Ford,49 D 14
Ford;river,99 E 1
Forde,54 F 2
Fordham,51 E 13
Fordingbridge,50 J 8
Fordland,102 G 3
Fordsville,104 C 1
Fordville,98 B 2
Fordyce,103 K 3
Forecariach,75 K 3
Foreman,97 K 13
Foremost,88 L 8
Forest,90 M 4
Forest Acres,104 H 7
Forest Beach,105 K 7
Forest City,98 I 5
Forest Hill,101 L 9
Forest Lake,98 F 6
Forest Lawn,88 K 7
Forest Park,105 I 3
Forest River,98 B 2
Forester,103 J 2
Forestville,91 I 9
Forgan,97 I 9
Forion,107 H 10
Fork;river,104 E 7
Forked River,95 E 9
Forks,94 B 2
Forli;prov,38 E 6
Forman,98 E 2
Formby,49 L 2
Formigas;isl.,36 M 3
Formosa,116 F 8
Formosa;isl.,75 J 1
Formosa;prov,116 E 7
Fornham Saint Martin, 51 E 14
Forres,116 G 5
Forrest,99 L 10
Forrest City,103 I 5
Forreston,105 I 8
Forsand,55 I 2
Forsayth,83 D 10
Forsmark,53 O 4
Forsmo,54 D 10
Forsnas,52 G 5
Forssa,53 N 8
Forsyth,105 I 4
Fort Adams,103 N 4
Fort Albany,90 G 5
Fort Assiniboine,88 I 7
Fort Atkinson,99 I 10
Fort Augustus,46 G 7
Fort Barnwell,104 F 11
Fort Benning,105 J 2
Fort Benton,96 B 4
Fort Bragg,104 G 9
Fort Chipewyan,88 F 8
Fort City,100 J 6
Fort Collins,96 G 7
Fort Coulonge,90 K 6
Fort Covington,101 D 10
Fort Crampel,76 K 3
Fort Davis,105 J 2
Fort Dodge,98 J 5
Fort Drum,105 P 7
Fort Dundas,82 B 6
Fort Eustis,104 D 12
Fort Fairfield,101 A 16
Fort Fisher,104 H 11
Fort Fitzgerald,86 H 6
Fort Fitzperald,88 E 8
Fort Franklin,88 A 5
Fort Fraser,88 H 3
Fort Fremont,105 J 7
Fort Gaines,105 K 2
Fort Gardel,74 E 11
Fort Gay,100 M 2
Fort George,90 F 6
Fort George;river,90 F 6
Fort Good Hope,86 E 5
Fort Grahame,88 G 3
Fort Hall,78 C 11
Fort Hancock,97 M 7
Fort Hope,90 F 2
Fort Kent,88 I 8
Fort Kent,101 A 15
Fort Khunnek,68 D 8
Fort Knox,104 C 2
Fort Lauderdale,105 N 13
Fort Lawn,104 G 7
Fort Lee,95 A 9
Fort Liard,88 D 4
Fort Liberta,109 G 11
Fort Lockhart,68 D 4
Fort Lupton,96 G 7
Fort Mac Kay,88 G 8
Fort Madison,98 M 7
Fort Matanzas National M- onument,105 N 7
Fort Mc Murray,86 I 6
Fort McKenzie,91 D 9
Fort McLeod,88 L 7
Fort McPherson,86 D 4
Fort Meade,95 G 2
Fort Mill,104 G 7
Fort Motte,105 I 7
Fort Munro,68 F 3
Fort Myers,105 N 10
Fort Nelson,88 E 4
Fort Nelson;river,88 E 4
Fort Norman,88 A 4
Fort Ogden,105 M 10
Fort Oglethorpe,104 G 2
Fort Payne,104 H 2
Fort Peck,96 B 6
Fort Pierce,105 M 12
Fort Pierre,96 D 9
Fort Plain,101 G 10
Fort Portal,78 C 8
Fort Providence,88 D 6
Fort Ranson,98 D 2
Fort Reliance,88 C 9
Fort Resolution,88 D 7
Fort Ripley,98 E 5
Fort Rupert,90 G 6
Fort Saint James,88 H 3
Fort Saint John,88 G 5
Fort Sandemen,71 H 13
Fort Saskatchewan,88 I 7
Fort Scott,102 F 1
Fort Seikirk,40 A 1
Fort Severn,90 D 3
Fort Sibut,76 L 3
Fort Simpson,88 C 5
Fort Smith,88 E 8
Fort Summer,97 K 7
Fort Valley,105 J 4
Fort Vermilion,88 F 6
Fort Walton Beach, 103 O 10
Fort Wayne,99 L 13
Fort White,105 M 5
Fort William,46 G 6
Fort Worth,97 L 11
Fort Yates,96 C 9
Fort Yucon,86 D 3
Fort de Possel,76 L 3
Fort-Dauphin,79 P 12
Fort-Garnot,79 O 12
Fort-de-France,109 M 15
Fortaleza,115 G 11
Fortaleza Santa Teresa, 117 K 11
Fortaleza de Ituxi, 115 E 11
Fortaleza;river,114 G 4
Fortar,46 G 8
Forteau,91 G 14
Fortescue;river,82 F 2
Forth Bridge;point, 49 B 11
Forth;river,49 B 9
Fortin Castre,118 E 6
Fortin Lavalle,116 F 7
Fortin Mutum-,115 M 16
Fortin Olmos,116 H 7
Fortin Paredes,115 M 15
Fortin Ravelo,115 M 14
Fortin Uno,118 D 6
Fortuna,102 E 3
Fortune,91 J 16
Fortville,99 M 13
Forty Mile,86 E 3
Forward,89 L 11
Fossil,94 E 4
Fossil Downs,82 D 5
Fosston,98 C 4
Foster,102 E 1
Foster City,99 E 11
Foster;river,89 G 10
Fosterdele,101 I 10
Fostervilie,104 F 1
Fostoria,99 L 15
Fougeres,44 D 6
Fouke,103 L 1
Foula;isl.,46 C 9
Foulness;isl.,51 G 15
Foulney;isl.,49 J 12
Foulpointe,79 M 13
Foumban,75 M 12
Fountain,103 N 9
Fountain City,98 H 7
Fountain Hill,103 L 4
Fountain Inn,104 G 6
Fountains Abbey,49 J 15
Four Roads,108 K 2
Fourche la Fave;river, 103 J 1
Fourche;isl.,109 B 11
Fourchu,91 K 14
Fovant,50 J 8
Fowey,50 M 1
Fowey;river,50 L 2
Fowler,99 M 11
Fowlers Bay,82 I 6
Fowlerville,100 G 1
Fowlstown,105 L 3
Fox Bay,91 I 13
Fox Islands;isl.,99 F 12
Fox Lake,99 I 10
Fox Valley,89 K 9
Fox;river,99 K 10
Foxboro,98 E 7
Foxdale,48 J 8
Foxe Peninsula;pen., 87 F 11
Foxe;pen.,84 E 9
Foxe;sea,84 D 8
Foxhome,98 E 3
Foxworth,103 N 6
Foz do Gregorio,114 D 8
Foz do Iguacu,117 F 10
Foz do Jatai,115 B 10
Foz do Jordao,114 F 7
Fracis Harbour,91 F 15
Frackville,95 B 3
Fraga,116 K 5
Fraguas,107 K 16
Fraile Muerto,117 J 10
Frailes;range,115 M 11
Framingham,101 H 13
Framlingham,51 E 15
Francavilla Fontana, 39 J 12
France,117 J 9
Frances Lake,88 C 3
Franceville,78 C 1
Francis,89 L 11
Francisco Ignacio Madero, 106 G 7
Francisco Madero,116 M 6
Francisco de Orellana, 114 B 7
Francois,91 I 15
Francois Le-,109 M 16
Frankford,104 C 8
Frankfurt,98 F 1
Franklin,96 G 10
Franklin Delano Roosevelt Lake,94 B 5
Franklin Mountains;range,

Franklin Mountains;range, 86 E 5
Franklin;isl.,33 G 2
Franklinton,103 O 6
Franklinville,100 H 7
Frankrike,53 J 2
Frankton,99 M 13
Frankville,103 N 8
Frant,51 J 13
Franz,90 I 3
Fraser,82 I 4
Fraser;isl.,80 J 5
Fraser;river,88 K 4
Fraserburgh,46 F 9
Fraserdale,90 H 5
Frater,90 J 3
Fray Bentos,116 K 8
Fray Luis Beltran,118 E 6
Fray Marco,117 L 10
Frazee,98 D 4
Frazer,95 D 5
Fred,103 O 1
Freda,99 D 10
Frederic,98 F 6
Frederica,95 H 5
Fredericia,55 N 5
Frederick,98 E 1
Fredericklown,99 L 16
Fredericksburg,97 M 11
Fredericktown,100 J 3
Fredericton,91 K 11
Fredericton Junction, 91 K 11
Frederikshab,87 E 15
Frederikshavn,55 L 5
Frederikssund,55 N 6
Frederiksted,109 B 9
Frederiksvaerk,55 M 5
Fredonia,96 H 12
Fredrikstad,55 I 6
Freeburg,102 E 4
Freedom,102 D 9
Freehold,95 C 8
Freeland,95 A 4
Freeland Park,99 M 11
Freeman,98 I 2
Freeport,99 J 9
Freetown,102 D 10
Freetown,75 K 3
Freirina,116 H 1
Fremantle,82 I 2
Fremont,98 K 2
French Camp,103 L 7
French Guyana,113 F 4
French Lick,102 E 10
French Polynesia,83
Frenchburg,104 C 4
Frenchman;river,96 A 6
Frenchtown,95 B 6
Freshwater,51 K 9
Fresia,118 F 2
Fresnillo,106 H 7
Fressingfield,51 D 15
Frette,54 H 2
Freyburg,101 E 14
Fria,75 K 3
Friars Point,103 J 5
Frias,114 C 2
Friday Harbor,94 B 3
Friedensburg,95 B 3
Friedrichshafen,42 K 5
Friend,98 L 2
Friendship,99 H 9
Fries,104 E 7
Frinton,51 F 15
Frio;river,112 D 7
Friona,97 J 8
Frisco City,103 N 9
Frobisher Bay,87 F 12
Frogmore,103 N 4
Frome,50 I 7
Frome Downs,37 C 4
Front Renge;range,96 G 7
Front Royal,100 M 7
Frontenac,102 F 1
Frontera,107 K 14
Fronteras,106 B 4
Frosinone;prov,38 H 8
Frostburg,100 L 6
Frostproof,105 P 6
Frountain,99 H 12
Froutino,112 E 4
Frovi,53 P 3
Froya;isl.,54 C 4
Fruita,96 H 5
Fruitdale,103 N 8
Fruithurst,104 H 2
Fruitport,99 I 12
Frutillar,118 F 2
Frydlant,43 F 10
Fudai,62 D 13
Fuente Bulnes,119 N 4
Fuerte;river,106 E 4
Fuga;isl.,65 A 11
Fuglasker;isl.,52 C 1
Fuglo;isl.,54 A 2
Fujin,61 B 15
Fujisaki,62 C 12
Fukangawa,62 C 3
Fukaura,62 C 12
Fukuma,63 L 1
Fukuoka,62 D 13
Fukushima,62 F 2
Fukuyama,63 O 2
Fulaerjii,61 C 13
Fulda,98 H 3
Fulford,49 J 16
Fullarton,108 M 1
Fullerton,98 K 1
Fullinas,53 M 1
Fulojpszallas,43 L 13
Fulton,98 K 8
Fumav,45 B 11
Funafuti;isl.,83 E 14
Funchal,74 B 3
Fundacion,112 C 5
Funtua,75 J 11
Furano,62 D 4
Furman,105 J 7
Furneaux Group;isl., 83 L 11
Furneaux;isl.,80 L 6
Furstenfeld,43 L 11
Furukawa,62 F 13
Furukuchi,62 F 12
Fushan,61 F 13
Fusong,61 D 14
Fussen,42 K 6
Futa Ruin,118 F 4
Futaleufu,118 G 3
Futatsui,62 D 12
Futunxi;river,61 J 13
Fuwa,77 A 10
Fuya,62 F 11
Fuyu,61 C 13
Fuyuan,61 B 15
Fuyun,60 C 6
Fuzesabony,43 J 14
Fyresdal,55 I 4
Fyresdal,55 I 4
Fyzabad,108 M 2
Fyzabad,108 M 2
Gaan,77 I 14
Gabbs,94 I 6
Gabela,78 G 2
Gabes,74 B 12
Gabon;c.,73 I 6
Gabras,76 J 6
Gabrovo,40 G 9
Gacheltville,95 E 3
Gacko,40 G 3
Gadaisu,83 D 14
Gadake,60 H 3
Gadarwera,68 K 8
Gaddele,53 I 3
Gadra,68 I 3
Gadsden,102 H 7
Gaduki,66 B 9
Gadzi,76 L 2
Gaerwen,50 A 2
Gaferunt;isl.,80 D 5
Gaffney,104 G 6
Gag;isl.,65 I 15
Gagnoa,75 L 6
Gagnon,91 G 10
Gago Coutinho,78 H 4
Gaiab;river,79 M 4
Gail,97 L 9
Gailey,50 C 7
Gaillimh,47 L 3
Gaima,83 C 12
Gaiman,118 G 6
Gain,43 D 13
Gainesboro,104 E 2
Gainesville,100 M 8
Gairioch,46 F 6
Gaithersburg,100 L 8
Gaithersbury,95 G 1
Gaiwal,67 M 4
Gajle,67 P 6
Galan,53 M 7
Galarza,117 G 9
Galashiels,49 D 12
Galati,40 D 12
Galatia,96 H 10
Galatina,39 J 13
Galax,104 E 7
Galeana,106 C 5
Galegu,77 I 9
Galela,65 H 14
Galena,86 C 2
Galesburg,98 L 8
Galesville,98 H 7
Galeton,90 G 5
Galgate,49 J 13
Galion,100 J 3
Gallabat,77 I 9
Gallatin,96 B 4
Gallatin-Gateway,96 C 4
Gallatin;river,92 E 5
Gallegos;river,119 M 4
Galliakot,68 K 5
Gallipoli,39 J 12
Gallipolis,100 L 3
Gallon,99 L 15
Galloo;isl.,101 F 9
Gallows Corner,51 G 13
Gallup,97 J 5
Galston,49 D 9
Galt,90 M 5
Galveston Bay;river, 93 J 9
Gam;isl.,65 I 15
Gama;isl.,118 E 8
Gamaliel,104 E 2
Gamarra,112 D 6
Gambaga,75 K 8
Gambela,77 K 9
Gambele,86 A 1
Gambie;river,75 J 3
Gambier,99 M 16
Gambier;isl.,81 J 16
Gamboma,78 C 2
Gamboula,76 M 2
Gamlaba,65 I 14
Gamvik,52 A 10
Gan Gan,118 G 5
Ganag,77 B 10
Gananoque,90 L 7
Gancedo,116 G 6
Gandak;river,69 I 11
Gandava,68 G 2
Gander,91 H 16
Gandhinagar,68 K 4
Ganetli,76 G 7
Ganga;river,69 I 13
Ganganagar,68 G 5
Gangapur,68 I 7
Gangcha,60 G 8
Gangoh,68 G 7
Gangran,68 K 6
Gangtok,69 H 13
Ganhe;river,61 B 13
Gankui,61 B 13
Gannvalley,96 D 10
Ganongga,83 C 16
Ganqika,61 D 13
Gansdisishan;range,60 G 3
Gansen,60 G 6
Gansusheng;prov, 34 164 E 7
Ganzhuermiao,61 C 12
Ganzur,77 D 11
Gao,74 H 8
Gaona,116 F 5
Gaotai,60 F 8
Gaoua,75 K 7
Gaoual,75 J 3
Gap,95 D 4
Gap Mills,104 D 8
Gapan,65 B 11
Garachine,112 E 3
Garamba;river,78 A 8
Garantah,65 M 9
Garba Tula,78 C 11
Garberville,94 H 2
Garboldisham,51 D 14
Garbyang,69 F 9
Gardby,55 M 10
Garden,99 E 12
Garden City,96 H 9
Garden Peninsula;pen., 99 E 12
Garden River,99 D 14
Garden;isl.,99 E 13
Gardenton,89 L 14
Gardey,118 C 10
Gardez,71 G 13
Gardiner,90 I 5
Gardiners;isl.,101 I 13
Gardner,96 H 7
Gardner Pinnacles;isl., 81 A 12
Gardner;isl.,81 F 11
Gardo,77 J 14
Garelochhead,48 B 8
Garesnica,38 C 11
Garhakota,68 K 8
Garhi Khair,68 H 2
Gariamanga,112 M 2
Garibaldi,88 K 3
Garissa,78 C 2
Garland,103 K 2
Garmab,71 G 12
Garmischartenkirchen, 42 K 7
Garner,98 I 5
Garnet,96 H 12
Garoe,77 K 14
Garonne,75 K 13
Garot,68 J 6
Garoua,76 J 1
Garret,99 K 13
Garrett Park,95 H 1
Garrison,48 H 1
Garruchos,117 G 10
Garsdale,49 I 13
Garsen,78 D 12
Garstang,49 K 13
Garston,49 M 12
Gartempe;river,44 G 8
Garthmy,50 C 5
Gartield,105 J 6
Garvagh,48 F 4
Garvellachs;isl.,48 A 6
Garwolin,43 E 15
Gary,99 K 11
Garza,116 H 5
Garza Garcia,107 F 9
Gas City,99 M 13
Gasa Dzong,69 H 14
Gasconade;river,102 E 4
Gascoyne Junction,82 G 2
Gascoyne;river,82 G 1
Gashaka,75 L 13
Gashua,75 J 12
Gaskins,105 M 2
Gaspar,117 G 14
Gaspar Grande;isl., 108 K 1
Gasparillo,108 L 2
Gaspe,91 I 12
Gaspe Peninsula;pen., 91 I 11
Gassaway,100 L 4
Gassol,75 K 12
Gaston,99 M 13
Gastonia,104 G 7
Gastre,118 G 4
Gastsjon,53 K 3
Gate City,104 E 5
Gatehouse of Fleet,49 G 9
Gateshead,49 F 15
Gatesville,104 E 12
Gateway,90 K 5
Gatico,116 D 1
Gatineau,90 K 7
Gatineau;river,90 K 7
Gatlingurg,104 F 4
Gattman,103 K 8
Gauley;river,100 M 4
Gaurihar,69 J 9
Gavdhos;isl.,41 P 1
Gave de Pau;river,44 K 6
Gaviotas,118 D 7
Gavle,53 N 4
Gay,99 D 10
Gay Mills,98 I 8
Gaya,75 J 9
Gaydah,83 G 12
Gaydon,51 E 9
Gaylord,98 G 5
Gaza,76 L 2
Gazamn,75 I 12
Gbanga,75 L 4
Gbarnga,75 L 4
Gdansk,43 A 12
Gdynia,43 A 12
Geamartaluiul;river, 40 E 8
Geary,97 J 11
Geba;river,75 J 2
Geban,66 G 10
Gebe;isl.,65 I 14
Gebeit,77 E 9
Gebel Iweibid,77 F 15
Gebo,96 E 5
Gedaref,77 H 9
Gedid Ras el Fil,76 I 6
Gediz;river,41 M 12
Gehua,83 D 14
Geidam,75 J 13
Geigertown,95 D 4
Geikie;river,89 G 10
Geili El-,76 H 8
Geilo,54 G 4
Geisill,48 M 2
Geita,78 D 9
Gel;river,76 K 7
Gela,39 N 9
Geladi,77 K 13
Gelib,78 C 13
Gelibolu,41 J 11
Gelligaer,50 G 5
Gemena,78 B 4
Gemlik korfezi;gulf, 41 J 13
Gemmell,98 B 5
Gemu,60 G 4
Genale;river,77 L 11
General Acha,118 C 7
General Alvarado,116 E 4
General Alvear,116 M 7
General Alvear,116 L 4
General Aquino,117 E 9
General Arenales,116 L 7
General Artigas,117 G 10
General Belgrano,116 M 8
General Bravo,107 F 10
General Cepeda,106 F 8
General Conesa,118 E 7
General Deheza,116 K 5
General Delgado,117 G 9
General Diaz,116 G 8
General Enrique Martinez, 117 K 11
General Enrique Mosconi, 116 D 5
General Galarza,116 K 8
General Guido,117 M 9
General Jose de San Mart-in,116 F 8
General Juan Madariaga, 117 M 9
General La Madrid,118 C 8
General Lagos,116 A 1
General Las Hems,116 L 8
General Lavalle,116 L 5
General Lazaro Cardenas, 106 D 6
General Martin M. de Guemes,116 E 5
General O'Brien,116 L 7
General Obligado,116 G 8
General Pando;ftin.', 116 B 8
General Paz,116 L 8
General Pico,116 M 6
General Pinedo,116 G 7
General Pinto,116 L 7
General Piran,118 C 10
General Plaza,112 L 3
General Racedo;valley, 118 G 5
General Roca,118 D 5
General Saavedra,115 K 13
General San Martin, 118 D 7
General Santos,65 F 13
General Simon Bolivar, 106 G 7
General Teran,107 F 9
General Tosevo,40 F 12
General Trevino,107 F 9
General Trias,106 D 6
General Vargas,117 I 10
General Viamonte,116 L 7
General Vicente Guerrera, 106 G 7
General Villegas,116 L 6
General Vintter,118 E 7
Genesee,94 C 6
Genesee;river,100 G 7
Geneseo,98 K 8
Geneva,96 G 11
Genhe,61 B 12
Genoa,96 F 11
Genola,98 E 5
Genova;prov,38 E 3
Genovesa;isl.,112 G 2
Gentryville,102 F 9
George,98 I 3
George Town,109 F 9
George;river,91 C 11
Georgetown,75 I 2
Georgia;state,93 I 11
Georgiana,103 N 10
Georgina;river,82 E 8
Gerald,102 E 5
Geraldton,82 H 2
Gerardmer,45 D 13
Gering,96 F 8
Germansen Landing,88 G 3
Germantown,100 K 1
Germany East-,76
Germany West-,76
Germfask,99 E 12
Gerona;prov,37 C 14
Gerrard,88 K 5
Gerufa,79 J 6
Gestro;river,77 L 12
Geta,53 O 6
Gethsemani,91 H 13
Gettysburg,96 D 9
Getulio Vargas,117 G 12
Ghabat el Arab,76 J 7
Ghadai,66 D 6
Ghaggar;river,68 G 6
Ghaghra;river,69 10
Ghaham,105 I 2
Ghana;c.,72 H 3
Ghandi,68 J 6
Ghanpokhara,69 G 11
Ghanzi,79 K 5
Ghaomukb,66 F 9
Gharbiya El-;prov,77 C 11
Gharo,68 J 1
Gharyan,76 A 1
Ghat,69 K 14
Ghatampur,69 I 9
Ghazali El-,77 C 14
Ghaziabad,68 G 7
Ghazipur,69 I 11
Ghazni,71 G 13
Gheen,98 B 6
Gheorgheni,40 C 9
Ghilarza,39 J 3
Ghizac,71 G 12
Ghogra;river,67 I 6
Ghost River,90 G 1
Ghotki,68 H 3
Ghudamis,74 C 12
Giamama,78 C 13
Giarre,39 M 10
Giarso,77 L 10
Gibb River,82 D 5
Gibb;river,82 C 5
Gibbon,96 G 10
Gibbs;isl.,33 C 7
Gibe;river,77 K 10
Gibert;isl.,119 O 4
Gibertsville,95 C 5
Gibraltar,112 D 7
Gibroy,94 J 4
Gibsland,103 M 2
Gibson,105 I 5
Gibson City,99 M 10
Gibson Island,95 H 3
Gibson;desert,80 J 2
Gibson;isl.,95 H 3
Gibsonburg,99 K 15
Gibsons,88 L 3
Gibsonton,105 P 5
Gibsonville,104 F 9
Gidar,68 G 1
Gide alv;river,53 K 5
Gideon,102 H 6
Gidole,77 L 10
Gifford,105 P 8
Gigante,112 H 4
Gigha;isl.,48 C 6
Giibertown,103 M 8
Gila Bend,97 K 3
Gila Desert,97 K 2
Gila;river,97 K 2
Gilbana,77 C 16
Gilbert,98 C 6
Gilbert Plains,89 K 12
Gilbert;isl.,81 E 9
Gilberton,83 D 10
Gildeskal,52 F 3
Gildford,96 B 5
Gile,79 I 11
Gilford,48 I 5
Gilgat;river,68 B 5
Gilgil,78 C 11
Gilgit,66 B 9
Gilgit;river,66 B 8
Gilkey,104 F 6
Gill,103 M 1
Gillam,89 G 14
Gillet,103 K 4
Gillette,96 D 7
Gillham,103 J 1
Gillingham,51 H 14
Gilman,98 F 8
Gilmanton,98 G 7
Gilo,77 K 9
Gilortul;river,40 E 8
Gilsland,49 F 13
Gilwern,50 G 5
Gimbi,77 J 9
Gimli,89 K 14
Gingin,82 I 2
Gingoog,65 E 13
Ginir,77 K 1
Ginosa,39 J 11
Giohar,77 M 13
Giralia,82 F 1
Giraltovce,43 H 15
Girard,96 H 13
Girardot,112 G 5
Girishk,71 G 12
Giron,112 L 2
Girvan,48 E 8
Girvan;river,49 E 9
Giscome,88 I 4
Gitega,78 D 8
Giuba;river,78 B 13
Giurgeni,40 E 12
Giurgiu,40 F 10
Givena,83 C 13
Gizai,77 D 11
Giziga,57 D 15
Gizycko,43 B 15
Gjerstad,55 J 4
Gjevilvashyta,54 E 5
Gjirokaster,41 K 4
Gjoa Haven,87 E 9
Gjovdal,55 J 4
Gjovik,54 G 5
Gla,98 L 8
Glace Bay,91 J 14
Glacier,88 K 5
Gladewater,97 K 13
Gladstone,83 F 12
Gladwin,99 H 14
Gladys,104 D 9
Glama;river,54 G 6
Glan,65 G 13
Glanamman,50 G 3
Glasan,48 L 1
Glasford,99 M 9
Glasgow,49 C 9
Glasloch,48 I 3
Glaslyn,89 J 9
Glassboro,95 E 6
Glassdrumond,48 J 6
Glastonbury,50 I 6
Gleason,99 F 9
Gleichen,88 K 7
Gleisdorf,43 L 10
Glen Arbor,99 G 12
Glen Burnie,95 G 2
Glen Rock,95 E 2
Glen Ullin,96 C 9
Glenariff;river,48 F 5
Glenarm,48 F 6
Glenavy,48 H 5
Glenbarr,48 D 6
Glenboro,89 L 13
Glencarse,49 A 11
Glencoe,104 H 1
Glendale,97 K 3
Glendale Cove,88 K 2
Glendive,96 C 7
Glendon,88 I 8
Glendora,103 K 5
Glendun;river,48 F 5
Glenfall,105 L 1
Glenford,99 M 16
Gleniron,95 A 1
Glenluce,48 G 8
Glenmont,99 L 16
Glenmoore,95 D 4
Glenmora,103 N 3
Glennallen,86 E 2
Glenns Ferry,94 F 7
Glennville,105 K 6
Glenora,88 E 1
Glenrock,96 E 7
Glenrothes,49 A 11
Glenroy,82 D 5
Glens Falls,101 F 11
Glenshesk;river,48 F 5
Glentrool,48 F 8
Glentworth,89 L 10
Glenville,100 L 4
Glenwood,91 H 16
Glenwood City,98 F 7
Glenwood Springs,96 G 6
Glidden,98 E 8
Glina,38 D 10
Glissjoberg,53 L 2
Globe,97 K 4
Glogow,43 E 11
Glomfjord,52 F 3
Glommerstrak,52 H 6
Gloria,106 E 7
Gloria La-,112 D 6
Glossop,49 M 14
Gloster,103 N 5
Glote,54 F 7
Gloucester,50 F 7
Gloucester City,95 E 6
Gloversville,101 G 11
Glowno,43 E 13
Gluchstadt,42 B 5
Glyn Neath,50 G 4
Glyncorrwg,50 G 4
Glyndon,98 D 3
Glynn,48 G 6
Glyntawe,50 F 4
Gmund,43 L 9
Gmunden,43 J 9
Gniew,43 B 12
Gniewkowo,43 C 12
Gniezno,43 D 12
Gnjilane,40 H 5
Gnosall,50 C 7
Gnurock,48 B 8
Goageb,79 M 3
Goaso,75 L 7
Goba,77 K 11
Gobabis,79 K 4
Gobernador Civil,116 K 3
Gobernador Costa,118 H 4
Gobernador Crespo,116 I 7
Gobernador Duval,118 D 6
Gobernador Galvez,116 K 7
Gobernador Garmendia, 116 G 5
Gobernador Georges, 119 K 4
Gobernador Mayer,119 L 4
Gobernador Moyano,119 J 5
Gobernador Racedo,116 J 7
Gobles,99 J 12
Gobuin,78 C 13
Godalming,51 I 11
Godbout,91 I 10
Goderich,90 L 4
Godhra,68 K 5
Godmanchester,51 E 12
Godmanstone,50 K 6
Gods Lake,89 H 14
Gods;river,89 G 15
Godstone,51 I 12
Godthaab,87 E 14
Godwin,104 G 10
Goeree Overflakkee;isl., 42 E 1
Goetzville,99 E 14
Goffstown,101 G 13
Gogama,90 J 4
Gogebic Station,99 E 9
Goginan,50 D 3
Gogrial,76 K 7
Gogunda,68 J 5
Gohad,68 I 8
Gohi,117 K 9
Goicoechea;isl.,119 L 8
Goiden,88 K 6
Goiden Meadow,103 P 5
Gojeb;river,77 K 10
Gojome,62 D 12
Gojra,68 F 5
Goka,63 I 5
Gokase;river,63 M 2
Gokurt,68 G 2
Gol,54 G 4
Golcicha,56 D 8
Golconda,102 F 8
Gold Bar,88 G 4
Gold Beach,94 F 2
Gold Hiil,96 G 2
Goldap,43 A 15
Golden,98 M 8
Golden Prairie,89 L 9
Goldendale,94 D 4
Goldenville,91 K 13
Goldfiel,89 E 9
Goldfields,86 H 7
Goldonna,103 M 2
Goldsboro,104 F 10
Goldsmith,97 L 8
Goldston,104 F 9
Goldthwaite,97 L 11
Goldville,104 H 6
Goleniaw,43 C 10
Goljama Kamcija;river, 40 G 11
Golspie,46 F 7
Golub-Dobrzyn,43 C 13
Goma,78 C 8
Gomare,79 J 5
Gombe,75 K 12
Gomez Farias,107 H 10
Gomez Palacio,106 F 7
Gomez Renson,112 L 1
Gommerstrack,54 A 11
Gonabad,71 F 10
Gonave;isl.,109 H 10

Gonave;isl.,

Gonaves, 109 H 11
Gonda, 69 I 10
Gondal, 68 L 3
Gonder, 77 I 10
Gonen;river, 41 K 12
Gongola;river, 75 K 12
Gonohe, 62 C 13
Gonoura, 63 L 1
Gonzales, 97 N 12
Gonzales;river, 116 D 8
Gonzalez Chaves, 118 D 9
Gonzalez Moreno, 116 L 6
Gonzalez Suarez, 114 B 5
Gonzanama, 112 M 2
Goochland, 104 C 10
Good Hope, 98 M 8
Good Thunder, 98 H 5
Goodenough;isl., 83 C 14
Gooderham, 90 L 6
Goodeve, 89 K 11
Gooding, 94 F 7
Goodland, 96 G 9
Goodlettsville, 104 E 1
Goodman, 99 F 10
Goodrich, 50 F 6
Goodsoil, 89 I 9
Goodville, 95 D 4
Goodwater, 105 I 4
Goombalie, 37 B 7
Goose Bay, 88 J 1
Goose Creek, 105 J 8
Goosport, 103 O 2
Gopaiganj, 69 K 14
Gopalganj, 69 I 11
Gor. Oryakhovitsa, 40 G 10
Gora, 43 E 11
Goran El-, 77 L 12
Gorazde, 40 F 3
Gorbea, 118 E 2
Gordo, 103 L 8
Gordon, 86 C 4
Gordon Downs, 82 D 5
Gordon;isl., 119 O 5
Gordonsburg, 103 I 9
Gordonsville, 104 C 10
Gordonton, 104 F 8
Gordonvale, 83 D 11
Gore, 77 K 9
Gore Bay, 90 K 4
Goreda, 65 C 14
Gorgona;isl., 112 H 3
Gorgor, 114 G 4
Gorgora, 77 I 10
Gorham, 101 E 13
Gorica, 38 C 8
Gorin, 98 M 7
Gorizia, 38 C 8
Gorizia;prov, 38 C 8
Gorj;prov, 40 E 7
Gorleston, 51 C 16
Gorlice, 43 H 15
Goro, 77 K 11
Goroka, 83 B 13
Gorong;isl., 65 K 15
Gorontalo, 65 I 12
Gortin, 48 G 3
Gorzow Wielkopolski, 43 D 9
Gosberton, 51 B 12
Gose, 62 G 11
Gosforth, 49 F 15
Goshen, 99 K 13
Goshogawara, 62 C 12
Gosier, 109 M 12
Gosport, 51 K 10
Goss, 103 N 6
Gossen;isl., 54 D 3
Gossinga, 76 K 6
Gostivar, 41 I 5
Gostynin, 43 D 13
Gota, 77 J 11
Gota alv;river, 55 K 6
Gotaru, 68 H 3
Gote, 54 A 1
Goteborg, 55 K 6
Gothenburg, 96 F 10
Gotland;isl., 55 K 11
Gottshalls, 95 2
Goubera, 76 L 6
Goudiry, 75 I 3
Goudreau, 99 B 13
Gouessant;river, 44 D 5
Gough, 105 I 6
Gough;isl., 32 K 6
Goul City, 99 E 12
Goulais River, 99 D 13
Goulburn Islands;isl., 82 B 7
Gould, 103 K 4
Goulds, 105 O 12
Goumenissa, 41 J 6
Goundam, 74 H 7
Gourcy, 75 J 7
Goure, 75 I 12
Gouro, 76 F 3
Gouverneur, 101 E 9
Govan, 89 K 10
Gove, 96 H 9
Govi;desert, 60 E 7
Gowan, 98 D 6
Gowanda, 100 H 6
Gower, 102 C 1
Gowganda, 99 C 16
Gowrie, 98 J 5
Goya, 112 E 3
Goyave, 109 M 12
Goyaves;river, 109 M 11
Goyllarisquizga, 114 G 5
Gracefield, 90 K 7
Graceton, 98 A 4
Graceville, 98 F 3
Gracey, 102 G 9
Gracias, 108 J 4
Gradule, 38 A 9
Grady, 103 K 4
Grafenau, 42 I 8
Grafton, 99 K 16
Grafton;isl., 119 N 3
Graham, 104 F 9
Graham, 98 A 8
Graham;isl., 88 H 1
Grahamville, 105 J 7
Grain, 51 H 14
Grainseach an Disirt, 48 K 5
Grajewo, 43 B 15
Gramalote, 112 E 6
Grambling, 103 M 3
Gramby, 102 G 1
Grampian, 100 J 6
Grampian Mountains;range, 46 H 7
Gramshi, 41 J 4
Gran Abaco;isl., 85 K 10
Gran Bajo de San Julian, 119 K 4
Gran Guardia, 116 F 8
Gran Inagua;isl., 85 K 10
Gran Malvina;isl., 119 M 8
Gran Morelos, 106 D 6
Gran Park, 99 L 11
Granada, 108 K 5
Granadines;isl., 85 L 13
Granard, 48 K 2
Granbury, 97 L 11
Granby, 90 K 8
Grand Bank, 91 I 15
Grand Bassam, 75 M 6
Grand Bay, 103 O 8
Grand Bend, 99 I 16
Grand Bourg, 109 M 12
Grand Bruit, 91 I 14
Grand Canyon, 97 I 3
Grand Canyon National Park, 97 I 3
Grand Cayman;isl., 108 H 7
Grand Chain, 102 G 7
Grand Chenier, 103 P 2
Grand Comore island, 79 K 11
Grand Coulee, 94 C 5
Grand Cul-de-Sac Marin, 109 L 11
Grand Falls, 91 J 10
Grand Forks, 88 L 5
Grand Haven, 99 I 12
Grand Island, 98 L 1
Grand Isle, 101 A 16
Grand Junction, 96 H 5
Grand Lahou, 75 M 6
Grand Marais, 98 C 8
Grand Marsh, 99 H 9
Grand Meadow, 98 H 6
Grand Mere, 90 K 8
Grand Portage, 99 C 9
Grand Rapids, 89 I 12
Grand Ridge, 105 L 2
Grand Teton National Park, 96 E 4
Grand Turk, 109 G 11
Grand flet;isl., 109 M 12
Grand;isl., 99 D 11
Grand;river, 96 C 9
Grande Anse, 109 L 13
Grande Baleine;river, 90 E 7
Grande Entree, 91 J 13
Grande Prairie, 88 H 5
Grande Riviere, 109 L 15
Grande Riviere, 91 I 12
Grande Ronde, 94 D 3
Grande Ronde;river, 94 D 6
Grande Terre Islands;isl., 103 P 6
Grande Terre;isl., 109 L 12
Grande Vallee, 91 I 11
Grande de Lipez;river, 114 L 2
Grande de Tarija;river, 116 D 5
Grande de Tierra del Fuego;isl., 119 N 5
Grande o Grijalva;river, 107 L 14
Grande;river, 97 O 10
Grandes Bergeronnes, 91 I 10
Grandfalls, 97 M 8
Grandfield, 97 K 11
Grandin, 102 G 5
Grandois, 91 G 15
Grandview, 89 K 12
Grandy, 104 E 13
Grange, 49 I 12
Grangemouth, 49 B 10
Grangeville, 94 D 6
Granite, 95 G 1
Granite City, 102 E 6
Granite Falls, 98 G 4
Graniteville, 105 I 6
Granna, 55 K 8
Grannis, 103 J 1
Gransha, 48 G 6
Grant, 96 F 9
Grant, 90 H 3
Grant City, 98 M 4
Grantham, 95 D 1
Grantham, 51 B 11
Grants, 97 J 5
Grants Pass, 94 F 3
Grantsburg, 98 F 6
Grantsville, 96 G 3
Grantville, 105 I 3
Granun, 88 L 7
Granville, 86 E 3
Graso;isl., 53 O 5
Grasonville, 95 H 3
Grass;river, 89 H 12
Grassrange, 96 C 5
Gratz, 95 B 2
Gravatana;river, 112 K 8
Gravelbourg, 89 L 10
Gravenhurst, 90 L 5
Gravesend, 51 H 13
Gravette, 102 G 1
Gravina in Puglia, 39 I 1
Gravois Mills, 102 E 3
Grawn, 99 G 12
Gray, 105 I 4
Grayling, 100 D 1
Grayrigg, 49 I 13
Grayson, 100 M 2
Grayton Beach, 105 M 1
Grayville, 102 E 8
Graz, 43 L 10
Great Abaco;isl., 109 D 9
Great Ashby, 49 H 13
Great Barrington, 101 H 11
Great Bear;river, 88 A 5
Great Bend, 96 H 10
Great Camanoe;isl., 109 A 9
Great Cumbrae;isl., 48 C 8
Great Dividing Range;range, 83 D 10
Great Dunmow, 51 F 13
Great Egg Harbor;river, 95 F 7
Great Exuma;isl., 109 F 9
Great Falls, 96 B 4
Great Harwood, 49 K 13
Great Inagua;isl., 109 G 10
Great Malvern, 50 E 7
Great Miami;river, 100 L 1
Great Missenden, 51 G 11
Great Ponton, 51 B 11
Great Ruaha;river, 78 F 11
Great Salked, 49 G 13
Great Salt Lake Desert;desert, 94 G 8
Great Sandy Desert;desert, 82 E 3
Great Smeaton, 49 H 15
Great Torrington, 50 J 2
Great Victoria Desert;desert, 82 H 5
Great Wass;isl., 101 E 16
Great Witley, 50 E 7
Great Yarmouth, 51 C 16
Greater Antilles;arch., 108 G 6
Greater Leech Lake Indian Reserve, 98 C 5
Grebeoun Mont-;mount, 74 G 12
Greco, 117 K 9
Greedmoor, 104 E 10
Greeley, 96 G 7
Greeleyville, 105 I 8
Green, 90 G 2
Green Bay, 99 G 10
Green Brier, 104 E 1
Green Camp, 99 L 15
Green Castle, 98 M 6
Green Cove Springs, 105 M 6
Green Hammerton, 49 J 16
Green Haven, 95 G 2
Green Hill, 108 K 2
Green Lake, 89 9
Green Lane, 95 C 5
Green Point, 105 M 2
Green Pond, 105 J 7
Green Ridge, 102 E 2
Green River, 96 F 5
Green Springs, 99 K 15
Green-Forest, 102 H 2
Green;river, 85 I 5
Greenbelt, 95 H 1
Greenbrier;river, 104 D 7
Greenbush, 98 A 3
Greencastle, 48 J 5
Greene, 98 I 6
Greeneville, 104 E 5
Greenfield, 98 L 4
Greengates, 48 K 2
Greenhaugh, 49 E 13
Greenhead, 49 F 13
Greening, 90 J 7
Greenland, 99 D 9
Greenlaw, 49 C 13
Greenock, 48 B 8
Greenodd, 49 I 12
Greenough;river, 82 H 2
Greenport, 101 I 12
Greensboro, 103 L 9
Greensburg, 96 I 10
Greenspond, 91 H 16
Greentown, 99 M 12
Greenup, 100 M 2
Greenvale, 83 D 10
Greenville, 75 M 4
Greenwich, 99 L 16
Greenwood, 88 L 5
Greer, 104 G 6
Gregorio;river, 114 E 8
Gregory, 96 E 10
Gregory Downs, 83 D 9
Gregory;river, 82 D 8
Greinton, 50 I 6
Grena, 55 M 5
Grenada, 103 K 6
Grenada;c., 85 L 13
Grenada;isl., 109 D 14
Grenade, 44 J 8
Grenbrier, 95 B 2
Grenivik, 52 A 3
Grennbrier;river, 100 M 5
Grenville, 98 F 2
Grervy;isl., 119 O 6
Gresham, 94 D 3
Gresham, 104 H 9
Greta;river, 49 H 14
Gretna, 98 A 2
Grevena, 41 K 5
Grewelthorpe, 49 J 15
Grey Abbey, 48 H 6
Grey De-;river, 82 E 3
Grey Islands Harbour, 91 G 15
Greybull, 96 D 6
Greystoke, 49 H 12
Grianfort An-, 48 J 5
Gribingui;river, 76 K 3
Gridley, 99 L 10
Griekwastad, 79 N 5
Griffin, 89 L 11
Grifton, 104 F 11
Griggsville, 102 C 5
Grijalva;river, 107 K 14
Grim Cape;, 83 L 10
Grimari, 76 L 3
Grimesland, 104 F 11
Grimsby, 47 K 10
Grimsey;isl., 52 A 3
Grimshaw, 88 G 6
Grimsstadhir, 52 A 4
Grimstad, 55 J 4
Grindavik, 52 C 2
Grindsted, 55 N 4
Grinnell, 98 K 6
Grinnell Peninsula;pen., 87 C 9
Griquet, 91 F 15
Grise Fiord, 87 C 10
Griswold, 98 L 4
Groais;isl., 91 G 15
Grobming, 43 K 9
Grodkow, 43 G 12
Grodzisk Mazowiecki, 43 D 14
Groenland;isl., 84 B 9
Groenland;sea, 84 B 9
Groix;isl., 44 E 4
Grojec, 43 E 14
Gromite Bay, 88 K 2
Gronlid, 89 J 10
Groomsport, 48 H 6
Groot Laagte;river, 79 K 4
Grootfontein, 79 J 3
Gros Morne, 109 L 15
Gross, 105 L 6
Grosse Tete, 103 O 4
Grosseto;prov, 38 G 5
Groton, 101 I 13
Groton, 98 F 1
Grottaglie, 39 J 12
Groundhog;river, 90 I 4
Grove, 102 G 1
Grove City, 99 M 15
Grove-Hill, 103 N 8
Groveland, 105 O 6
Grover, 96 F 7
Groves, 103 P 1
Groveton, 101 E 13
Grovetown, 105 I 6
Grudovo, 40 H 11
Grudziadz, 43 C 12
Grumeti;river, 78 C 10
Grundy, 104 D 6
Grundy Center, 98 J 6
Grunidora, 106 G 8
Grutnes, 46 C 9
Gruver, 97 I 9
Gryfe;river, 48 C 8
Gryfice, 43 B 10
Grygla, 98 B 4
Grytoyo;isl., 52 D 4
Guacamaya, 112 H 5
Guacara, 113 C 9
Guacayaco;river, 112 J 7
Guachaqui, 116 F 3
Guachara, 113 E 9
Guachipas, 116 F 4
Guachipas;river, 116 F 4
Guachiria;river, 112 F 7
Guachochic, 106 E 5
Guadalajara, 106 I 7
Guadalcanal;isl., 80 G 7
Guadalcazar, 107 H 9
Guadales, 116 L 3
Guadaloupe, 106 C 3
Guadalquivir, 36 I 6
Guadalupe Bravos, 97 M 6
Guadalupe Calvo, 106 F 5
Guadalupe Victoria, 106 G 7
Guadalupe de Bravo, 106 B 6
Guadalupe de los Reyes, 106 G 5
Guadeloupe, 109 M 12
Guafo;isl., 118 H 2
Guagramano, 114 B 5
Guaiba, 117 I 12
Guaico, 108 K 3
Guaillabamba;river, 112 J 2
Guainia;river, 113 H 10
Guaira, 117 E 11
Guaira;prov, 117 F 9
Guam, 80 C 5
Guamas, 106 D 3
Guanaceri, 106 F 6
Guanachacabides;gulf, 108 F 6
Guanaja;isl., 108 I 5
Guanajuato, 106 I 8
Guanajuato;state, 106 I 8
Guanape;isl., 114 F 3
Guanapo, 108 K 3
Guanare, 112 D 8
Guanare Viejo, 112 D 8
Guanare;river, 113 D 9
Guanarito, 112 D 8
Guandacol, 116 I 3
Guane, 108 F 6
Guangxizhuangzizhigu, 61 K 10
Guanico, 112 E 1
Guanipa;river, 113 C 12
Guano, 112 K 2
Guanoco, 113 C 12
Guanta, 113 C 11
Guantanamo, 109 G 10
Guapi, 112 H 3
Guapiles, 108 L 6
Guapore;river, 115 I 14
Guaqui, 115 K 9
Guaranda, 112 K 2
Guarapuava, 117 F 12
Guaraquito;river, 113 D 10
Guaratuba, 117 F 14
Guaraunos, 113 C 12
Guarayos;river, 115 J 13
Guardal;river, 37 I 9
Guarenas, 113 C 10
Guarico;river, 113 D 9
Guarico;state, 113 D 10
Guaritico;river, 112 E 8
Guarne, 112 F 5
Guarrojo;river, 112 G 6
Guasare;river, 112 B 6
Guasave, 106 F 4
Guasdalito, 112 E 7
Guasipati, 113 E 13
Guatemala, 108 J 4
Guatemala;c., 85 M 7
Guateque, 112 G 6
Guatimozin, 116 K 6
Guatire, 113 C 10
Guatisimina, 113 G 12
Guatrache, 118 D 7
Guaviare;river, 113 G 9
Guayabal, 113 E 10
Guayabero;river, 112 H 6
Guayaguayare, 108 M 3
Guayameo, 106 K 8
Guayane;river, 113 H 10
Guayaneco;archp., 119 J 1
Guayaouil, 112 L 2
Guayaouil;gulf, 110 F 3
Guayape;river, 108 J 5
Guayaramerin, 115 G 11
Guayas;prov, 112 L 1
Guayas;river, 112 I 5
Guaycuru;river, 116 F 8
Guayeleyo;river, 107 G 10
Guaymas;river, 106 D 3
Guayouiraro, 116 I 8
Guayouiraro;river, 116 I 8
Guayuriba;river, 112 G 6
Guazapares, 106 E 5
Guba, 77 J 9
Guchab, 79 J 3
Gudjon, 102 G 3
Gue Du-;river, 91 D 9
Guebwiller, 45 E 13
Guejar;river, 112 H 6
Guekedou, 75 K 4
Guelengdeng, 76 J 2
Guelph, 90 L 5
Guelta Zemmur, 74 E 4
Guemes, 107 G 10
Gueoue;river, 112 B 8
Gueppi, 114 A 5
Guer;river, 44 D 4
Guera, 74 G 2
Guerban angeer, 60 G 7
Guere;river, 113 D 11
Guereda, 76 H 4
Guernsey;isl., 47 P 9
Guerrero;state, 107 K 9
Gueydan, 103 P 3
Guguan;isl., 80 B 5
Guichon, 117 J 9
Guide Rock, 98 M 1
Guider, 76 J 1
Guidiguir, 75 J 12
Guiglo, 75 L 5
Guigue, 113 C 9
Guildford, 51 I 11
Guildhall, 101 E 13
Guilford, 101 D 15
Guilford College, 104 F 8
Guimaras;isl., 65 D 11
Guimbalete, 106 E 7
Guina, 108 J 6
Guindulman, 65 E 12
Guinea Bissau;c., 75 J 1
Guinea;c., 75 J 2
Guinea;gulf, 72 H 4
Guines, 108 D 7
Guingamp, 44 D 4
Guinguineo, 75 I 2
Guinguo, 76 M 2
Guion, 102 H 4
Guiria, 113 C 13
Guisborough, 49 H 16
Guiseley, 49 K 15
Guist, 51 B 14
Guita Koulouba, 76 L 5
Guiuan, 65 D 13
Guizhousheng;prov, 61 J 9
Gujar Khan, 66 F 8
Gujarat;state, 68 K 3
Gujranwala, 68 E 5
Gujrat, 68 E 5
Gulabgarh, 66 F 12
Gulang, 61 G 9
Gulf, 104 F 9
Gulfport, 105 P 5
Gulihe;river, 61 A 13
Gulkana, 86 E 2
Gull Lake, 89 L 9
Gulliver, 99 E 12
Gulsvik, 54 H 5
Gultari, 66 D 11
Gulu, 78 B 9
Guluogongba, 60 G 4
Gulwe, 78 E 11
Gum Neck, 104 F 12
Gum Spring, 104 C 10
Guma, 60 F 3
Gumel, 75 J 12
Gummi, 75 J 10
Gumti;river, 69 I 9
Gumuru, 76 K 8
Gumzai, 65 C 14
Guna, 68 J 7
Gungo, 78 G 1
Gungu, 78 E 3
Gunisao;river, 89 I 13
Gunnar, 89 E 9
Gunnerside, 49 I 14
Gunnison, 96 G 3
Gunpowder Falls;river, 95 F 2
Guntersville, 104 H 1
Gunungapi;isl., 65 L 13
Gunzburg, 42 J 6
Guoxian, 61 F 11
Gupis, 66 B 9
Gurais, 66 D 10
Gurara;river, 75 K 11
Gurba;river, 78 A 7
Gurdaspur, 68 E 6
Gurdon, 103 K 2
Gurgaon, 68 H 7
Gurha, 68 J 4
Gurkha, 69 H 11
Gurley, 104 G 1
Gurney, 98 E 8
Gurun, 70 C 4
Gurup;river, 110 F 9
Gurur, 70 D 5
Gurvan Dzagal, 61 C 11
Gusau, 75 J 11
Gusilike, 60 E 5
Guspini, 39 K 3
Gussing, 43 L 11
Gustavia, 109 11
Gustavo Sotelo, 106 B 2
Gustavus, 86 G 2
Guthrie, 97 J 11
Guthrie Center, 98 K 4
Gutierrez, 115 M 12
Gutierrez Zamora, 107 I 11
Guttenberg, 98 J 7
Guyana;c., 113 F 14
Guyandot;river, 100 M 3
Guyang, 61 E 10
Guyhirnc, 51 C 12
Guymon, 97 I 9
Guynandot;river, 104 D 6
Guysborough, 91 K 13
Guyton, 105 J 7
Guyuan, 61 E 12
Guzman, 106 B 5
Gvaliyar, 67 i 4
Gwadabawa, 75 J 10
Gwai, 79 J 7
Gwai;river, 79 J 7
Gwalchmal, 49 M 9
Gwandu, 75 J 10
Gwane, 78 A 6
Gwinn, 99 E 11
Gwinner, 98 E 2
Gwyddelwern, 50 F 2
Gydanskij Poluostrov;pen., 56 E 8
Gyoma, 43 K 15
Gyonk, 43 L 13
Gypsumvilie, 89 J 13
Gypsumville, 89 J 13
Gyula, 43 L 16
Gyula, 43 L 16
Haapai Group;isl., 83 F 15
Haapai;isl., 81 I 11
Haapajarvi, 53 J 9
Haapavesi, 53 J 9
Haasts Bluff, 82 F 6
Habana;prov, 108 F 7
Habarana, 67 P 6
Habaswein, 78 B 12
Habay, 88 F 5
Habban, 70 M 5
Haboro, 62 B 3
Hacha, 112 J 4
Hachijb;isl., 63 M 12
Hachimori, 62 D 12
Hachinohe, 62 C 13
Hackberry, 103 P 2
Hackensack, 98 D 5
Hackensack;river, 95 A 9
Hackettstown, 95 A 7
Hackleburg, 103 J 8
Haddenham, 51 D 13
Haddington, 49 B 12
Haddock, 105 I 4
Haddonfield, 95 E 6
Hadejia, 75 J 12
Hadejia;river, 75 J 12
Haderslev, 55 N 4
Hadham, 51 F 12
Hadleigh, 51 F 15
Hadley, 104 D 1
Hadsel, 52 D 3
Hadseloy;isl., 52 D 3
Hadsund, 55 L 5
Hafizabad, 68 E 5
Hafnarjordhur, 52 C 2
Hafun, 77 J 15
Hag Abdulla, 76 H 8
Hagadera, 78 C 12
Hagerman, 97 K 7
Hagerstown, 99 M 13
Haggerston, 49 C 4
Hagley, 50 D 7
Hagood, 104 H 8
Hague, 101 F 11
Haguenau, 45 C 13
Hahira, 105 L 4
Hahnville, 103 P 5
Hahwell, 50 M 3
Hahwhistle, 49 F 13
Haig Lake, 88 G 6
Hailakandi, 69 J 16
Hailar, 61 B 12
Hailesjo, 53 K 4
Hailey, 94 E 7
Haileybury, 90 J 5
Haileyville, 118 A 7
Hailiutu, 61 E 9
Hailsham, 51 J 13
Hailuoto, 53 I 9
Hailuoto;isl., 53 I 8
Hainan;isl., 59 K 11
Haines, 86 G 3
Haines City, 105 P 6
Haines Junction, 86 F 3
Hainesburg, 95 A 6
Haiti;c., 109 H 10
Haivorgate, 89 K 10
Haiya, 77 G 9
Hajaudorog, 43 J 15
Hajduboszormeny, 43 J 15
Hajduhadhaz, 43 J 15
Hajduszoboszlo, 43 K 15
Hajfabad, 71 H 9
Hajnowka, 43 C 16
Hakodate, 62 F 2
Hala, 68 I 2
Halaib, 77 E 9
Halbert, 103 M 1
Halberton, 50 J 4
Halden, 55 I 6
Haldersvig, 54 A 1
Haldwani, 68 G 8
Hale, 102 D 2
Hale Center, 97 K 9
Hale;river, 82 F 7
Halesowen, 50 D 7
Halesworth, 51 D 16
Halethorpe, 95 G 2
Haleyville, 103 J 9
Halford, 50 F 8
Halfway;river, 88 G 4
Hali, 70 K 3
Haliburton, 90 L 6

Jaipur,

Jaisalmer, 68 H 3
Jajarkot, 69 G 10
Jajpur, 69 M 12
Jakhal, 68 G 6
Jakhan, 68 K 2
Jakin, 105 L 2
Jakobshavn, 87 C 13
Jakobstad, 53 J 7
Jakutsk, 57 F 12
Jal, 97 L 8
Jala, 106 I 7
Jalaca de Ledesma, 107 I 10
Jalalabad, 71 F 13
Jalapa, 107 K 14
Jalaun, 68 I 8
Jalesar, 68 H 8
Jalingo, 75 K 13
Jalisco;state, 106 I 7
Jalkot, 66 C 8
Jalon;river, 37 D 10
Jalor, 68 J 4
Jalostotitlen, 106 I 8
Jalpa de Mendez, 107 K 13
Jaltenango, 107 L 14
Jaltepec;river, 107 L 12
Jaltipan de Morelos, 107 K 12
Jaltocan, 106 I 10
Jama, 112 J 1
Jamada, 63 L 1
Jamaica, 95 B 10
Jamaica;c., 109 H 9
Jamari, 115 F 13
Jamas, 65 B 16
Jamboi, 40 H 11
Jambongan;isl., 65 F 9
James Bay, 90 F 5
James Craik, 116 J 6
James;isl., 118 H 2
James;river, 96 C 10
Jamesburg, 95 C 8
Jamestown, 90 I 3
Jamesville, 104 F 12
Jamison, 95 C 6
Jammu, 68 E 6
Jamnagar, 68 K 3
Jampur, 68 G 3
Jamrao Ki Nahar;river, 68 I 2
Jamtland;lan., 54 D 7
Jan Mayen;isl., 58 B 11
Jana;river, 57 D 12
Janaperi;river, 113 K 13
Janauaca, 113 L 14
Jand, 68 D 4
Jandiatuba, 115 D 9
Jandiatuba;river, 113 M 8
Jandola, 68 E 3
Janesville, 98 H 5
Jangal, 68 H 1
Jani Khel, 68 D 3
Janira;river, 40 G 10
Janos, 106 B 5
Janoshalma, 43 L 14
Janoshaza, 43 K 12
Janow-Lubelski, 43 F 16
Janskij Zaliv;gulf, 57 D 12
Japan;c., 65 B 14
Japero, 65 C 15
Japiim, 114 E 7
Japura, 113 K 10
Japura;river, 113 L 11
Jaque, 112 E 3
Jaqui, 114 J 6
Jaquirana;falls, 115 F 13
Jaralpur, 68 E 5
Jaramillo, 119 J 6
Jaranwala, 68 F 5
Jarcevo, 56 G 8
Jarocin, 43 E 12
Jaromel, 43 G 10
Jaroslaw, 43 G 16
Jarpen, 53 K 2
Jarratt, 104 E 11
Jarrettsville, 95 F 3
Jarrow, 49 F 15
Jaru;river, 115 G 13
Jaruma, 115 L 9
Jarvis;isl., 81 E 13
Jarwa, 69 H 10
Jasfo, 43 H 15
Jask, 71 I 9
Jason;pen., 33 C 7
Jasonville, 102 D 9
Jasper, 88 J 5
Jati, 68 J 2
Jatupu, 113 K 15
Jatupu;river, 115 A 16
Jau;river, 115 A 13
Jauja, 114 H 5
Jaumave, 107 G 9
Jauru;river, 115 K 16
Jauya, 108 J 6
Java;isl., 59 O 11
Java;sea, 59 O 12
Javier;isl., 119 J 2
Jawor, 43 F 11
Jay, 101 E 11
Jayanca, 114 D 2
Jayapura, 65 B 16
Jayess, 103 N 6
Jazminal, 106 F 8
Jean Marie River, 88 D 5
Jeanerete, 103 P 4
Jeannette, 100 K 5
Jeater Houses, 49 I 16
Jebba, 75 K 10
Jebelein El-, 76 I 8
Jeberos, 114 C 5
Jebri, 68 H 1
Jedburgh, 49 D 13
Jedrzejow, 43 F 14
Jefferson, 97 K 13
Jefferson City, 102 E 4
Jefferson Heights, 103 P 6
Jefferson;river, 96 D 3
Jeffersontown, 102 E 11
Jeffersonville, 105 J 4
Jega, 75 J 10
Jelenia Gora., 43 F 10
Jellico, 104 E 4
Jellicoe, 90 H 2
Jemison, 103 L 10
Jena, 103 N 3
Jenbach, 42 K 7
Jenkinjones, 104 D 6
Jenkins, 104 D 5
Jenkinsville, 104 H 7
Jenkintown, 95 D 6
Jenner, 88 K 8
Jennings, 102 E 6
Jens Munks;iala, 87 C 15
Jensen Beech, 105 M 12
Jequetepeque;river, 114 E 3
Jerecuaro, 107 J 9
Jeremie, 109 H 10
Jerico, 112 F 4
Jerome, 96 E 2
Jersey City, 95 B 9
Jersey Shore, 100 I 8
Jersey;isl., 47 P 9
Jerseyville, 102 D 5
Jerusalem, 69 I 3
Jervaulx Abbey, 49 I 15
Jesenice, 38 B 8
Jesup, 98 J 7
Jesus, 117 G 10
Jesus Carranza, 107 K 12
Jesus Mana, 116 I 5
Jesus de Machaca, 115 K 9
Jetait, 89 G 11
Jetersville, 104 D 10
Jetmore, 96 H 10
Jetpur, 68 L 3
Jewell, 96 E 10
Jhainti, 69 I 14
Jhajjar, 68 G 7
Jhal, 68 G 1
Jhang Maghiana, 68 F 4
Jhansi, 68 J 8
Jhawani, 69 H 11
Jhelum, 68 D 5
Jhelum;river, 68 E 4
Jhil Manchhar, 68 H 1
Jhunjhunu, 68 H 6
Jiachazong, 60 I 6
Jialingjiang;river, 61 H 11
Jiamusi, 61 C 14
Jiangjunmiao, 60 D 6
Jiangsusheng;prov, 61 G 12
Jiangxisheng;prov, 61 I 12
Jiangzi, 60 I 5
Jiayin, 61 B 14
Jiboia, 112 I 8
Jicamorachic, 106 D 4
Jicaron;isl., 108 M 7
Jico, 107 J 11
Jiddah, 70 J 3
Jiesjokka;river, 52 C 8
Jiexiu, 61 G 11
Jihocesky;prov, 43 I 9
Jihomoravsky;prov, 43 I 10
Jijiga, 77 J 12
Jilantai, 61 F 9
Jilemutu, 61 A 12
Jilibulake, 60 H 6
Jilinsheng;prov, 61 D 13
Jim Thorpe, 95 A 5
Jimani, 109 H 11
Jimbolia, 40 D 5
Jimenez, 107 D 9
Jimma, 77 K 10
Jimmi;river, 83 B 13
Jimunai, 60 C 5
Jin, 61 D 12
Jind, 68 G 6
Jingbian, 61 G 10
Jinghe, 60 D 4
Jinhe, 61 A 12
Jinhongshan, 60 F 6
Jinja, 78 C 9
Jinlonggou, 61 A 13
Jinotega, 108 J 5
Jinta, 60 F 8
Jiparana;river, 115 F 13
Jipijapa, 112 K 1
Jirofc, 71 H 9
Jiul;river, 40 E 8
Jiulongjiang;river, 61 K 13
Jiwani, 71 J 10
Jiwuer, 61 B 13
Jixi, 61 C 15
Joacaba, 117 G 12
Joaquin, 103 M 1
Joaquin Amaro, 107 M 13
Joaquin V. Gonzalez, 116 E 5
Jocoli, 116 K 3
Jodhpur, 68 I 5
Jodiya Bandar, 68 K 3
Joensuu, 53 L 12
Joggins, 91 K 12
Johan, 68 G 1
Johangiribad, 68 H 7
Johi, 68 H 2
John Day, 94 E 5
John Day;river, 94 E 4
Johnsburg, 101 F 11
Johnson City, 97 I 9
Johnson;isl., 90 D 6
Johnsonburg, 100 I 6
Johnsons Crossing, 88 C 1
Johnsonville, 104 H 9
Johnston, 104 H 6
Johnstone, 49 C 9
Johnstonebridge, 49 F 11
Johnstown, 100 K 3
Johnswood, 100 C 2
Johor Baharu, 64 H 4
Joinville, 117 F 14
Joinville;isl., 33 C 7
Jokkmokk, 52 G 6
Joliet, 99 K 10
Joliette, 90 K 8
Jolo, 65 F 11
Jolo;isl., 65 G 11
Jolo;sea, 59 L 13
Jomala, 53 O 6
Jombo;river, 78 G 3
Jones, 103 L 10
Jonesboro, 102 F 7
Jonesport, 101 D 16
Jonestown, 103 J 5
Jonesville, 99 K 14
Jonkoping, 55 K 8
Jonouiere, 91 J 9
Jonuta, 107 K 14
Joplin, 102 G 1
Joppa, 95 F 3
Jora, 68 I 7
Jordan, 96 B 6
Jordan Valley, 94 F 6
Jordan;c., 70 E 3
Jordao;river, 117 F 12
Jorge;isl., 119 M 9
Jorhat, 66 H 10
Jorpeland, 55 I 2
Jorge Montt;isl., 119 L 2
Jos, 75 K 11
Josa B. Casas, 118 E 8
Josavka, 40 E 2
Jose Abad Santos, 65 F 13
Jose Enrique Rodo, 117 K 9
Jose Maria, 113 I 9
Jose Maria Blanco, 116 M 6
Jose Panganiban, 65 C 12
Jose Pedro Varela, 117 K 10
Jose de San Martin, 118 H 4
Joseph, 94 D 6
Joseph Bonaparte Gulf, 82 C 5
Joshimath, 68 F 8
Jostvan Dyke;isl., 109 A 9
Joutla de Juarez, 107 K 1
Joutla de Juarez, 107 K 10
Jowai, 69 I 15
Joy, 98 L 8
Joynes, 104 E 7
Jozankei, 62 D 3
Ju;isl., 65 I 14
Juaguari, 117 H 10
Juami;river, 113 K 9
Juan Aldama, 106 G 7
Juan Bautista Arruabarrena, 116 I 8
Juan E. Barra, 118 D 9
Juan Eugenio, 106 F 7
Juan Fernandez;archp., 119 P 1
Juan G. Bazan, 116 E 7
Juan Garat, 116 I 8
Juan Jose Castelli, 116 F 7
Juan Jose Paso, 116 M 6
Juan N Fernandez, 118 D 10
Juan Stuven;isl., 119 J 2
Juan de Garay, 118 D 7
Juan de Mena, 117 E 9
Juana Ramirez;isl., 107 H 11
Juancho, 118 C 11
Juangriego, 113 C 12
Juanjui, 114 E 4
Juankoski, 53 K 11
Juarez, 118 D 9
Juarez, 106 B 2
Juba, 76 L 8
Jubones;river, 112 L 2
Juca de Oro, 116 G 2
Jucara;isl., 115 B 11
Jucaro, 108 G 8
Juchipila, 106 I 7
Juchipila;river, 106 I 7
Juchitan de Zaragoza, 107 L 12
Judenburg, 43 K 10
Judoma;river, 57 F 13
Juguarao;river, 117 J 11
Juigalpa, 108 K 6
Juina;river, 115 I 15
Jujuy;prov, 116 D 4
Julaca, 114 L 2
Julesburg, 96 F 8
Juli, 115 K 9
Julia Creek, 83 E 9
Juliaca, 114 J 8
Julianehab, 87 E 15
Juliette, 105 I 4
Julijske Alpe;range, 38 C 8
Julio de Castilhos, 117 H 11
Jullundur, 68 F 6
Juma;river, 115 D 15
Jumari;river, 115 F 13
Jumbilla, 114 D 4
Jumeje, 78 G 4
Jumla, 69 G 10
Jumna;river, 59 K 6
Jump River, 98 F 8
Junagadh, 68 L 3
Junction, 96 H 3
Junction City, 94 E 3
Jundah, 83 F 10
Juneau, 86 G 3
Jungshahi, 68 J 1
Juniata;river, 100 J 8
Junin, 112 I 3
Junin de los Andes, 118 E 3
Juniper, 91 K 11
Juntura, 94 E 5
Jupiter, 105 M 12
Juquia, 117 E 15
Juquila Mixes, 107 L 12
Jur;river, 76 K 7
Jura;isl., 48 C 5
Jurado, 112 E 3
Juramento;river, 116 F 5
Jurby, 49 I 9
Jurua, 115 B 11
Jurua;river, 115 B 11
Juruena, 115 I 15
Juruena;river, 115 E 16
Juruezinho;river, 115 D 9
Jurupari;river, 114 F 8
Jurva, 53 L 7
Justiniano Posse, 116 K 6
Justo Daraet, 116 K 5
Jutai, 115 C 9
Jutai;river, 115 B 10
Jutiapa, 108 J 4
Juticalpa, 108 J 5
Jutiyaco, 116 H 7
Jutland;pen., 34 F 7
Juuka, 53 K 11
Juva, 53 M 11
Juventud;isl., 108 F 6
Juwain, 71 G 11
Juz-Bug;river, 56 E 1
Jyvaskyla, 53 L 9
Jyvaskyla, 53 L 9
Jzan, 70 L 3
Ka;river, 75 J 10
Kaabong, 78 A 10
Kaavi, 53 K 11
Kaba, 43 K 16
Kaba, 75 L 10
Kabaena;isl., 65 L 11
Kabala, 75 K 3
Kabale, 78 C 8
Kabasha, 78 C 8
Kabia;isl., 65 L 11
Kabo, 76 K 3
Kabrit, 77 E 16
Kabugao, 65 A 11
Kabul, 71 F 13
Kabul;river, 71 F 13
Kabunda, 78 G 8
Kaburuang;isl., 65 G 13
Kabuzhuo, 60 I 4
Kacanik, 40 H 5
Kachia, 75 K 11
Kachin;state, 66 H 11
Kacincbarcika, 43 J 14
Kadarkut, 43 M 12
Kadaura, 68 I 8
Kadei;river, 76 M 2
Kaden;isl., 67 M 13
Kadoka, 96 E 9
Kadugli, 76 J 7
Kaduna, 75 K 11
Kaduna;river, 75 K 11
Kaedi, 74 H 3
Kaele, 76 J 1
Kaf, 70 F 3
Kafanchan, 75 K 11
Kafia Kingi, 76 J 5
Kafr Shukr, 77 D 12
Kafr el Battikh, 77 A 13
Kafr el Garayda, 77 A 12
Kafr saqr, 77 C 13
Kagan, 66 D 9
Kagelike, 60 F 5
Kagera;river, 78 C 8
Kagmar, 76 H 7
Kahajan;river, 64 J 8
Kahama, 78 D 9
Kahan, 71 H 13
Kahe, 78 D 11
Kahntah, 86 H 5
Kahoka, 98 M 7
Kahomba, 78 F 3
Kahror, 68 G 4
Kahuava, 53 K 8
Kahuta, 66 F 8
Kahutah, 88 F 5
Kai Besar;isl., 65 L 16
Kai Ketjil;isl., 65 L 16
Kai;isl., 80 F 3
Kaia, 60 F 4
Kaiama, 75 K 10
Kaibeab, 65 D 16
Kaiblechonggu, 60 H 4
Kailo, 78 D 6
Kaim;river, 83 B 12
Kaimana, 65 C 13
Kaimer;isl., 65 K 16
Kainantu, 83 B 13
Kaira, 68 K 4
Kairana, 68 G 7
Kairiru;isl., 83 A 12
Kaithal, 68 G 6
Kaitong, 61 D 13
Kaitum alv;river, 52 E 6
Kajaani, 53 J 10
Kajaapu, 64 L 3
Kajan;river, 65 H 9
Kajang, 64 H 3
Kajiado, 78 C 11
Kajiki, 63 N 1
Kajoa;isl., 65 I 13
Kajuagung, 64 K 4
Kaka, 76 J 8
Kakabeka Falls, 99 B 9
Kakamega, 78 C 10
Kakata, 75 L 4
Kakenge, 78 E 5
Kakeshili, 60 G 6
Kakia, 79 L 5
Kakkar, 68 H 2
Kakuda, 62 G 13
Kakumaa, 78 A 10
Kakunodate, 62 E 12
Kakuto, 63 N 2
Kal, 43 J 14
Kala, 60 E 5
Kala Naga, 69 J 16
Kalabagh, 68 D 4
Kalabahi, 65 M 12
Kalabakan, 65 G 9
Kaladan;river, 69 K 16
Kaladar, 90 L 6
Kalahesti, 67 N 5
Kalajoki, 53 J 8
Kalajoki;river, 53 J 9
Kalakan, 57 H 12
Kalam, 66 C 7
Kalamai, 41 O 6
Kalamazoo, 99 J 13
Kalamazoo;river, 99 J 12
Kalamulunhe;river, 60 F 4
Kalang;river, 69 H 16
Kalannie, 82 I 2
Kalao;isl., 65 L 11
Kalaotoa;isl., 65 L 11
Kalashaerhe;river, 60 E 5
Kalat, 68 G 1
Kalatse, 66 E 13
Kalayat, 68 H 5
Kalb De-, 101 E 10
Kalbarri, 82 H 2
Kaldakvisl;river, 52 3
Kaldrananes, 52 A 2
Kale;river, 49 D 13
Kaledupa;isl., 65 L 12
Kalehe, 78 D 7
Kalemyo, 67 I 10
Kaletwa, 67 J 10
Kaleva, 99 G 12
Kalewa, 67 I 10
Kalfafell, 52 C 3
Kalfafellsstadhur, 52 B 4
Kalgoorlie, 82 I 4
Kali Bay, 83 A 13
Kali Gandak;river, 69 H 11
Kali;river, 69 G 9
Kalibo, 65 D 11
Kalima, 78 D 7
Kalimantan Barat;prov, 64 I 6
Kalimantan Selatan;prov, 64 J 8
Kalimantan Tengah;prov, 64 J 7
Kalimantan Timur;prov, 65 I 9
Kalimantan;isl., 64 I 6
Kalimnos, 41 O 11
Kalimnos;isl., 41 O 1
Kalimpong, 69 H 13
Kalinovik, 40 G 2
Kalispell, 94 B 7
Kalisz, 43 E 12
Kalisz Pomorski, 43 C 10
Kalix, 52 H 8
Kalix alv;river, 52 F 7
Kalixfors, 52 E 6
Kalkaska, 99 G 13
Kalkfeld, 79 K 2
Kalkfontein, 79 K 4
Kalkidhiki;pen., 41 J 7
Kalkuni, 113 F 16
Kallsjon;lake, 53 J 2
Kalmar, 55 L 10
Kalmar;plain, 55 L 9
Kalni;river, 69 J 15
Kalo, 83 D 14
Kalocsa, 43 L 30
Kalofer, 40 H 9
Kalona, 98 K 7
Kalong, 60 I 5
Kalpa, 68 E 8
Kalso;isl., 54 B 2
Kalundborg, 55 N 6
Kalur Kot, 68 E 4
Kama, 99 A 10
Kamadugu Yobe;river, 72 G 5
Kamaishi, 62 E 13
Kamakusa, 113 F 14
Kamalia, 68 F 4
Kamalu, 75 K 3
Kamamaung, 67 K 12
Kaman, 68 H 7
Kamanjab, 79 J 2
Kamaran;isl., 70 M 3
Kamarod, 71 I 11
Kamassa, 76 K 1
Kambhayat Ki Khadi;gulf, 67 K 2
Kambia, 75 K 3
Kamdesh, 66 C 6
Kameda, 62 E 12
Kamen, 65 A 14
Kami, 66 B 8
Kami-Shihovo, 62 D 5
Kamien Krajenski, 43 C 12
Kamien Pomorski, 43 B 9
Kamienna Gora, 43 G 11
Kamiiso, 62 F 2
Kamikoshiki;isl., 63 N 1
Kamileroi, 83 D 9
Kaministikwia, 99 B 9
Kaminokuni, 62 F 1
Kaminoyama, 62 F 12
Kamisha;river, 78 D 3
Kamkoram;range, 68 B 6
Kamloops, 88 K 4
Kamnik, 38 B 9
Kamo, 62 F 12
Kamoija;river, 40 G 12
Kamorta;isl., 67 O 11
Kampala, 78 C 9
Kampar, 64 G 3
Kampar Kanan;river, 64 I 3
Kampar;river, 64 I 3
Kamparkiri;river, 64 I 2
Kampene, 78 D 7
Kamsack, 89 K 12
Kamuli, 78 B 9
Kamundan;river, 65 B 13
Kanaaupscow, 90 E 7
Kanaaupscow;river, 90 E 7
Kanab, 97 I 3
Kananga, 78 E 5
Kanarraville, 96 H 3
Kanatak, 86 E 1
Kanawha;river, 93 G 12
Kanbalu, 67 I 11
Kanczuga, 43 G 16
Kandale, 78 E 3
Kandangan, 64 J 9
Kandersteg, 42 L 3
Kandhkot, 68 G 2
Kandhura, 66 D 6
Kandi, 75 J 9
Kandla, 68 K 3
Kandreho, 79 M 12
Kane, 102 D 5
Kanegasaki, 62 E 13
Kaneyama, 62 E 12
Kangaba, 75 J 5
Kangaruma, 113 F 15
Kangasala, 53 M 9
Kangasniemi, 53 L 10
Kangba, 60 I 5
Kangean;isl., 65 L 9
Kangen;river, 76 L 8
Kangma, 69 G 14
Kangmaer, 60 I 4
Kaniet Islands;isl., 83 A 13
Kanin Poluostrov-;pen., 56 D 6
Kanjiza, 40 C 4
Kankaanpaa, 53 M 7
Kankakee, 99 L 11
Kankakee;river, 99 L 10
Kankan, 75 K 4
Kannanj, 68 I 8
Kannapolis, 104 F 7
Kannus, 53 J 8
Kano, 75 J 11
Kano;river, 75 K 11
Kanonga, 78 G 10
Kanosh, 96 H 3
Kanowit, 64 H 7
Kanoya, 63 O 2
Kanpur, 69 I 9
Kansas, 102 D 8
Kansas City, 102 D 1
Kansas;river, 92 G 8
Kansas;state, 92 G 7
Kansk, 57 H 9
Kantchari, 75 J 9
Kantishna, 86 D 2
Kantunil, 107 I 16
Kantunil Kin, 107 I 16
Kaolack, 75 I 2
Kaongeshi;river, 78 F 5
Kapagere, 83 D 14
Kapalu, 66 C 12
Kapasan, 68 J 5
Kapcagaj, 56 J 6
Kapchorwa, 78 B 10
Kapfenberg, 43 K 10
Kapili;river, 69 I 16
Kapiskau, 90 F 5
Kapit, 64 H 7
Kapiti;isl., 83 K 15
Kaplan, 103 P 3
Kapoata, 76 L 8
Kapoavar, 43 L 12
Kappar, 71 J 11
Kapuas;river, 64 I 6
Kapurthala, 68 F 6
Kapuskasing, 90 H 4
Kapuskasing;river, 90 H 4
Kaputur, 78 B 10
Kar Nicobar;isl., 67 O 11
Kara;river, 71 E 9
Karacabey, 41 K 13
Karace, 57 D 12
Karachi, 68 J 1
Karadeniz Daglari;range, 70 B 4
Karaginskiy;isl., 57 D 16
Karaguinskii;isl., 58 B 13
Karakelong;isl., 65 G 13
Karakum;desert, 56 J 3
Karakuwisa, 79 J 4
Karama;river, 65 J 10
Karambu, 65 K 9
Karamian;isl., 64 K 8
Karasjok, 52 C 9
Karasjokka;river, 52 C 9
Karatj, 52 G 5
Karatsu, 63 L 1
Karau, 83 A 13
Karaul, 56 E 8
Karauli, 67 I 4
Karavanke;range, 38 B 8
Karbole, 54 F 9
Kardhitsa, 41 L 6
Kardzali, 41 I 10
Kareima, 76 F 8
Karesuando, 52 D 7
Karevandar, 71 I 10
Kargasok, 56 G 7
Kargi, 78 B 11
Kargil, 66 D 12
Karhula, 53 N 10
Kari, 75 J 12
Kariasniemi, 52 C 9
Kariava, 83 C 13
Karibib, 79 K 2
Karimata;isle, 64 J 6
Karimganj, 69 J 13
Karin, 77 J 13
Kariovac, 38 C 10
Kariovo, 40 H 9
Kariovy Vary, 42 G 8
Karkag, 43 K 15
Karkar;isl., 83 B 13
Karkh, 68 H 1
Karkkila, 53 N 8
Karkola, 53 N 9
Karkossa, 74 H 4
Karlshamn, 55 M 8
Karlskoga, 55 I 8
Karlskrona, 55 M 9
Karlstad, 55 I 7
Karluk, 86 E 1
Karnack, 103 L 1
Karnal, 68 G 7
Karnali;river, 69 G 9
Karnes City, 97 N 1
Karnobat, 40 H 11
Karompa;isl., 65 L 11
Karonie, 82 I 4
Karora, 77 G 10
Karotho, 77 L 9
Karpathos;isl., 41 P 12
Karpaty;range, 43 I 13
Karskoje More;sea, 56 D 7
Kartovik, 86 C 4
Karttula, 53 K 10
Kartuni, 113 F 14
Kartuzy, 43 B 12
Karufa, 65 C 13
Karumba, 83 D 9
Karungi, 52 G 8
Karungie, 82 C 5
Karungu, 78 C 10
Karunki, 52 G 8
Karvatn, 54 E 4
Karvia, 53 L 8
Karwi, 69 J 9
Karzok, 68 D 7
Kasaan, 86 H 2
Kasai;river, 78 D 3
Kasasa, 63 O 1
Kaseda, 63 O 1
Kasganj, 68 H 8

Manuel Rodriguez;isl.,
119 M 3
Manuel Urbano,115 F 9
Manui;isl.,65 K 12
Manuk;isl.,65 L 15
Manuripi;river,115 H 10
Manus;isl.,83 A 13
Manville,95 B 7
Many,103 N 2
Manyane,79 L 5
Manyonga;river,78 D 9
Manyoni,78 E 10
Manzai,68 E 3
Manzanillo,106 J 7
Maoke;cord.,80 F 3
Maopora;isl.,65 M 14
Maothail,48 J 1
Maowo,60 E 8
Mapastepec,107 M 14
Mapi;river,65 D 16
Mapia;isl.,80 E 3
Mapiri,115 J 10
Mapiri;river,115 G 11
Maple Creek,89 L 9
Maple Hill,104 G 11
Maple Lake,98 F 5
Maple View,101 F 9
Maple;river,98 C 2
Maplesville,103 L 9
Mapleton,98 H 5
Maplewood,99 G 11
Mapoon,83 B 9
Maporillai,112 E 8
Mappi Post,65 D 16
Maprik,83 A 12
Mapuera;river,113 K 15
Mapur;isl.,64 I 4
Maqha,70 G 2
Maquanhe;river,60 I 4
Maquena,117 J 9
Maquinchao,118 F 5
Maquinchao;river,118 F 5
Maquoketa,98 K 8
Maquoketa;river,98 J 7
Maquon,98 L 8
Mar de Ajo,117 M 9
Mar del Plata,118 D 10
Mara,65 K 11
Mara;river,78 C 10
Maraa,113 K 11
Marabuma,113 D 14
Maraca;isl.,110 E 9
Maracaibo,112 C 7
Maracay,113 C 9
Maradah,76 B 3
Maradi,75 J 11
Marais des Cygnes;river,
96 H 13
Maral,68 C 6
Maralal,78 B 11
Maralinga,82 H 6
Marampa,75 K 3
Maranoa;river,83 G 11
Maranon rio,114 C 4
Maranon;river,110 F 4
Marapa;river,116 G 4
Marasesti,40 D 11
Marathon,90 I 3
Maratua;isl.,65 H 10
Marauau;river,113 J 14
Marauia;river,113 J 11
Maravari,83 C 16
Maravatio de Ocampo,
107 J 9
Marawi,65 F 12
Marayes,116 J 4
Marble Bar,82 E 3
Marblehead,100 I 3
Marbu,54 H 4
Marcabal,114 E 4
Marcal;river,43 K 12
Marcapata;river,114 I 8
Marcapomacocha,114 H 5
Marceline,102 C 3
Marcellus,94 C 5
March,51 D 12
Marcha;river,57 F 11
Marche,42 G 2
Marchena;isl.,112 G 1
Marco,105 O 10
Marcoma,114 J 6
Marcos Juarez,116 K 6
Marcos Paz,116 L 8
Marcus Hook,95 E 5
Marda,82 H 3
Mardan,66 E 7
Mardie,82 F 2
Mare;isl.,83 F 13
Mareanna,103 J 5
Mareeba,83 D 10
Marengo,98 K 7
Marenisco,99 E 9
Marfa,97 M 8
Margam,50 H 3
Margaret Bay,88 J 2
Margaretville,101 H 10
Margarita,116 H 7
Margarita Belen,116 G 8
Margarita;isl.,112 D 5
Margate,51 H 16
Margate City,95 G 8
Margento,112 E 5
Margos,114 G 4
Margosatubic,65 F 12
Marguerite,88 J 4
Marguerite;river,91 G 11
Maria,91 I 11
Maria Cleofas;isl.,
106 I 5
Maria Juana,116 J 7
Maria Madre;isl.,106 I 5
Maria Magdalena;isl.,
106 I 5
Maria Theresa,116 K 6
Maria la Baja,112 C 5
Maria-Elena,116 D 2
Maria;isl.,82 C 8
Maria;isl.,81 I 14
Marianas;isl.,80 B 5
Marianna,105 L 2
Marianske Lazne,42 H 8
Marib,70 M 4
Maribel,99 H 11
Maribo,55 O 6
Maribor,38 B 10
Marica;river,41 I 10
Maridi;river,76 L 7
Marie Galante;isl.,
109 M 13
Marie;river,113 K 9
Marienville,100 I 6
Mariestad,55 J 8
Marieta;river,113 F 10
Marietta,97 K 12
Marimba,78 F 3
Marinduque;isl.,65 C 11
Marine City,100 G 3
Marine on Saint Croix,
98 F 6
Marinette,99 F 11
Maringouin,103 O 4
Marinque,79 J 10
Marion,94 B 7
Marion Downs,83 F 9
Marion Junction,103 M 9
Mariow,51 H 10
Maripa,113 E 11
Mariposas,116 M 2
Mariquina,118 E 2
Mariquita,112 G 5
Marisa,65 I 11
Marissa,102 E 6
Mariusa;isl.,113 C 13
Markali,43 L 15
Markanyol,55 M 8
Markerwaard;isl.,42 D 2
Markesan,99 H 9
Market Deeping,51 C 12
Market Drayton,50 B 6
Market Harborough,51 D 10
Markethill,48 I 4
Markham,98 C 7
Markham Moor,51 A 10
Markham;river,83 B 13
Markinch,49 A 12
Marks,103 J 5
Marks Tey,51 F 14
Marksbury,50 I 6
Marksville,103 N 3
Marktredwitz,42 H 7
Markville,98 E 6
Marlboro,101 G 12
Marlborough,50 H 8
Marlette,100 F 2
Marmaduke,102 H 5
Marmara;isl.,41 J 12
Marmath,96 C 8
Marmaton;river,102 F 1
Marmelos,115 D 14
Marmelos;river,115 E 14
Marmora,90 L 6
Marne,98 K 4
Maroa,99 M 10
Maroantsetra,79 L 13
Maroob,65 D 16
Maros,65 K 10
Maroth,68 H 6
Maroua,76 J 1
Marovoay,79 M 12
Marquand,102 F 6
Marquesas Keys;isl.,
105 P 9
Marquete,99 D 11
Marquete;isl.,99 E 13
Marquetta;isl.,100 C 2
Marquez;river,115 M 10
Marrawah,83 L 10
Marromeu,79 J 10
Marrupa,78 H 11
Mars Hill,101 B 16
Mars Le-,98 I 3
Marsabit,78 B 11
Marsafa,77 E 12
Marsat Uwayja,76 B 3
Marsden,49 L 14
Marseille,99 L 10
Marsh Harbour,109 D 9
Marsh;isl.,103 P 4
Marshal,100 H 1
Marshall,86 C 1
Marshall Bennett Islands;
isl.,83 C 15
Marshall;isl.,81 C 9
Marshallberg,104 G 12
Marshalls Creek,95 A 6
Marshallton,95 E 5
Marshalltown,98 K 6
Marshallville,105 J 4
Marshfield,50 H 7
Marshville,104 G 8
Marsqui,91 I 11
Marston,102 H 6
Martaban,67 L 12
Martapura,64 K 4
Marten,88 H 7
Martha's Vineyard;isl.,
101 I 14
Marthasville,102 E 5
Marthaville,103 M 2
Martigny,42 M 3
Martin,96 E 9
Martin de Loyola,116 M 4
Martin;pen.,33 E 3
Martina Franca,39 I 12
Martinez,94 I 3
Martinica;isl.,85 L 13
Martinicus;isl.,101 E 16
Martinique;isl.,109 L 15
Martins Creek,69 B 6
Martins Ferry,100 K 4
Martinsburg,100 K 7
Martinsville,102 D 8
Martley,50 E 7
Martli,68 J 2
Marton,49 H 16
Marttila,53 N 8
Marua,83 C 14
Marudi,64 G 8
Marumori,62 G 13
Maruseppu,62 C 5
Marvell,103 J 5
Marwar Junction,68 I 5
Marwayne,88 I 8
Mary,56 J 3
Mary,68 D 4
Mary;river,82 B 6
Maryborough,83 G 12
Marydel,95 H 4
Maryfield,89 L 12
Maryland Line,95 E 2
Maryland;state,100 L 6
Maryport,49 G 11
Marys Harbour,91 F 15
Marystown,91 I 16
Marysvale,96 H 3
Marysville,96 G 11
Maryville,96 G 12
Marzafal,74 H 8
Masaka,78 C 9
Masamba,65 J 11
Masara,77 A 12
Masaramu,112 K 4
Masardis,101 B 16
Masaryktown,105 O 5
Masasi,78 G 12
Masaya,108 K 5
Masbate,65 D 12
Masbate;isl.,65 D 12
Mascarenas;isl.,79 P 9
Mascasin,116 J 4
Mascateasqatascot,104 F 4
Mascota,106 I 6
Masefield,89 L 9
Masela;isl.,65 M 15
Mashala,77 C 12
Masham,49 I 15
Mashhad,71 E 10
Mashike,62 C 3
Mashoes,104 F 13
Mashtul es Suq,77 E 12
Masi Manimba,78 E 3
Masimbu,65 J 10
Masindi,78 B 9
Masindi Port,78 B 9
Masinloc,65 B 10
Masirah;isl.,71 L 9
Masisea,114 F 6
Masisi,78 C 8
Masjut,68 B 5
Masoller,117 J 10
Masomeloka,79 N 13
Mason,99 J 14
Mason City,98 I 6
Mason Creek,88 F 4
Masparro;river,112 D 8
Masqat,71 J 9
Massa Carrara;prov,38 E 4
Massachusetts;state,
101 H 12
Massafra,39 J 1
Massaguet,76 I 2
Massakory,76 I 2
Massena,101 D 10
Massenya,76 J 2
Masset,88 H 1
Massey,99 E 15
Massillon,100 J 4
Massinga,79 L 10
Masson;isl.,83 K 4
Mastic Beach,95 B 12
Mastuj,66 B 7
Mastung,71 H 12
Mata Mata,79 M 4
Matador,89 K 9
Matagalpa,108 J 5
Matagami,90 H 6
Matak;isl.,64 H 5
Matam,75 I 3
Matamoros,106 F 7
Matandu;river,78 F 11
Matane,91 I 10
Matanui,66 E 6
Matanzas,108 F 7
Matanzas;prov,108 F 7
Matapalo,114 C 2
Mataquito;river,116 M 2
Matara,116 G 6
Mataram,65 M 9
Matarani,114 K 7
Mataranka,82 C 7
Matasiri;isl.,65 K 9
Matauro;river,115 D 14
Matawan,95 C 8
Matbul,77 B 11
Matechewan,90 I 5
Mategua,115 I 13
Matehuala,107 G 9
Matelot,108 K 3
Matera,39 I 11
Matera;prov,39 J 11
Mateszalka,43 J 16
Matewan,104 D 6
Matheson,90 I 5
Matheson Island,89 J 13
Mathew Town,109 G 10
Mathews,104 D 12
Mathiston,103 K 7
Mathura,68 H 7
Mati,65 F 13
Mati;river,41 I 3
Matias Romero,107 L 12
Matinga;river,78 B 4
Matlmanyane,79 J 6
Matlock,51 A 9
Mato Grosso,115 J 15
Mato Grosso;state,
115 G 14
Matoaca,104 D 11
Matoaka,104 D 7
Matong,83 B 15
Matos;river,115 J 11
Matrei,42 L 8
Matruh,76 A 6
Matsel,66 F 12
Matsubase,63 M 1
Matsumae,62 F 1
Matsushima,62 F 13
Matsuura,63 L 1
Mattagami;river,90 H 4
Mattaponi;river,104 C 11
Mattawa,90 K 6
Mattawamkeag,101 C 16
Mattawamkeag;river,
101 C 16
Mattawitchewan;river,
99 A 13
Matthew;isl.,83 G 13
Matthews,99 M 13
Mattice,90 H 4
Mattituck,95 A 13
Mattoon,102 D 8
Matu,64 H 7
Matua,64 J 7
Matucana,114 H 5
Matura,108 K 3
Matura;river,108 K 3
Maturin,113 C 12
Maturuca,113 G 14
Mau,69 I 9
Mau Summit,78 C 10
Maua,78 H 11
Mauari;river,115 A 9
Mauas,115 B 16
Mauchline,49 D 9
Mauckport,102 F 10
Maud,103 K 1
Maudaha,69 I 9
Maues;river,115 B 16
Maug;isl.,80 B 5
Mauhan,67 I 11
Maule,116 M 2
Maule;prov,116 M 1
Maullin,118 F 2
Maumee,99 K 15
Maumee;river,99 K 15
Maumere,65 M 11
Maun,79 J 5
Maunie,102 F 8
Maupiti;isl.,81 H 14
Mauri;river,115 L 9
Maurice,103 P 3
Maurice;river,95 F 6
Mauricetown,95 G 6
Mauriceville,103 O 1
Mauritania;c.,74 G 2
Mauritius,79 O 10
Mauritius;isl.,73 L 13
Maury,104 F 11
Mauston,98 H 8
Mauterndf,43 K 9
Mava,83 C 12
Mavaca;river,113 I 11
Mavinga,79 I 4
Mawer,89 K 10
Max,96 B 9
Max Meadows,104 D 7
Maxaila,79 K 9
Maxcami,107 I 15
Maxeys,104 H 4
Maxton,49 D 13
Maxville,105 M 6
Maxwell,97 I 7
May Pen,109 H 9
May;isl.,49 A 13
May;river,83 B 12
Maya;isl.,64 J 6
Maya;river,58 D 11
Maya;river,57 F 13
Mayaguana;isl.,109 F 10
Mayama,78 D 2
Maybell,96 G 5
Mayberry,95 F 1
Mayboie,48 E 8
Maydi,70 L 3
Mayenne,44 D 7
Mayenne;river,44 E 6
Mayer,97 J 3
Mayersville,103 L 5
Mayerthorpe,88 I 6
Mayfair,89 J 9
Mayfield,51 J 13
Mayfield Heights,100 I 4
Mayflower,103 N 1
Mayhew,103 K 7
Mayland,104 F 2
Maymyo,67 J 11
Mayo,75 K 13
Mayo Mayo,115 H 11
Mayor;river,106 E 4
Mayor Buratovich,118 E 8
Mayotte;isl.,73 K 11
Mayport,105 M 7
Mayreau;isl.,109 C 15
Mays Landing,95 F 7
Maysville,98 M 4
Maytiguid;isl.,65 D 10
Maytown,95 D 2
Mayville,98 C 2
Maza,116 M 6
Mazan,114 B 7
Mazan;river,114 B 6
Mazapa de Madero,107 M 14
Mazara del Vallo,39 M 7
Mazarredo,119 I 6
Mazaruni;river,113 F 15
Mazatan,106 C 3
Mazatenango,108 I 3
Mazatlan,106 H 5
Mazeppa,95 A 1
Mazha,60 F 2
Mazo Cruz,115 K 9
Mazocahui,106 C 4
Mazoe,79 I 8
Mazomanie,99 I 9
Mazon,99 L 10
Mazrub,76 H 7
Mazzarino,39 N 9
Mbaere;river,76 L 2
Mbaiki,76 M 3
Mbale,78 B 10
Mbalmayo,75 M 13
Mbamba Bay,78 G 10
Mbandaka,78 C 3
Mbanza-Ngungu,78 E 2
Mbarara,78 C 8
Mbe,78 D 2
Mbemkuru;river,78 F 12
Mbere;river,76 K 2
Mbinda,78 D 1
Mbini;river,75 N 12
Mboi,78 E 5
Mboma;river,72 H 8
Mbres,76 K 3
Mbulu,78 D 10
Mburucucuya,116 G 8
Mbuyapey,117 F 9
Mc Alpen,105 M 5
Mc Dame,88 E 2
Mc Farlane;river,89 F 9
Mc Grath,98 E 6
Mc Gregor,98 I 7
Mc Gregor,98 D 6
Mc Gregor,96 C 13
Mc Henry,98 C 1
Mc Intosh,96 C 9
Mc Kee,104 D 4
Mc Keesport,100 K 5
Mc Kenney,104 D 11
Mc Kinley Park,86 D 2
Mc Kinney,97 K 12
Mc Laughlin,96 C 9
Mc Lean,97 J 9
Mc Lennan,88 H 6
Mc Leod Lake,88 H 4
Mc Leod;river,88 I 6
Mc Millan,99 E 12
Mc Minnville,94 D 3
Mc Murray,88 G 8
Mc Pherson,96 H 11
Mc Roberts,104 D 5
Mc Veigh,104 D 6
Mc Vicar Arm;gulf,88 A 6
Mc Ville,98 B 2
McAdam,91 K 11
McAdoo,95 A 4
McAlexander,102 H 7
McAlisterville,95 B 1
McArthur,100 L 3
McArthur;river,82 C 8
McBain,99 H 13
McClure,100 J 8
McClusky,96 B 9
McColl,104 G 9
McComb,99 L 4
McConnellsburg,100 K 7
McConnelsville,100 K 3
McCook,96 G 9
McCool,103 L 7
McCormick,104 H 6
McCreary,89 K 13
McCrory,103 I 4
McCullough,103 N 9
McDavid,103 O 9
McDonough,105 I 3
McEwen,102 H 9
McGehee,103 K 4
McGivney,91 K 11
McHenry,103 O 7
McKamie,103 L 2
McKenzie,102 H 8
McKirdy,99 A 10
McLain,103 N 7
McLeansboro,102 F 8
McLeod,98 D 2
McMichael,95 A 5
McNeill,103 O 6
McRae,105 K 5
McSherrystown,95 E 1
McWilliams,103 M 9
Mchinja,78 F 12
Mckague,89 J 11
Mckenzie,103 N 10
McIntire,98 I 6
McIntosh,103 N 8
Mdhampur,68 E 6
Mdina,39 P 9
Meachan,89 J 10
Meade,97 I 10
Meade;river,86 3
Meadok Lake,86 J 6
Meadow,104 C 7
Meadow Lake,89 I 9
Meadow Lake Provincial P-
ark,88 I 8
Meadow Valley Wash;river,
94 J 7
Meadowlands,98 D 6
Meadowview,104 E 6
Meadville,103 N 5
Meaford,90 L 5
Mealsgate,49 G 11
Meander River,88 E 6
Measham,51 C 9
Meathas Troim,48 L 2
Mebane,104 F 9
Mecatihe Little;isl.,
91 G 14
Mecaya;river,112 I 6
Mechanics Grove,95 E 3
Mechanicsburg,95 D 1
Mechanicsville,98 K 7
Mechengue,112 H 3
Meconta,78 H 12
Mecufi,78 H 12
Mecula,78 G 11
Medan,64 H 2
Medang;isl.,65 M 9
Medanito,116 I 4
Medano,106 D 3
Medano Amarillo,106 F 2
Medano;isth.,112 B 8
Medanos,106 A 1
Medaryville,99 L 11
Medellin,112 F 4
Medemblik,42 C 2
Medenine,74 B 12
Mederdra,74 H 2
Medford,94 F 3
Medfra,86 D 1
Medgidia,40 E 12
Media,95 E 5
Media Agua,116 J 3
Mediapolis,98 L 7
Medias,40 C 8
Medias Aguas,107 K 12
Medical Lake,94 C 6
Medicine Bow,96 F 6
Medicine Bow Range,96 F 6
Medicine Hat,86 K 6
Medicine Lodge,97 I 11
Medina,99 L 16
Mediodia,112 K 6
Medora,96 C 8
Medstead,89 J 9
Medstugan,53 J 4
Medway;river,51 H 4
Medway;river,51 H 14
Medzilaborce,43 H 15
Medzilaborce,43 H 15
Meeberrie,82 G 2
Meeberrie,82 G 2
Meekatharra,82 G 3
Meeker,96 G 5
Meerut,68 G 7
Mega,77 L 10
Mega;isl.,64 K 2
Megantic,91 K 9
Megha;river,69 J 15
Mehakit,65 J 9
Mehar,71 I 13
Meherrin,104 D 10
Meherrin;river,104 E 11
Mehetia;isl.,81 I 15
Mehsana,68 K 4
Meifod,50 C 5
Meiganga,76 L 1
Meighen;isl.,86 A 8
Meigs,105 L 3
Meikhitila,67 J 11
Meir,50 B 7
Meirose,103 N 2
Mejanodas,65 M 15
Mejillones,116 E 1
Mekalina,99 D 14
Mekambo,78 B 1
Mekatina,90 J 3
Mekdela,77 I 11
Mekele,77 I 11
Mekhtar,68 F 3
Mekrou;river,75 J 9
Melaka,64 H 3
Melalap,65 G 9
Melber,102 G 7
Melborne,99 I 16
Melbourn,51 F 12
Melbourne,98 K 6
Melbourne Beach,105 P 8
Melcher,98 L 5
Melchor Ocampo,106 G 8
Melchor;isl.,118 H 2
Meldrim,105 K 7
Meldrum Bay,90 K 4
Meleden,77 J 14
Melfa,104 C 14
Melfi,76 J 3
Melfort,89 J 10
Melincue,116 K 7
Melipilla,116 L 2
Melita,89 L 12
Melito,39 M 10
Melitota,95 G 3
Melksham,50 H 7
Mellette,98 F 1
Mellit,76 H 5
Mellwood,103 J 5
Melmerby,49 G 13
Melo,117 J 11
Meloy,52 F 2
Melrose,49 D 12
Melrose,105 N 6
Meltaus,52 F 9
Meltham,49 L 15
Melton Mowbray,51 C 10
Meltosjarvi,52 G 9
Melville,89 K 11
Melville Peninsula,
87 E 10
Melville;isl.,86 C 7
Melville;pen.,84 D 8
Melvine,104 F 2
Melykut,43 M 14
Memala,64 J 7
Meman;isl.,83 A 13
Memba,78 H 12
Memboro,65 M 10
Memo;river,113 D 10
Mempakul,64 G 8
Memphis,98 M 6
Memuro,62 D 4
Mena,97 J 13
Mena;river,77 L 11
Menahga,98 D 4
Menai Bridge,50 A 3
Menaka,75 I 9
Menanga Laku,65 M 10
Menard,97 M 10
Menasha,99 H 10
Menate,64 I 7
Menchari;river,114 4
Mencue,118 E 4
Mendanau;isl.,64 J 5
Mendawai,64 K 8
Mendawai;river,64 J 8
Mendenhall,103 M 6
Mendez,112 L 3
Mendham,95 B 7
Mendhar,66 F 9
Mendi,77 J 9
Mendieta La-,116 E 5
Mendon,98 M 7
Mendong,60 H 4
Mendota,99 K 9
Mendoza,114 D 4
Mendoza;prov,116 M 3
Mendoza;river,116 K 3
Mendung,64 I 3
Mene Grande,112 C 7
Mene de Mauroa,112 C 7
Meneo Majagua;river,
112 D 7
Menggala,64 K 4
Menghe,43 C 10
Menguinzi,41 I 2
Mengyong,60 K 8
Menihek,91 G 10
Menlo,104 G 2
Menna;river,77 I 10
Menno,98 I 2
Menominee,99 F 11
Menominee;river,99 F 10
Menomonee Falls,99 I 10
Menomonie,98 G 7
Mentana,83 C 9
Menton,45 J 14
Mentone,97 L 8
Menzies,82 H 4
Meon;river,51 J 10
Meoqui,106 D 6
Mequon,99 I 10
Mer Rouge,103 L 4
Meraker,54 D 6
Meramec;river,102 F 4
Meranggau,64 I 6

Morgan City,

Morganfield, 102 F 8
Morganton, 104 F 6
Morgantown, 100 L 5
Morganza, 103 O 4
Mori, 62 E 2
Moriarty, 97 J 6
Moricetown, 88 H 2
Morichal Largo; river, 113 D 12
Moriches, 95 A 12
Moriches Bay, 95 B 12
Morillo, 116 D 6
Morin Creek, 89 I 9
Morioka, 62 D 13
Moris, 106 D 4
Morjarv, 52 G 7
Morkret, 53 M 1
Morlaix, 44 D 4
Morley, 49 K 15
Morley River, 88 D 2
Morne Rouge, 109 L 15
Morne-a-L'eau, 109 L 12
Mornington; isl., 83 D 9
Moro, 103 J 5
Moro Gulf; gulf, 65 F 12
Morobay, 103 L 3
Morobe, 83 C 14
Morocco, 99 L 11
Morochata, 115 L 11
Morogoro, 78 E 11
Moromaho; isl., 65 L 12
Morombe, 79 O 11
Moromoro, 115 L 12
Moron, 108 F 8
Morona, 112 L 3
Morona; river, 114 B 4
Morondava, 79 N 11
Moroni, 79 K 10
Moronna-Santiago; prov, 112 L 3
Morotai; isl., 65 H 14
Morowali, 65 J 11
Morrilton, 103 I 3
Morris, 89 L 14
Morrison, 99 K 9
Morrisonville, 102 D 7
Morriston, 50 G 3
Morristown, 101 E 9
Morrisville, 101 E 12
Morrope, 114 D 2
Morropon, 114 C 2
Morrosquillo; gulf, 112 D 4
Morrow, 100 L 1
Morrumbala, 79 I 10
Morrumbene, 79 L 10
Morson, 98 A 5
Mortagne, 44 E 6
Morteros, 116 I 6
Mortimers Cross, 50 E 6
Mortlock; isl., 80 D 6
Morton, 94 C 3
Moruga, 108 M 2
Moruga; river, 108 M 3
Morven, 104 G 8
Morvi, 68 K 3
Moscow, 56 E 3
Moscow, 103 I 7
Mosel; river, 42 G 3
Moselle, 103 N 7
Moselle; river, 45 C 12
Moses Lake, 94 C 5
Moses Point, 86 B 1
Mosher, 99 A 13
Mosinee, 99 G 9
Mosjoen, 52 G 2
Moskenesoy; isl., 52 E 2
Mosmota, 116 K 4
Mosquera, 112 H 3
Mosquero, 97 J 7
Mosquitos; gulf, 85 M 10
Moss, 55 I 5
Moss Bluff, 105 N 6
Moss Point, 103 O 8
Mossaka, 78 C 2
Mossala, 76 K 3
Mossbank, 89 L 10
Mossendjo, 78 D 1
Mossman, 83 D 10
Mossuril, 78 H 12
Mossy Head, 105 L 1
Most, 43 G 9
Mostardas, 117 J 12
Mosterton, 50 K 6
Mostyn, 49 M 11
Motaba; river, 78 B 3
Motacucito, 115 L 14
Motagua; river, 108 I 4
Motala, 55 J 8
Motatan, 112 D 7
Motatan; river, 112 C 7
Motherby, 49 H 12
Motherwell, 49 C 10
Moti; isl., 65 I 13
Motley, 98 E 4
Motozintla de Mendoza, 107 M 14
Motrul; river, 40 E 7
Mott, 96 C 8
Motul, 107 I 15
Motupe, 114 D 2
Mouali, 78 C 2
Mouchoir Bank, 109 G 11
Moudhros, 41 K 10
Moudjeria, 74 H 3
Mouka, 76 K 4
Mould Bay, 86 B 7
Moulmein, 67 L 12
Moulton, 103 J 9
Moultonboro, 101 F 13
Moultrie, 105 L 4
Mound Bayou, 103 K 5
Mound City, 96 C 9
Moundou, 76 K 2
Mounds, 102 G 7
Moundsville, 100 K 4
Mount Aetna, 95 C 3
Mount Airy, 95 F 1
Mount Arlington, 95 A 7
Mount Augustus, 82 G 2
Mount Ayr, 96 F 13
Mount Barnett, 82 D 5
Mount Brydges, 99 I 16
Mount Carmel, 101 J 9
Mount Carroll, 99 K 9
Mount Clemens, 100 G 3
Mount Coolon, 83 E 11
Mount Croghan, 104 G 8
Mount Cuthbert, 83 E 9
Mount Davies, 82 G 6
Mount Desert; isl., 101 E 16
Mount Dora, 105 O 6
Mount Doreen, 82 E 6
Mount Douglas, 83 E 11
Mount Erie, 102 E 8
Mount Forest, 90 L 5
Mount Gilead, 100 J 2
Mount Hagen, 83 B 13
Mount Holly, 103 L 3
Mount Holly Springs, 95 D 1
Mount Hope, 98 I 8
Mount Horeb, 99 I 9
Mount Ida, 97 J 13
Mount Isa, 83 E 9
Mount Jewett, 100 I 6
Mount Joy, 95 D 2
Mount Magnet, 82 H 3
Mount Mc-Kinley National Park, 86 D 1
Mount Morgan, 83 F 12
Mount Morris, 100 G 7
Mount Nebo, 95 E 3
Mount Olive, 102 D 6
Mount Olivet, 100 M 1
Mount Orab, 100 L 1
Mount Oxide, 83 D 9
Mount Penn, 95 C 4
Mount Perry, 83 G 12
Mount Pleasant, 96 G 4
Mount Pleasant Mills, 95 B 1
Mount Pocono, 101 J 10
Mount Pulaski, 99 M 9
Mount Savage, 100 K 6
Mount Sterling, 98 M 8
Mount Stewart, 91 K 12
Mount Storm, 100 L 6
Mount Union, 100 K 7
Mount Vernon, 82 F 2
Mount Victory, 104 D 3
Mount Willoughby, 82 G 7
Mount Wolf, 95 D 2
Mount Zion, 105 I 2
Mountain, 99 F 10
Mountain Ash, 50 G 4
Mountain Brook, 103 K 10
Mountain City, 94 G 7
Mountain Fork; river, 103 J 1
Mountain Grove, 102 G 3
Mountain Home, 94 F 7
Mountain Inn, 98 C 6
Mountain Lake, 98 H 4
Mountain Pine, 103 J 2
Mountain View, 102 G 4
Mountain Village, 86 C 1
Mountainair, 97 K 6
Mountville, 104 H 6
Mour; river, 48 G 2
Moura, 113 K 13
Mourdiah, 75 I 5
Moussoro, 76 I 2
Moutanita, 112 I 4
Mouyondzi, 78 D 1
Mouzon, 105 I 8
Movas, 106 D 4
Moville, 98 J 3
Moy, 48 H 4
Moya, 60 F 3
Moyahua, 106 I 7
Moyale, 78 A 12
Moyamba, 75 L 3
Moyie, 88 L 6
Moyo, 78 A 9
Moyobamba, 114 D 4
Moyock, 104 E 12
Moyowosi; river, 78 D 8
Moyto, 76 I 2
Mozambique; c., 73 M 9
Mraeso, 75 L 8
Mrarabenangin, 64 J 8
Muaratuhup, 64 I 8
Muaratunan, 65 J 9
Muarawahau, 65 I 9
Mubi, 76 J 1
Mubrani, 65 A 13
Mucajai; river, 113 H 13
Much Birch, 50 F 6
Much Wenlock, 50 C 6
Muchalat, 88 K 2
Muchiri, 115 M 12
Muckalee Creek; river, 105 K 3
Muco; river, 112 G 7
Mucojo, 78 G 12
Mucuchies, 112 D 7
Mucusso, 79 J 4
Mudanya, 41 J 13
Mueda, 78 G 12
Muene, 78 G 5
Mugayd, 70 I 7
Mugeba, 79 I 11
Mugi, 60 G 4
Mugu, 69 G 10
Mugu Karnali; river, 69 G 10
Muhamdi, 69 H 9
Muhammad Qol, 77 F 10
Muhammadgarh, 68 K 7
Muhos, 53 I 9
Muhulu, 78 C 7
Muhuwesi; river, 78 G 11
Muinchille, 48 J 3
Muineachan, 48 I 3
Muirkirk, 49 D 10
Muisne, 112 J 2
Muite, 78 H 12
Mujeres; isl., 107 I 16
Mukaehama, 63 P 1
Mukah, 64 H 7
Muker, 49 I 14
Muktsar, 68 F 6
Mukur, 71 G 13
Mukwonago, 99 J 10
Mula; river, 68 G 1
Mulalo, 112 K 3
Mulan, 61 C 14
Mulanay, 65 C 11
Mulatas; archp., 112 C 2
Mulato, 106 C 7
Mulatos, 112 D 4
Mulbehk, 66 D 12
Mulberry, 99 M 12
103 K 10
Mulberry; river, 102 H 2
Mulchen, 118 D 3
Mulege, 106 E 2
Muleshoe, 97 K 8
Mulgrave, 91 K 13
Mulgrave; isl., 83 A 9
Mulhouse, 45 E 13
Muligudje, 79 I 12
Mulinghe; river, 61 C 15
Mull; isl., 46 H 6
Mullach, 48 K 3
Mullach ide, 48 M 5
Mullen, 96 F 9
Mullens, 104 D 7
Muller; river, 82 F 7
Mullewa, 82 H 2
Mullica Hill, 95 E 6
Mullica; river, 95 F 7
Mulligan; river, 80 J 4
Multan, 68 F 4
Multia, 53 L 9
Muluwusuhe; river, 60 H 5
Muluya; river, 72 D 4
Mulvane, 97 I 11
Mumeng, 83 C 13
Mun, 65 L 16
Muna, 107 I 15
Muna; isl., 65 K 11
Muna; river, 57 E 11
Munburra, 83 C 10
Munchberg, 42 G 7
Muncho Lake, 86 H 4
Muncil, 99 M 13
Muncy, 100 I 8
Mundelein, 99 J 10
Mundesley, 51 B 16
Mundford, 51 D 14
Mundiwindi, 82 F 3
Mundra, 68 K 3
Mundrabilla, 82 I 5
Munducurus, 113 M 15
Mundwa, 68 H 5
Munford, 103 I 6
Mungallala; river, 83 H 11
Mungana, 83 D 10
Mungbere, 78 B 7
Mungeli, 69 L 9
Munich, 98 A 1
Muniches, 114 D 5
Munising, 99 E 11
Munoho Lake, 88 E 3
Munoz Gamero; pen., 119 M 3
Munslow, 50 D 6
Muntok, 64 J 4
Muonio, 52 E 8
Mupa, 79 I 2
Muqaynimah, 70 J 6
Mur; river, 43 L 10
Murakami, 62 F 11
Murakeresztuz, 43 L 11
Muramvya, 78 D 8
Murat, 45 I 9
Murau, 43 L 9
Muraukba, 65 C 14
Muravera, 39 K 4
Murayama, 62 F 12
Murchinson; river, 80 K 1
Murchison; river, 82 H 2
Murdale, 88 G 5
Murder; river, 103 N 9
Murderkill; river, 95 H 4
Murdo, 96 E 9
Murdochville, 91 I 11
Murdock, 105 M 10
Muresul; river, 40 D 6
Murfreesboro, 104 E 11
Murgha Kibzai, 68 F 3
Murgon, 83 G 12
Murieje, 78 G 4
Murjek, 52 G 7
Murlata; peak, 77 J 12
Murle, 77 L 10
Murphy, 96 E 1
Murphysboro, 102 F 7
Murray, 96 F 3
Murray Harbour, 91 K 13
Murray; river, 83 J 9
Murrayville, 102 D 6
Murree, 66 E 8
Murri; river, 112 F 4
Murska Sobota, 38 B 10
Muru; river, 114 F 8
Murugan, 65 D 16
Murui; river, 64 J 8
Murupu, 113 H 14
Murutinga, 115 B 15
Murware, 69 K 9
Murzuq, 76 D 1
Murzuq Edeien El-; desert, 76 D 1
Musa Khel Bazar, 68 F 3
Musa Qala, 71 G 12
Musa; river, 83 C 14
Musala; isl., 64 H 1
Muscatine, 98 K 8
Muscoda, 98 I 8
Musconetcong; river, 95 B 6
Musgrave, 83 C 10
Mushalagan; river, 91 G 9
Mushie, 78 D 2
Musi; river, 64 J 4
Muskegon, 99 I 12
Muskegon Heights, 99 I 12
Muskegon; river, 93 E 11
Muskingum; river, 100 K 3
Musko; isl., 55 I 11
Muskogee, 97 I 12
O'Higgins, 116 E 1
O'Higgins; prov, 116 L 2
Muskwa, 88 E 4
Muskwa; river, 88 E 4
Musmar, 77 G 9
Musoma, 78 C 10
Musquodoboit, 91 K 12
Mussau; isl., 83 A 14
Mussel Fork; river, 102 C 3
Musselburgh, 49 B 12
Musselshel; river, 96 B 6
Mussende, 78 G 2
Mustai, 77 D 12
Mustajidda, 70 H 4
Mustang, 69 G 11
Mustasaari, 53 K 7
Mustique; isl., 109 C 15
Mut, 76 D 6
Mutano, 79 I 2
Mutata, 112 E 4
Muthill, 49 A 10
Muting, 65 D 16
Mutsamudu, 79 K 11
Mutshatsha, 78 G 6
Mutsu, 62 B 13
Muttaburra, 83 F 10
Mutton Bay, 91 G 14
Mutubis, 77 A 10
Mutum; river, 113 M 9
Muyinga, 78 D 8
Muymanu; river, 115 H 9
Muyuri, 115 J 10
Muzaffarabad, 66 E 9
Muzo, 112 F 5
Mvolo, 76 L 7
Mvomero, 78 E 11
Mvoung; river, 78 B 1
Mwanza, 78 D 9
Mweka, 78 E 4
Mwenga, 78 D 7
Mwingi, 78 C 11
Myakka City, 105 M 10
Myakka; river, 105 M 10
Myaungmya, 67 L 11
Myeik Kyunzu; archp., 64 E 1
Myerstown, 95 C 3
Myggenaes, 54 1
Myingyan, 67 J 11
Myitkyina, 66 H 11
Myitta, 67 L 13
Mymensingh, 69 J 14
Myrdal, 54 G 3
Myrnan, 88 I 8
Myrtle, 102 H 3
Myrtle Beach, 105 I 9
Myrtle Grove, 103 O 9
Myrtle Point, 94 F 2
Mys Zelanija, 56 C 8
Mysen, 55 I 6
N Dali, 75 K 9
N Djamena, 76 I 2
N Guigmi, 75 I 13
N'Gouri, 76 I 2
Naarunoro; river, 83 K 16
Nabaroh, 77 B 12
Nabesna, 86 E 2
Nabeul, 74 A 12
Nabha, 68 F 6
Nabim, 65 B 14
Nabuga, 112 F 3
Nacajuca, 107 K 13
Nacala, 78 H 12
Nacaome, 108 J 5
Nacebe, 115 G 10
Nachana, 68 H 4
Naches, 94 C 4
Nachiland, 66 F 11
Nachingwea, 78 G 11
Nachitai, 60 G 6
Nachokoda, 57 J 15
Nacimiento, 118 D 2
Nacogdoches, 97 L 13
Nacori Chico, 106 C 4
Nacori Grande, 106 C 4
Nacozari de Garcia, 106 C 4
Nadeau, 99 F 11
Nadir, 77 D 11
Naerung, 66 E 13
Naes, 54 A 2
Naestved, 55 N 6
Nafada, 75 J 12
Nafun, 71 L 9
Nag, 71 I 12
Naga, 65 C 12
Naga; isl., 63 N 1
Nagai, 60 H 6
Nagala, 75 J 5
Nagaland; state, 66 H 10
Naganuma, 62 G 12
Nagaoka, 62 G 11
Nagar, 66 B 10
Nagar Parkar, 68 J 3
Nagasaki, 63 M 1
Nagata, 63 P 1
Nagina, 68 G 8
Nagishot, 76 L 8
Nagor, 67 I 3
Nagpur, 68 M 8
Nagrota, 68 E 6
Nagua, 109 12
Nagykallo, 43 J 16
Nagykanizsa, 43 L 12
Nagykata, 43 K 14
Nagykoros, 43 K 14
Nagyleta, 43 K 16
Nahan, 68 F 7
Nahanni Butte, 88 D 5
Nahar; river, 68 H 2
Nahma, 99 E 11
Nahuel Huapi, 118 F 3
Nahunta, 105 L 6
Naica, 106 D 6
Naicam, 89 J 11
Naikliu, 65 M 12
Nailswoth, 50 G 7
Naim, 46 F 7
Nain, 91 D 12
Nainital, 68 G 8
Nairn, 99 D 16
Nairobi, 78 C 11
Naivasha, 78 C 11
Najibabad, 68 G 8
Naka-Satsunai, 62 D 5
Naka-Tombetsu, 62 B 4
Nakajo, 62 G 11
Nakanno, 57 G 10
Nakanosawa, 62 D 3
Nakasato, 62 C 12
Nakashibetsu, 62 C 6
Nakatane, 63 P 2
Nakayama, 62 D 13
Nakechake, 60 I 3
Nakfa, 77 G 10
Nakhtarana, 68 K 2
Nakina, 88 D 1
Nakkila, 53 M 7
Naklo, 43 C 12
Nakskov, 55 O 6
Naktong; river, 61 F 15
Nakuru, 78 C 10
Nakusp, 88 K 5
Nal, 68 H 1
Naleczow, 43 E 15
Nalige, 60 D 4
Nalinhe; river, 60 E 8
Nalut, 74 C 13
Namacurra, 79 I 11
Namakzar; lake, 71 F 10
Namangan, 56 K 5
Namarroi, 79 I 11
Namasagali, 78 B 9
Namatanai, 83 A 15
Nambour, 83 G 12
Namche Bazar, 69 H 12
Nameh, 65 H 9
Namekagon; river, 98 E 7
Nametil, 79 I 12
Namibia, 73 M 7
Namie, 62 G 13
Namies, 79 N 4
Namiquipa, 106 C 5
Namleo, 65 K 13
Nammchek, 67 K 12
Namoerhe; river, 61 B 14
Nampa, 88 H 6
Nampala, 75 I 6
Nampula, 78 H 12
Nampur, 69 L 9
Namsen; river, 54 B 7
Namsos, 54 C 6
Namsskogan, 52 H 2
Namuel Mapa, 116 L 4
Namwala, 79 I 7
Namyslow, 43 F 12
Nana Candungo, 78 G 6
Nana; river, 76 L 2
Nanaimo, 88 L 3
Nananjiao; bay, 64 G 7
Nanay; river, 114 B 6
Nancha, 61 C 14
Nancorainza, 114 L 4
Nancy, 103 N 1
Nandewar Range, 79 K 9
Nanga Eboko, 75 M 13
Nangapinoh, 64 I 7
Nangaraun, 64 I 8
Nangataiap, 64 J 7
Nanhai; sea, 61 L 12
Nankana Sahib, 68 E 5
Nannine, 82 G 3
Nannup, 82 J 2
Nanortalik, 87 E 16
Nansio, 78 D 9
Nantes, 44 F 6
Nanticoke, 101 I 9
Nanticoke; river, 101 M 9
Nanton, 88 L 7
Nantucket, 101 I 15
Nantucket; isl., 101 I 15
Nantulo, 78 G 12
Nantwich, 50 A 6
Nanty-Glo, 100 J 6
Nanumanga; isl., 81 F 9
Nanumea; isl., 81 F 10
Nanweidao; isl., 64 E 7
Napier Mount-, 82 D 6
Naples, 100 G 8
Napo; prov, 112 J 4
Napo; river, 114 B 7
Napoleon, 96 C 10
Napoleonville, 103 P 5
Napoli; prov, 39 I 9
Nappanee, 99 K 12
Napperby, 82 F 7
Nar; river, 51 C 14
Nara, 68 K 2
Nara Visa, 97 J 8
Nara; river, 63 I 10
Narachic, 106 E 5
Naraime; river, 114 B 4
Naraina, 68 I 6
Naranja, 105 O 12
Naranjal, 114 C 5
Naranjo, 112 H 3
Naranjos, 107 I 11
Narbonne, 45 K 10
Narborough, 51 C 14
Nardo, 39 J 12
Nare, 116 I 7
Narendra Nagar, 68 F 8
Naretha, 82 I 5
Narew; river, 43 C 14
Nari; river, 68 G 2
Naricual, 113 C 11
Narino, 112 F 5
Narken, 52 F 7
Narnaul, 68 H 6
Narok, 78 C 10
Narowal, 68 E 6
Narpio, 53 L 7
Narpur, 68 E 4
Narraguagus; river, 101 D 16
Narrogin, 82 I 3
Narrows, 104 D 7
Narsinghgarh, 68 K 7
Narsinghpur, 68 K 8
Narssaq, 87 E 15
Narssarssuaq, 87 E 16
Narugo, 62 F 13
Narvacan, 65 A 11
Narvik, 52 D 5
Narwana, 68 G 6
Narwar, 68 I 7
Nasaker, 54 D 9
Nasarawa, 75 K 11
Naschel, 116 K 5
Nashart Bahr-; river, 77 B 10
Nashua, 98 I 5
Nashville, 97 K 13
Nashwauk, 98 C 6
Nasice, 40 D 2
Nasielsk, 43 D 14
Nasir, 76 K 8
Nasirabad, 68 G 2
Naskaupi; river, 91 E 12
Nassau, 101 H 11
Nassau; gulf, 119 O 5
Nassau; isl., 83 E 16
Nassauwadox, 104 D 13
Nassjo, 55 K 8
Nastapoka Islands; archp., 90 D 7
Nastapoka; river, 90 D 7
Natagaima, 112 H 5
Natal, 64 I 2
Natashquan, 91 H 12
Natashquan; river, 91 H 13
Natchez, 103 N 4
Natchitoches, 103 M 2
Nathan, 99 F 11
Nathia Gali, 66 E 8
Natitingou, 75 K 8
Natividad, 115 H 11
Natividad; isl., 106 D 1
Natsrat, 70 E 3
Natuna Besar; isl., 64 G 5
Natuna; isl., 59 N 11
Naturita, 96 H 5
Naubinway, 99 E 13
Nauchas, 79 L 3
Nauders, 42 L 6
Naudville, 91 I 9
Naugatuck, 101 I 12
Nauroz Kalat, 68 G 1
Nauru, 80 E 8
Naushahro Firoz, 68 H 2
Nauta, 114 C 6
Nautanwa, 69 H 10
Nauteyri, 52 A 2
Nautla, 107 J 11
Nauvoo, 98 M 7
Nava; river, 78 B 7
Navarino; isl., 119 O 5
Navarra; prov, 37 B 9
Navarre, 100 J 4
Navarro, 114 D 5
Navassa; isl., 109 H 10
Navenby, 51 A 11
Navet; river, 108 L 3
Navia, 116 L 4
Navibandar, 68 L 2
Navidad, 116 L 1
Navojoa, 106 E 4
Navolato, 106 F 5
Navrongo, 75 K 7
Nawabshah, 68 I 2
Nawakot, 69 H 11
Nawalgarh, 68 H 6
Nawasharh, 68 F 6
Naxahachie, 97 L 12
Naxos, 41 O 10
Naxos; isl., 41 O 10
Naya, 112 H 3
Naya; river, 112 H 3
Nayagarh, 69 M 12
Nayant; state, 106 H 6
Nayland, 51 F 14
Naylor, 105 L 5
Nayoro, 62 B 4
Naytahwaush, 98 C 4
Nazacara, 115 K 9
Nazareno, 114 L 3
Nazaret, 115 H 11
Nazareth, 95 B 6
Nazas, 106 F 7
Nazas; river, 106 F 7
Nazca, 114 J 6
Nazca; river, 114 J 5
Nazilli, 41 M 13
Nazko, 88 I 3
Nazret, 77 K 11
Nazwa, 71 K 9
Ncheu, 78 H 10
Ndao; isl., 65 M 11
Ndele, 76 K 4
Ndeni; isl., 80 G 8
Ndikinimeki, 75 M 12
Ndonga; river, 79 K 3
Nea; river, 54 D 6
Neales; river, 82 G 8
Neamt; prov, 40 B 10
Neapolis, 41 O 7
Neath, 50 G 3
Neath; river, 50 G 4
Nebine; river, 83 H 11
Nebit Dag, 56 I 2
Nebo, 83 E 11
Nebraska City, 98 L 3
Nebraska; state, 92 F 7
Nebster City, 96 E 13
Necedah, 98 H 8
Nechako; river, 88 I 3
Neches; river, 103 O 1
Nechi; river, 112 E 6
Necker; isl., 81 A 12
Necochea, 118 D 10
Nederlands, 103 P 1
Needham Market, 51 E 15
Needles, 94 K 8
Neembucu; prov, 116 G 8
Neemuch, 68 J 5
Neenah, 99 H 10
Neepawa, 89 K 13
Neeses, 105 I 7
Neffs, 95 B 5
Neffsville, 95 D 3
Negage, 78 F 2
Negai, 62 F 10
Negaunee, 99 D 11
Negaunee, 99 D 11
Negele, 77 L 11
Negotin, 40 F 7

Negra La-, 118 D 10
Negra; range, 114 F 4
Negreiros, 116 B 1
Negrillos, 115 M 9
Negritos, 114 C 1
Negro Baia del-, 77 K 15
Negro Muerto, 118 E 6
Negro; river, 112 G 6
Negros Los-, 106 G 6
Negrossla, 65 E 11
Negru-Voda, 40 F 12
Nehe, 61 B 13
Neiba, 109 H 11
Neidin, 47 M 3
Neidpath, 89 L 9
Neilburg, 89 J 9
Neillsville, 98 G 8
Neisse; river, 43 F 9
Neiva, 112 H 5
Nejo, 77 J 9
Nekemte, 77 J 10
Nekoosa, 99 H 9
Neligh, 98 J 1
Nelkan, 57 G 13
Nelluru, 67 M 5
Nelson, 49 K 14
Nelson Forks, 88 E 4
Nelson House, 89 H 12
Nelson area, 39 F 4
Nelson; river, 89 G 13
Nelsonville, 100 L 2
Nema, 74 H 5
Nemawar, 68 L 7
Nembe, 75 M 10
Nemegos, 99 C 14
Nemegose; river, 99 B 15
Nemuro, 62 D 7
Nemuro-Hanto; pen., 62 D 7
Nenana, 86 D 2
Nene; river, 51 C 13
Nenggiri; river, 64 G 3
Nenthead, 49 G 13
Neodesha, 96 H 12
Neoga, 102 D 8
Neola, 98 K 3
Neopit, 99 G 10
Neosho, 102 G 1
Neosho; river, 96 H 12
Neoskweskau, 90 G 7
Nepal; c., 69 G 9
Nepalganj, 69 H 9
Nepena, 114 F 3
Nephi, 96 G 3
Nepomuk, 43 H 9
Neptune, 89 L 11
Neptune Beach, 105 M 7
Nercinsk, 57 I 12
Neretva; river, 40 F 2
Neriquinha, 79 I 4
Nerresundby, 55 L 5
Nescopeck, 95 A 3
Neshaminy, 95 D 6
Neshkoro, 99 H 9
Neshoba, 103 L 7
Neskaupstadhur, 52 B 4
Nesmith, 105 I 9
Nesna, 52 G 2
Nesoya; isl., 52 F 2
Nesquehoning, 95 A 4
Ness City, 96 H 10
Nestawkanow; river, 90 I 8
Nesterville, 99 E 14
Neston, 49 M 12
Nestor, 108 L 3
Nestor Falls, 98 A 5
Nestos; river, 41 J 9
Nesttun, 54 H 2
Net Lake-, 98 B 6
Netcong, 95 A 7
Nether Stowey, 50 I 5
Netherdale, 83 E 11
Netherlands, 42 C 3
Netherwiton, 49 E 14
Nettlebed, 51 H 10
Nettleton, 102 H 5
Neubrandenburg, 42 C
Neudorf, 89 K 11
Neufchateau, 42 G 2
Neumarkt, 42 J 8
Neuquen, 118 D 5
Neuquen; prov, 118 D 4
Neuquen; river, 118 D 5
Neurara, 116 E 2
Neuse; river, 104 G 12
Neustadt, 42 J 4
Neuville, 103 N 1
Nevada, 98 K 5
Nevada, 102 F 1
Nevada; state, 92 F 3
Nevelsk, 57 H 15
Neversink; river, 101 I 10
Nevesinje, 40 G 2
Nevils, 105 J 6
Nevin, 50 B 2
Nevis, 98 D 4
Nevis; isl., 109 C 12
New Abbey, 49 F 10
New Albany, 102 E 10
New Albin, 98 I 7
New Amsterdam, 113 F 16
New Athens, 102 E 6
New Auburn, 98 F 7
New Augusta, 103 N 7
New Bedford, 101 I 14
New Berlin, 101 G 10
New Bern, 104 G 11
New Caledonia, 80 I 8
New Caledonia, 80 I 8
New Castle, 99 M 13
New Castleton, 49 E 12
New City, 101 I 11
New Columbia, 95 A 2
New Concord, 100 K 3
New Cumberland, 100 J 5
New Cumnock, 49 E 9
New Delhi, 68 G 6
New Delhi, 68 G 7
New Denver, 88 L 5
New Edinburg, 103 K 3
New Egypt, 95 D 8
New Ellenton, 105 I 6
New England, 96 C 8
New Florence, 102 D 4
New Franken, 99 G 10
New Freedom, 95 E 2
New Galloway, 49 F 10
New Germany, 91 L 12
New Glarus, 99 J 9
New Glasgow, 91 K 13
New Grant, 108 M 2
New Gretna, 95 F 8
New Guinea; isl., 80 F 3
New Hampshire; state, 93 E 14
New Hampton, 98 I 6
New Hanover; isl., 83 A 14
New Hartford, 98 J 6
New Haven, 99 L 13
New Hazelton, 88 G 2
New Hebrides; isl., 83 F 13
New Holland, 104 F 13
New Holstein, 99 H 10
New Hope, 104 G 1
New Iberia, 103 P 4
New Inn Green, 51 J 15
New Ireland; isl., 83 A 15
New Jersey; state, 93 F 13
New Kensington, 100 J 5
New Kent, 104 D 11
New Kingston, 95 C 1
New Laredo, 107 E 9
New Leipzig, 96 C 8
New Lexington, 99 M 16
New Lisbon, 98 H 8
New Liskeard, 90 J 5
New London, 98 L 7
New Madrid, 102 G 6
New Manchester, 99 L 13
New Market, 100 M 6
New Martinsville, 100 K 4
New Matamoras, 100 L 4
New Meadows, 94 D 6
New Mexico; state, 92 I 5
New Milford, 101 I 9
New Mills, 49 M 14
New Orleans, 103 P 6
New Oxford, 95 E 1
New Pekin, 102 E 10
New Philadelphia, 100 J 4
New Port Richey, 105 O 5
New Prague, 98 G 5
New Providence; isl., 109 E 9
New Quay, 50 E 2
New Radnor, 50 E 5
New Richland, 98 H 5
New Richmond, 91 I 11
New Ringgold, 95 B 4
New Rochelle, 95 A 10
New Rockford, 98 B 1
New Romney, 51 J 15
New Salem, 95 E 2
New Sharon, 98 K 6
New South Wales; state, 83 I 10
New Straitsville, 100 K 3
New Tazewell, 104 E 4
New Town, 96 B 8
New Tripoli, 95 B 4
New Ulm, 98 G 4
New Underwood, 96 D 8
New Vienna, 100 L 1
New Village, 95 B 6
New Washington, 99 L 15
New Waterford, 91 J 14
New Westminster, 88 L 3
New Windsor, 95 F 1
New World; isl., 91 H 15
New York, 95 B 9
New York Mills, 98 D 4
New York; state, 93 E 12
New Zealand, 81 L 9
New; river, 104 E 7
Newala, 78 G 12
Newald, 99 F 10
Newark, 51 A 10
Newbern, 102 H 7
Newberry, 99 E 12
Newcastle, 91 J 11
Newcastle Emlyn, 50 F 2
Newcastle Water, 82 D 7
Newcastle under Lyme, 50 B 7
Newcastle upon Tyne, 49 F
Newcastle; river, 82 D 7
Newcomb, 97 I 5
Newcomerstown, 100 K 4
Newell, 104 G 7
Newellton, 103 M 4
Newent, 50 F 7
Newfane, 101 G 12
Newfield, 95 F 6
Newfoundland, 104 F 12
Newfoundland; isl., 84 G 11
Newfoundland; isl., 91 H 15
Newfoundland; prov., 84 G 11
Newhalem, 94 B 4
Newhalen, 86 D 1
Newhall, 98 K 7
Newhaven, 51 K 12
Newington, 105 J 6
Newkirk, 97 I 11
Newland, 104 F 6
Newman, 102 C 8
Newmanstown, 95 C 3
Newmarket, 51 E 13
Newmarket, 101 G 14
Newmilns, 49 D 9
Newnan, 105 I 3
Newport, 47 N 8
Newport News, 104 D 12
Newport Pagnell., 51 F 11
Newry, 48 J 5
Newton, 48 B 7
Newton Abbot, 50 L 4
Newton Aycliffe, 49 H 15
Newton Ferrers, 50 M 3
Newton Grove, 104 G 10
Newton Reigny, 49 G 12
Newton Saint Boswells, 47 I 8
Newton Stewart, 49 F 9
Newton le Willows, 49 L 13
Newtonville, 95 F 7
Newtown, 49 F 12
Newtown Butler, 48 J 2
Newtownabbey, 48 H 5
Newtownards, 47 J 6
Newtownstewart, 48 G 2
Newville, 100 K 8
Neyland, 50 G 1
Neyshabur, 71 E 10
Nezperce, 94 D 6
Nezpique Bayou; river, 103 O 3
Ngabang, 64 I 6
Ngama, 76 I 2
Ngambe, 75 L 13
Ngaoundere, 76 K 1
Ngerengere, 78 E 11
Ngoko; river, 78 B 2
Ngozi, 78 D 8
Ngulu; isl., 96 D 3
Nguru, 75 J 12
Nhambiquara, 115 I 15
Nhamunda, 113 L 16
Nhamunda; river, 113 K 16
Nhecolandia, 115 M 16
Nia Nia, 78 B 7
Niabo El-, 77 L 11
Niagara, 98 B 2
Niagara Falls, 100 G 6
Niagara; river, 100 F 6
Niagassola, 75 J 4
Niakaramandougou, 75 K 6
Niamey, 75 J 9
Niangare, 78 A 7
Niantic, 102 C 7
Niari; river, 78 D 1
Niarmada; river, 68 L 5
Nias; isl., 64 I 2
Nicaragua; c., 85 M 9
Nicastro, 39 L 11
Niceville, 105 L 1
Nichare; river, 113 E 11
Nicholasville, 104 C 3
Nicholls, 105 K 5
Nicholls Town, 108 E 8
Nichols, 98 K 7
Nicholson, 82 D 6
Nicholson; river, 82 D 8
Nicholville, 101 E 10
Nicman, 91 H 11
Nico Perez, 117 K 10
Nicobar Dvip; isl., 67 O 12
Nicobar; isl., 59 M 9
Nicocli, 112 D 4
Nicolas Bravo, 106 G 6
Nicolet, 90 K 8
Nicollet, 98 H 5
Nicosia, 39 M 9
Nicoya; gulf, 108 L 6
Nicoya; pen., 108 L 5
Nicut, 103 I 1
Nidd; river, 49 J 16
Nidelava; river, 55 J 4
Nido El-, 65 D 10
Nidzica, 43 C 14
Niemen; river, 58 D 2
Nierhe; river, 61 A 12
Nieves, 106 G 7
Nifsha, 77 D 15
Niger; c., 72 F 5
Niger; delta, 72 H 5
Niger; river, 75 K 10
Nigeria; c., 72 H 5
Niggins, 96 G 8
Nighasan, 69 H 9
Nigrita, 41 J 8
Nihoa; isl., 81 A 13
Nihommatsu, 62 G 12
Niigata, 62 G 11
Niikappu; river, 62 E 4
Niisama, 62 C 4
Niitsu, 62 G 11
Nikiniki, 65 M 12
Nikki, 75 K 9
Nikla el Inab, 77 C 10
Nikonga; river, 78 D 9
Nikopol, 40 F 9
Niksic, 40 G 3
Nil el Abyad; river, 76 I 8
Nil el Azraq; river, 77 I 9
Nila; isl., 65 L 14
Nile; delta, 72 D 9
Nile; river, 72 E 9
Niles, 99 K 12
Nilphamari, 69 I 14
Nilsia, 53 11
Niltepec, 107 L 13
Nilton, 103 L 9
Nimu, 68 D 7
Nimule, 76 M 8
Ninety Six, 104 H 6
Ninfield, 51 J 13
Ningwu, 61 F 11
Ninhue, 118 C 2
Ninigo Group; isl., 83 A 12
Ninnescah; river, 97 I 12
Ninock, 103 M 2
Nioro, 75 I 4
Niota, 104 F 3
Nipawin, 89 I 11
Nipigon, 99 A 10
Nipigon; river, 99 A 10
Nippon-Kai; sea, 63 I 3
Nipton, 94 K 7
Nirvana, 99 H 13
Nis, 40 6
Niscemi, 39 N 9
Nishibetsu, 62 D 7
Nishibetsu; river, 62 D 6
Nishinera, 63 N 2
Nishinoomote, 63 P 2
Nishwa, 77 D 13
Nisiros; isl., 41 O 12
Nissan; isl., 80 F 6
Nissan; river, 55 L 7
Nisswa, 98 D 5
Nitchequon, 91 F 9
Nith; river, 49 F 11
Niton, 51 L 9
Nitro, 99 D 16
Niuafoon; isl., 81 H 11
Niue; isl., 83 F 16
Niutao; isl., 81 F 10
Nivala, 53 J 9
Nivea; river, 114 C 4
Niying, 69 G 16
Nizn'aja Tunguska; river, 57 F 9
Nizne Udinsk, 57 I 9
Nizneangarsk, 57 H 11
Niznejansk, 57 D 12
Nkata Bay, 78 G 10
Nkongsamba, 75 M 12
Nmai Hka; river, 66 H 12
No Man's Land; isl., 101 I 14
Noanama, 112 G 4
Noatak, 86 B 2
Noatak; river, 86 B 2
Nobeoka, 63 M 3
Noble, 102 E 8
Noblesville, 99 M 12
Noboribetsu, 62 E 2
Nocatee, 105 M 10
Nochixtlan, 107 L 11
Nodaway; river, 98 M 4
Noel, 102 G 1
Noelville, 90 K 5
Noemi; river, 113 H 9
Noetinger, 116 J 6
Nogar Parkar, 71 J 13
Nogat; river, 43 B 13
Nogoya, 116 J 8
Nohar, 68 G 6
Noheji, 62 C 13
Noir; isl., 119 N 3
Noira, 101 D 10
Noirmoutier; isl., 44 F 5
Nok Kundi, 71 H 11
Nokia, 53 M 8
Nokomis, 89 K 10
Nola, 76 M 2
Nolan, 104 C 6
Nolichucky; river, 104 E 5
Nomal, 66 B 10
Nombre de Dios, 106 G 7
Nome, 86 B 2
Nomgon, 61 E 9
Nominingue, 90 K 7
Nonni; river, 58 F 11
Nonoava, 106 E 5
Nonouti; isl., 81 E 9
Nonza, 45 K 15
Noonkanbah, 82 D 4
Noose Jaw, 86 K 7
Nootka, 88 K 2
Noqui, 78 E 1
Nora, 53 O 2
Nora Springs, 98 I 6
Nora; isl., 77 G 11
Noranda, 90 I 5
Norborne, 102 D 2
Norco, 103 P 5
Norcross, 104 H 3
Nord Kavloy; isl., 52 B 5
Norderham, 42 C 5
Nordfold, 52 E 4
Nordhausen, 42 E 6
Nordhurfjordhur, 52 A 2
Nordhurishaf; sea, 52 A 2
Nordingre, 53 K 5
Nordkapp, 52 A 9
Nordkinnhalvoya; pen., 52 A 10
Nordli, 53 I 2
Nordmalings, 53 K 6
Nordreisa, 52 C 6
Nordvik, 57 D 10
Nore, 54 H 4
Norfolk, 98 J 2
Norfolk; isl., 80 J 8
Norfork, 102 H 3
Norham, 49 C 14
Noria La-, 106 D 8
Norily, 57 E 9
Norlina, 104 E 10
Norma, 95 F 6
Normal, 99 M 9
Norman, 103 J 2
Norman Cross, 51 D 12
Norman Park, 105 L 4
Norman Wells, 88 A 4
Norman; isl., 109 B 9
Norman; river, 83 D 9
Normanby; isl., 83 D 15
Normandie, region, 44 D 6
Normanton, 83 D 9
Nornalup, 82 J 3
Norra-Ny, 54 H 7
Norridgewock, 101 D 15
Norris, 96 C 4
Norris Arm, 91 H 15
Norris City, 102 F 8
Norristown, 95 D 6
Norrkoping, 55 J 9
Norrsundet, 54 G 10
Norrtalje, 54 H 11
Norskehavet; sea, 54 A 3
Nort Elmham, 51 C 14
Nort Uist; isl., 46 F 5
Nort Walsham, 51 B 15
North, 105 I 7
North, 49 D 15
North Adams, 101 G 12
North Anson, 101 D 14
North Atlanta, 104 H 3
North Augusta, 105 I 6
North Aulatsivik; isl., 91 A 11
North Baltimore, 100 I 2
North Battlefond, 89 J 9
North Bay, 90 K 5
North Bend, 94 F 2
North Berwick, 49 B 12
North Branch, 98 A 5
North Branch Raritan; river, 95 B
North Canadian; river, 97 I 10
North Carolina; state, 104 F 6
North Conway, 101 E 13
North Creek, 101 F 11
North East, 100 H 5
North English, 98 K 7
North Fabius; river, 98 M 7
North Fork Crow; river, 98 F 5
North Fork; river, 102 G 3
North Fork; river, 94 C 6
North Gayne; river, 90 A 8
North Haro, 101 D 12
North Holston; river, 104 E 6
North Horr, 78 A 11
North Islands; isl., 103 P 7
North Judson, 99 L 12
North Keeling; isl., 64 M 1
North Kentucky; river, 104 C 4
North Key Largo, 105 O 12
North Knife; river, 89 E 13
North Korea, 57 K 14
North Liberty, 99 K 12
North Licking; river, 104 C 4
North Loup; river, 96 F 9
North Manitou, 99 F 12
North Manitou; isl., 99 F 12
North McIntyre, 99 B 9
North Miami, 105 O 12
North Muskegon, 99 I 12
North Platte, 96 F 9
North Platte; river, 96 F 8
North Powder, 94 E 6
North Red; river, 102 G 9
North River, 89 E 13
North Rona; isl., 46 D 6
North Ronaldsay; isl., 46 D 9
North Sea; sea, 46 E 9
North Stratford, 101 D 13
North Sunderland, 49 D 15
North Sydney, 91 K 14
North Tawton, 50 K 3
North Thompson; river, 88 K 4
North Troy, 101 D 12
North Twin; isl., 90 F 5
North Tyne; river, 49 F 14
North Vancouver, 88 L 3
North Vermilion, 88 F 6
North Vernon, 102 D 11
North Wales, 95 D 6
North West River, 91 F 13
North Wilkesboro, 104 E 7
North York, 95 D 2
North; river, 91 E 14
Northallerton, 49 I 15
Northam, 50 J 2
Northampton, 51 E 10
Northbridge, 101 H 13
Northcliffe, 82 J 2
Northern Territory; state, 82 D 6
Northfield, 98 G 6
Northhampton, 95 B 5
Northleach, 50 G 8
Northome, 98 C 5
Northome, 98 C 5
Northport, 94 B 6
Northumberland, 95 A 2
Northview, 102 F 3
Northville, 101 G 11
Northway, 87 E 3
Northwest, 104 E 12
Northwest Frontier, 68 C 5
Northwich, 49 M 13
Northwood, 98 C 2
Norton, 50 E 8
Norton Bay, 86 C 1
Norton Sound; gulf, 86 B 1
Nortonville, 102 G 9
Norwalk, 98 H 8
Norway, 98 K 7
Norway House, 89 I 13
Norwich, 51 C 15
Norwood, 100 L 1
Noshim, 60 D 14
Nossi-Be; isl., 79 L 12
Nosy-Mitsio; isl., 79 L 12
Notasulga, 105 J 2
Notec; river, 43 D 10
Notia Rodhopi; range, 41 I 9
Notikewin; river, 88 G 6
Notingham; isl., 87 G 11
Noto, 39 N 10
Noto-Hanto; pen., 62 H 8
Notodden, 55 I 4
Notre Dame du Lac, 91 J 10
Notre-Dame-du Nord, 90 J
Nottaway; river, 90 H 6
Nottingham, 51 B 10
Nottingham Island, 87 G 11
Nottoway, 104 D 10
Nottoway; river, 104 E 11
Nouadhibou, 74 G 2
Nouakchott, 74 H 2
Noumea, 80 H 8
Nouna, 75 J 6
Nouvelle Anvers, 78 B 3
Nova Anadia, 65 M 13
Nova Chaves, 78 G 4
Nova Freixo, 78 H 11
Nova Gaia, 78 G 3
Nova Lamego, 75 J 2
Nova Lisboa, 78 H 2
Nova Luzitania, 79 J 10
Nova Olinda do Norte, 115 B
Nova Prata, 117 H 12
Nova Sagres, 65 M 13
Nova Scotia, 87 K 16
Nova Sofala, 79 J 10
Nova Vida, 115 G 13
Nova Zagora, 40 H 10
Novaja Sibir; isl., 57 C 12
Novalty, 98 M 6
Novara; prov, 38 C 3
Novi, 38 D 9
Novi Pazar, 40 G 4
Novi Sad, 40 D 4
Novillos Los-, 117 J 10
Novo Aripuana, 115 C 14
Novo Hamburgo, 117 I 12
Novo Mesto, 38 C 9
Novo Redondo, 78 G 1
Novokazalinsk, 56 I 4
Novorybnoje, 57 D 10
Novska, 40 D 1
Novyy Port, 56 E 7
Nowata, 97 I 12
Nowe Miasto, 43 E 14
Nowgong, 68 J 8
Nowogrod, 43 C 15
Nowshera, 66 E 6
Nowy Sacz, 43 H 14
Nowy Tomysl, 43 D 10
Noxapater, 103 L 7
Noxubee; river, 103 L 8
Noy-Varika, 79 N 13
Nozay, 44 E 6
Nsanje, 79 I 10
Nsok, 75 N 13
Nsukka, 75 L 11
Ntem; river, 75 N 12
Ntui, 75 M 13
Nu Garna, 68 G 1
Nuatja, 75 L 8
Nuba, 77 D 13
Nubaran, 70 E 7
Nubian; desert, 72 F 9
Nuboai, 65 B 14
Nucuray; river, 114 C 5
Nueva Alejandria, 114 D 6
Nueva Antioquia, 112 F 8
Nueva California, 116 K 3
Nueva Casas Grandes, 106 C 5
Nueva Galia, 116 L 5
Nueva Germania, 117 D 9
Nueva Gerona, 108 F 7
Nueva Imperial, 118 D 2
Nueva Lubecka, 118 H 4
Nueva Palmira, 116 K 8
Nueva Pompeya, 116 E 6
Nueva Rosita, 106 D 8
Nueva Vizcaya, 116 I 8
Nueva; isl., 119 O 6
Nueve de Julio, 116 L 7
Nuevitas, 109 G 9
Nuevo Ideal, 106 F 6
Nuevo Mamo, 113 D 12
Nuevo Morelos, 107 H 10
Nuevo Mundo, 115 G 10
Nuevo Necaxa, 107 J 10
Nuevo Rocafuerte, 112 K 4
Nuevo Tres Picos, 107 L 13
Nuevo; gulf, 118 G 7
Nugssuaq, 87 C 13
Nugssuaq Halvo; pen., 87 C 13
Nuguria; isl., 80 F 6
Nui; isl., 83 E 14
Nuits-Saint-Georges, 45 F 11
Nukehe; river, 60 I 6
Nukualofa, 83 F 15
Nukufetau; isl., 81 F 10
Nukuh, 83 B 14
Nukuhiva; isl., 81 H 16
Nukulailai; isl., 83 E 15
Nukumanu; isl., 80 F 7
Nukunau; isl., 81 E 10
Nukunono; isl., 83 E 15
Nukus, 56 I 3
Nukutavake; isl., 81 J 16
Nullagine, 82 F 3
Nullagine; river, 82 F 3
Nullarbor, 82 I 6
Numakawa, 62 A 3
Numan, 75 K 13
Numata, 62 C 3
Numatinna; river, 76 L 6
Numfoor; isl., 65 A 14
Numidia, 95 A 3
Nunaksaluk; isl., 91 D 13
Nunapitchuk, 86 C 1
Nunchia, 112 F 7
Nunda, 98 H 2
Nuneaton, 51 D 9
Nunez; isl., 119 N3
Nungo, 78 H 11
Nunivak; isl., 84 C 2
Nunkiang, 61 B 13
Nunkiang; river, 61 C 13
Nunnanen, 52 D 8
Nunnelly, 102 H 9
Nunney Catch, 50 I 7
Nunoa, 114 J 8
Nuominhe; river, 61 B 13
Nuorang, 69 H 16
Nuoro, 39 J 3
Nuoro; prov, 39 J 3
Nupani; isl., 80 G 8
Nuqui, 112 F 3
Nurakita; isl., 81 G 10
Nurba, 57 F 11
Nuremberg, 95 A 3
Nurmes, 53 J 12
Nurmijarvi, 53 O 9
Nusa Tenggara Barat; prov, 65 M 9
Nusa Tenggara Timur; prov, 64 M 11
Nushagak, 86 D 1
Nushagak; river, 86 D 1
Nushki, 68 F 1
Nutak, 91 C 12
Nutley, 51 J 12
Nutria La-, 107 F 10
Nutts Corner, 48 H 5
Nuvukjuak, 87 F 11
Nuwakot, 69 H 11
Nuyts Archipelago, 82 I 7
Nyaake, 75 M 5
Nyabing, 82 J 3
Nyack, 101 J 11
Nyakanazi, 78 D 8
Nyala, 76 I 5
Nyamlali, 76 J 6
Nyamtumbo, 78 G 10
Nyandge; river, 79 J 9
Nyanza, 78 D 8
Nyaunglebin, 67 K 12

Rumia,43 A 12
Rumoi,62 C 3
Rumsey,88 K 7
Rumson,95 C 9
Run Hats,98 J 6
Run;isl.,46 G 5
Runcorn,49 M 13
Rungan;river,64 J 8
Rungwa;river,78 E 9
Runtu,79 J 4
Ruoqiang,60 F 5
Ruoshi;river,60 F 8
Ruovesi,53 M 8
Rupa,38 C 8
Rupar,68 F 7
Rupat;isl.,64 H 3
Rupert,96 E 3
Rupert;river,90 G 6
Rupnagar,68 I 6
Rupununi;river,113 H 15
Rural Hall-,104 E 8
Rural Retreat,104 E 7
Rurmnabaque,115 J 10
Rurutu;isl.,81 J 14
Ruscaigh,48 K 1
Ruse,40 F 10
Rush,96 H 8
Rush City,98 F 6
Rushden,51 E 11
Rushford,98 H 7
Rushville,96 E 8
Ruskin,105 P 5
Russelkonda,69 M 11
Russell,89 K 12
Russell Springs,96 H 9
Russell;isl.,86 D 8
Russells Point,99 M 14
Russellville,100 L 1
Russelsheim,42 G 5
Russian Mission,86 C 1
Russkaya Gavar,56 C 8
Rustburg,104 D 9
Ruteng,65 M 10
Rutherfordton,104 F 6
Ruthin,50 A 5
Ruthsburg,95 H 4
Ruthven,98 I 4
Ruthwell,49 F 11
Rutland,98 E 2
Rutland Plains,83 C 9
Rutledge,104 E 4
Rutshuru,78 C 8
Ruvu;river,78 F 11
Ruvuma;river,78 G 12
Ruyigi,78 D 8
Rwanda,78 D 8
Rybtsu,62 G 10
Ryde,51 K 10
Rye,51 J 14
Rye Beach,101 G 14
Ryegate,96 C 5
Ryhope,49 G 16
Rylstone,49 J 14
Rymanow,43 H 15
Rypin,43 C 13
Ryter De-,101 G 9
Ryukyu;isl.,59 I 13
Rzepin,43 D 10
Rzeszow,43 G 15
Saadah,70 L 4
Saaminki,53 M 11
Saar;river,42 H 3
Saarijarvi,52 G 9
Saarikoski,52 D 7
Saariouis,42 H 3
Saarland,42 H 3
Saavedra,118 D 8
Saba;isl.,109 C 11
Sabac,40 E 4
Sabah;state,65 G 9
Sabana;archp.,108 F 8
Sabancuy,107 J 14
Sabang,65 I 10
Sabanilla,107 K 14
Sabano,112 K 6
Sabarei,78 A 11
Sabarmati;river,68 K 4
Sabaya,115 M 10
Saberania,65 B 15
Sabetha,98 M 3
Sabha,76 C 1
Sabinas Hidalgo;river, 107 E 9
Sabinas;river,107 E 9
Sabine Pass,103 P 1
Sabine;river,103 M 1
Sabinov,43 H 15
Sablayan,65 C 11
Sable River,91 L 12
Sable-sur-Sarthe,44 E 7
Sable;isl.,84 G 12
Sabya,70 L 3
Sabzevar,71 E 10
Sac City,98 J 4
Sac;river,102 F 2
Sacaba,115 K 11
Sacabchen,107 J 15
Sacaca,115 L 11
Sacanana,118 G 5
Sacendaga;river,101 F 11
Sach Khas,66 G 13
Sachalin;isl.,57 H 15
Sachalinskij Zaliv;gulf, 57 G 14
Sachayoj,116 F 6
Sachigo;river,90 D 2
Sachin,68 M 4
Sachs Harbour,86 D 6
Sackets Harbor,101 F 9
Sackville,91 K 12
Saco,96 B 6
Saco;river,101 F 14
Sacramento,106 F 8
Sacxan,107 K 16
Sada Bandeira,78 H 1
Sadani,78 E 11
Sadberge,49 H 15
Sadda,68 D 3
Sadhoowa,108 M 2
Sadiola,75 I 3
Sadiqabad,68 G 3
Sadiya,66 H 10
Sado;isl.,62 G 10
Sadowara,63 N 2
Sadri,68 J 5
Sadsburyville,95 D 4
Saenz Valiente,116 I 8
Safaniya,70 H 6
Safarikovo,43 I 14
Saffle,55 I 7
Safford,97 L 4
Saffron Walden,51 F 13
Saft Turab,77 C 11
Saft el 'Inab,77 C 12
Saft el Muluk,77 C 10
Sag Harbor,95 A 13
Saga,63 M 1
Sagae,62 F 12
Sagaing,67 J 11
Saginaw,100 F 2
Saginaw Bay,100 E 2
Sagua la Grande,108 F 8
Saguache,96 H 6
Saguenay;river,91 I 9
Sagwara,68 K 5
Sahaba,76 F 7
Sahagun,112 D 4
Saharanpur,68 G 7
Sahaswan,68 H 8
Sahe,60 I 4
Sahiwal,68 E 4
Sahra Es-;desert,74 D 7
Sahragt el Kubra,77 D 12
Sahuaripa,106 D 4
Sahuayo de Jose Maria Moreles,106 J 8
Sai;river,69 I 9
Saibai;isl.,83 C 12
Said Bundas,76 K 5
Saidor,83 B 13
Saidpur,69 I 13
Saidu,68 C 5
Saigo,63 I 5
Sailolof,65 J 15
Sain Alto,106 G 7
Saint Affrique,45 J 9
Saint Agustin-Saguenay, 87 I 16
Saint Alban's,91 I 15
Saint Albans,100 M 3
Saint Albans,51 G 11
Saint Amand Mont Rond, 45 F 9
Saint Andre de Cutzac, 44 I 6
Saint Andrew,109 B 16
Saint Andrew's,91 I 1
Saint Andrews,91 L 11
Saint Andrews,49 A 12
Saint Ann's Bay,109 H 9
Saint Anne,99 L 11
Saint Anthony,96 D 4
Saint Augustin;river, 91 G 14
Saint Augustine,105 M 7
Saint Austell,50 M 1
Saint Avold,45 C 12
Saint Barthelemy;isl., 109 B 11
Saint Bethlehem,102 G 9
Saint Blazey,50 L 1
Saint Boniface,89 K 4
Saint Brevin-lesPins, 44 F 5
Saint Camille,91 K 9
Saint Catharines,90 M 3
Saint Catherines;isl., 104 K 7
Saint Charles,98 H 7
Saint Christopher;isl., 109 C 11
Saint Clair,99 I 15
Saint Clair Shores, 100 G 3
Saint Clair;river,99 I 16
Saint Clairsville,100 K 4
Saint Claude,89 L 13
Saint Clears,50 F 2
Saint Cloud,96 C 12
Saint Croix Falls,98 F 6
Saint Croix;isl.,109 C 9
Saint Croix;river, 101 C 16
Saint David,108 K 3
Saint Devereux,50 F 6
Saint Elmo,102 D 7
Saint Eloy-les-Mines, 45 G 9
Saint Emilien,91 I 9
Saint Eustatius;isl., 109 C 11
Saint Felicien,90 I 8
Saint Flavien,91 K 9
Saint Florent,45 L 15
Saint Francis,96 G 9
Saint Francis;river, 103 I 5
Saint Francisville, 103 O 4
Saint Francois,109 M 13
Saint Gabriel,90 K 8
Saint Gedeon-de-Beauce, 91 K 9
Saint George,91 L 11
Saint George's,91 I 14
Saint Georges,91 K 9
Saint Germans,50 L 2
Saint Giles Islands;isl., 109 E 16
Saint Gilles-sur-Vie, 44 F 5
Saint Gowans Head,50 G 1
Saint Gregor,89 J 10
Saint Helen,99 G 14
Saint Helen's,51 K 10
Saint Helena;isl.,73 L 3
Saint Helens,47 K 8
Saint Helens,49 M 13
Saint Helier,47 P 9
Saint Hyacinthe,90 K 8
Saint Ignatius,113 H 14
Saint Ives,51 D 12
Saint James,98 H 4
Saint James City,105 N 10
Saint Jean,90 L 8
Saint Jean Baptiste, 89 L 14
Saint Jean de Dieu, 91 J 10
Saint Jean-Pied de-Port, 44 K 6
Saint Jean-de-Luz,44 K 5
Saint Jerome,90 K 8
Saint Joe,102 H 3
Saint John,91 K 11
Saint John's,109 C 12
Saint John's,91 I 16
Saint John;isl.,109 B 9
Saint John;river,91 K 10
Saint Johns,97 J 6
Saint Johns;river,105 M 6
Saint Johnsbury,101 E 13
Saint Jones;river,95 G 5
Saint Joseph,91 K 9
Saint Joseph;isl.,99 E 14
Saint Joseph;river, 99 K 12
Saint Jovite,90 K 7
Saint Kew Highway,50 L 1
Saint Kilda;isl.,46 F 4
Saint Kitts;isl.,85 K 2
Saint Landry,103 O 3
Saint Laurent,89 K 13
Saint Lawrence,83 F 12
Saint Lawrence;isl., 86 B 1
Saint Lawrence;river, 91 I 10
Saint Leonard,91 J 10
Saint Louis,74 H 2
Saint Louis;river,98 D 6
Saint Luce,109 M 15
Saint Lucia;isl.,109 A 15
Saint Maixent-L'Ecole, 44 G 7
Saint Malachie,91 K 9
Saint Malo,44 D 5
Saint Marc,90 K 8
Saint Maries,94 C 6
Saint Marks,105 M 3
Saint Marks;river,105 M 3
Saint Martin;isl.,99 F 11
Saint Martins,91 K 11
Saint Martinville,103 P 4
Saint Mary's,108 M 2
Saint Mary;isl.,112 H 1
Saint Mary;river,106 C 5
Saint Marys,83 L 11
Saint Marys;river,99 L 13
Saint Matthew;isl.,84 C 2
Saint Matthews,105 I 7
Saint Matthias Group;isl. ,83 A 14
Saint Maurice;river, 90 J 8
Saint Maximin-le Sainte-Baume,45 K 12
Saint Michael,49 K 12
Saint Michel des Saints, 90 K 8
Saint Monance,49 A 12
Saint Nazaire,44 E 5
Saint Neots,51 E 12
Saint Niklaas,42 E 1
Saint Norbert,89 K 14
Saint Omer,45 A 9
Saint Paris,99 M 14
Saint Pascal,91 J 9
Saint Paul,88 I 8
Saint Paul-Cap-de-Joux, 91 H 14
Saint Paul;isl.,91 J 13
Saint Paul;river,75 L 4
Saint Pauls,104 G 9
Saint Peter,98 G 5
Saint Peter Port,47 P 9
Saint Peters,91 K 13
Saint Petersburg,105 P 5
Saint Pierre,109 L 15
Saint Pierre,89 L 14
Saint Pierre,79 P 9
Saint Pierre et Miquelon, 91 J 15
Saint Pierre-le-Moutier, 45 F 10
Saint Pierre;isl.,91 J 15
Saint Quentin,91 J 10
Saint Raymond,91 J 9
Saint Regis Falls, 101 E 10
Saint Regis;river, 101 E 10
Saint Shotts,91 J 16
Saint Stephen,91 K 11
Saint Thomas,90 M 4
Saint Thomas;isl.,109 B 9
Saint Tite,90 K 8
Saint Truiden,42 F 2
Saint Tudwal's Isles;isl. ,50 B 2
Saint Ulric,91 I 10
Saint Valery-en-Caux, 44 B 7
Saint Vincent,65 A 11
Saint Vincent,98 A 6
Saint Vincent;gulf,80 L 4
Saint Vincent;isl., 85 L 13
Saint Vincent;isl., 105 N 2
Saint Walburg,89 I 9
Saint Yvon,91 I 11
Saint-Bruno-de-Montarville,87 L 14
Saint-Denis,79 P 9
Saint-Die,45 D 13
Saint-Julien-en Genevois, 45 G 12
Saint-Lo,44 C 6
Saint-Michaels,101 M 9
Sainte Anne,89 K 14
Sainte Anne de Beaupre, 91 J 9
Sainte Anne des Monts, 91 I 1
Sainte Germaine,91 K 9
Sainte Marie,91 K 9
Sainte Rose,109 L 11
Sainte Rose du Degelis, 91 J 10
Sainte Rose du Lac, 89 K 13
Sainte-Agathe des Monts, 90 K 7
Sainte-Marie;isl.,79 M 13
Saintfield,48 H 6
Saipan;isl.,80 C 5
Saipuru,115 M 13
Sairang,69 J 16
Saishihu,60 F 6
Saito,63 N 2
Saitula,60 F 2
Saivoumuotka,52 E 8
Sajama,115 L 9
Sajang;isl.,65 I 15
Sajnsand,61 D 10
Sajo;river,43 J 15
Sakal,75 I 2
Sakami;river,90 F 7
Sakar;isl.,83 B 14
Sakaraha,79 O 11
Sakata,62 E 12
Sakatonchee;river,103 K 7
Sakha;oasis,70 J 4
Sakhalin;isl.,58 E 13
Sakishima;archp.,80 A 2
Sakol Gopal,69 M 12
Sakrand,68 I 2
Sal Rei,75 P 2
Sala,54 H 10
Saladas,116 G 8
Saladillo,116 K 4
Saladillo;river,116 H 5
Salado El-,107 G 9
Salado;river,116 G 1
Salaga,75 K 8
Salaguolehe;river,60 F 7
Salal,76 H 2
Salala,75 L 4
Salalah,70 M 7
Salama,108 I 3
Salamanca,100 H 6
Salamat;river,76 J 3
Salamaua,83 C 13
Salamin,112 F 4
Salamun,77 B 13
Salanga;isl.,112 K 1
Salangen,52 D 4
Salaqi,61 F 10
Salaqui;river,112 E 3
Salar de Pocitos,116 E 3
Salas,106 D 6
Salaverry,114 E 3
Salavina,116 H 6
Salawati;isl.,65 J 15
Salawik,86 B 2
Salay Gomez;isl.,119 P 3
Salazar,78 F 2
Salcedo,109 G 11
Salces,45 L 9
Salcoats,47 I 7
Salcombe,50 M 3
Saldana;river,112 H 4
Saldungaray,118 D 8
Sale,49 M 14
Sale City,105 L 3
Sale Creek,104 F 2
Salebabu;isl.,65 G 13
Salem,94 E 3
Salem;river,95 F 5
Salemi,39 M 7
Salen,46 G 6
Salerno;prov,39 J 9
Salford,49 M 14
Sali;river,116 G 4
Saliaca;isl.,106 F 4
Salibea,108 K 3
Salida,96 H 6
Salihli,41 M 13
Salima,78 H 10
Salina,96 H 3
Salina Cruz,107 L 12
Salina;isl.,39 L 9
Salinas,112 L 1
Salinas Victoria,107 F 9
Salinas de Garcia Mendoza,115 M 10
Salinas del Penon Blanco, 106 H 8Saline,100 H 2
Saline;river,102 F 8
Salinitas,116 F 2
Salisbury,50 J 8
Salisbury;isl.,87 G 11
Salitral,114 D 2
Salix,98 J 3
Salkehatchie;river, 105 I 7
Salla,52 F 11
Salley,105 I 7
Salliquelo,116 M 6
Sallyana,69 H 10
Salmo,88 L 5
Salmon Arm,88 K 5
Salmon Gums,82 I 4Salo
Salmon;river,94 D 7
Saloinen,53 I 8
Salol,98 A 4
Salomon;isl.,80 F 6
Salomon;sea,80 F 6
Salonga;river,78 C 4
Salonta,40 B 6
Saloum;river,75 I 2
Salsacate,116 J 5
Salso;river,39 N 8
Salsuk,69 K 16
Salt Lake City,96 F 3
Salt;river,97 K 4
Salta,116 E 4
Salta;prov,116 E 5
Saltash,50 L 2
Saltcoats,48 D 8
Saltdal,52 F 4
Saltdalselva;river,52 F 4
Saltillo,103 I 8
Salto,116 L 7
Salto de Agua,107 K 14
Salto;prov,117 J 9
Saltoluokta,52 F 5
Saltsburg,100 J 6
Saltville,104 E 6
Saluda,104 D 12
Saluda;river,104 H 6
Salumbar,68 J 5
Salur,67 L 6
Salvador;isl.,85 K 10
Salvisa,104 C 3
Salvus,88 H 1
Salyersville,104 C 5
Salzach;river,42 J 8
Salzburg,42 K 8
Samabor,38 C 10
Samadun,77 E 11
Samahil,107 I 15
Samaipata,115 L 12
Samalayuca,106 B 6
Samana,109 H 12
Samangan,71 F 13
Samani,62 E 4
Samaniego,112 I 3
Samannud,77 B 12
Samar;isl.,65 D 13
Samarai,83 D 15
Samaro,68 I 2
Samba,78 E 6
Sambas,64 I 6
Sambas;river,64 H 6
Sambava,79 L 13
Sambhal,68 H 8
Sambhar,68 I 6
Sambu;river,112 E 3
Same,78 D 11
Samiria;river,114 C 5
Samo Alto,116 I 2
Samoa;isl.,81 G 11
Samokov,40 H 8
Samos;isl.,41 N 11
Samothraki;isl.,41 J 10
Sampa,75 K 7
Sampacho,116 K 5
Sampaga,65 J 10
Sampit,64 J 7
Sampit;river,64 J 8
Samso;isl.,55 N 5
Samson,105 L 1
Samthar,68 I 8
Samu,65 J 9
Samundri,68 F 5
San,75 J 6
San Agustin,112 I 4
San Agustin Chayuco, 107 L 11
San Agustin Loxicha, 107 L 11
San Agustin de Valle Fertil,116 I 3
San Agustine,103 N 1
San Alejo,107 D 9
San Ambrosio;isl.,119 O 1
San Andrea,39 L 11
San Andres,106 D 6
San Andres Tuxtla, 107 K 12
San Andres;isl.,112 A 1
San Andresy Providencia, 112 A 1
San Angel,112 C 5
San Antioco;isl.,39 L 2
San Antonio,75 N 11
San Antonio Este,118 F 7
San Antonio Oeste,118 E 7
San Antonio de Acucaray, 114 D 6
San Antonio de Caparo, 112 E 7
San Antonio de Esquilache,114 K 8
San Antonio de Lipez, 114 M 2
San Antonio de Maturin, 113 C 12
San Antonio de los Banos, 108 F 7
San Antonio de los Cobres,116 E 4
San Antonio del Golfo, 113 C 12
San Antonio del Parapeti, 115 M 13
San Augustine,97 L 13
San Bartolo,106 G 4
San Basilio,116 K 5
San Benito,106 F 5
San Bento,114 E 3
San Bernardo,106 E 4
San Bernardo;isl.,112 C 4
San Blas,106 E 8
San Blas Atempa,107 L 12
San Blas;gulf,112 D 2
San Blas;range,112 D 2
San Borja,115 J 10
San Buenaventura,115 J 10
San Camilo,116 E 7
San Carlos,65 D 12
San Carlos,115 L 12
San Carlos,115 I 11
San Carlos Centro,116 J 7
San Carlos Yautepec, 107 L 12
San Carlos de Bariloche, 118 F 3
San Carlos de Guaroa, 112 H 6
San Carlos de la Union, 108 J 5
San Carlos;river,113 D 9
San Cataldo,39 N 8
San Cayetano,118 D 9
San Celoni,117 H 14
San Cosme,116 G 8
San Criatobal,112 L 6
San Cristobal,114 L 2
San Cristobal,109 H 12
San Cristobal de Las Casas,107 L 14
San Cristobal;isl.,80 G 7
San Diego,114 L 4
San Diego de Cabrutica, 113 D 11
San Dimas,106 G 6
San Elizario,97 L 6
San Emeterio,106 B 2
San Enrique,116 M 7
San Estanislao,117 E 9
San Esteban,108 I 5
San Esteban;isl.,106 D 2
San Felipe,106 B 1
San Felix,116 H 2
San Felix;isl.,119 O 1
San Fernando,65 B 11
San Fernando de Apure, 113 E 9
San Fernando de Atabapo, 113 G 9
San Fernando de la Cal; river,115 L 15
San Fernando;river, 107 F 10
San Francique,108 M 1
San Francisco,112 I 4
San Francisco Gotera, 108 J 4
San Francisco de Belloc, 118 D 9
San Francisco de Borja, 106 D 5
San Francisco de Conchos, 106 E 6
San Francisco de Macoris, 109 G 12
San Francisco del Chanar, 116 I 5
San Francisco del Mezquital,106 G 7
San Francisco del Monte - de Oro,116 K 4
San Francisco del Oro, 106 E 6
San Francisco del Pampeti,114 K 4
San Francisco del Rincon, 106 I 8
San Francisco;river, 97 K 5
San Gabriel,112 J 3
San Gabriel Chilac, 107 K 11
San German,118 D 8
San Gil,112 F 6
San Gregorio,117 J 10
San Guillermo,116 I 6
San Gustavo,116 I 8
San Hilario,116 F 8
San Ignacio,106 E 2
San Ignacio;isl.,106 F 4
San Isidro,116 H 4
San Javier,116 M 2
San Javier,116 J 8
San Javier,115 K 13
San Javier;river,116 I 8
San Jeronimo,114 E 6
San Jeronimo Taviche, 107 L 11
San Joaquin,113 D 12
San Joaquin;river, 115 I 13
San Jon,97 J 8
San Jorge,112 J 3
San Jorge;gulf,119 I 5
San Jorge;river,112 D 5
San Josade Sisa,114 D 4
San Jose,65 B 11
San Jose,114 J 8
San Jose de Buenavista, 65 D 11
San Jose de Feliciano, 116 I 8
San Jose de Guaribe, 113 C 10
San Jose de Guaviare, 112 H 6
San Jose de la Noria, 107 G 10
San Jose de los Arce, 106 E 2
San Jose del Boqueron, 116 F 5
San Jose;gulf,118 F 7
San Jose;isl.,106 F 3
San Jose;prov,117 L 9
San Juan,65 E 13
San Juan Bautista, 113 C 12
San Juan Nepomuceno, 112 C 5
San Juan Sabinas,106 D 8
San Juan de Camarones, 106 F 6
San Juan de Galdonas, 113 C 12
San Juan de Guadalupe, 106 G 8
San Juan de Micay;river, 112 H 3
San Juan de Salvamento, 119 N 7
San Juan de la Costa, 118 F 2
San Juan de los Morros, 113 C 10
San Juan del Rio,106 G 7
San Juan del Sur,108 K 5
San Juan;prov,116 J 2
San Juan;river,97 I 4
San Juandelos Cayos, 113 B 9
San Juanico;isl.,106 I 5
San Julian,119 K 5
San Julio,106 C 1
San Justo,116 I 7
San Lope,112 F 7
San Lorenzo,114 L 3
San Lucas,106 H 3
San Luis,97 K 1
San Luis Acatlan,107 L 10
San Luis Potosi,107 H 9
San Luis Potosi;state, 106 H 8
San Luis Rio Colorado, 106 A 1
San Luis Valley,96 H 7
San Luis de la Loma, 107 L 9

San Luis;prov, 116 L 4
San Manuel, 97 L 4
San Marcial, 97 K 6
San Marcos, 114 E 3
San Marcos, 107 L 10
San Marcos, 113 H 14
San Marcos;isl., 106 E 2
San Martin, 112 H 6
San Martin Norte, 116 I 7
San Martin;isl., 106 B 1
San Martin;river, 115 I 13
San Mateo, 113 C 11
San Mateo del Mar, 107 L 12
San Matias, 115 K 16
San Matias;isl., 80 E 5
San Matias;gulf, 118 F 7
San Miguel, 106 D 1
San Miguel Chimalapa, 107 L 12
San Miguel Zapotitlan, 106 F 4
San Miguel de Allende, 107 I 9
San Miguel de Huachi, 115 J 10
San Miguel de Orcasitas, 106 C 3
San Miguel de Salcedo, 112 K 3
San Miguel de Tucuman, 116 G 4
San Miguel del Monte, 116 L 8
San Miguel el Grande, 107 L 11
San Miguel;gulf, 112 D 3
San Miguel;river, 106 E 5
San Miguelito, 106 B 4
San Narciso, 65 B 10
San Nicolas, 107 G 10
San Nicolas de Arriba, 106 F 6
San Nicolas de los Armyos, 116 K 7
San Nicolas de los Garza, 107 F 9
San Nicolas;river, 106 J 6
San Onofre, 112 C 4
San Pablo, 65 C 11
San Pablo Balleza, 106 E 6
San Pablo Huixtepec, 107 L 11
San Pablo de Loreto, 114 C 8
San Pascual, 65 C 12
San Patricio, 117 G 9
San Pedro, 106 C 6
San Pedro Martir, 106 B 1
San Pedro Martir;isl., 106 D 2
San Pedro Mixtepec, 107 M 11
San Pedro Nolasco;isl., 106 D 3
San Pedro Pochutla, 107 M 11
San Pedro Sula, 108 I 5
San Pedro de Arimena, 112 G 7
San Pedro de Arriba, 106 C 5
San Pedro de Atacama, 116 D 2
San Pedro de Buena Vista, 115 L 11
San Pedro de Chonta, 114 F 4
San Pedro de Lloc, 114 E 3
San Pedro de Macons, 109 H 12
San Pedro de las Colinas, 106 F 7
San Pedro del Parana, 117 F 10
San Pedro;prov, 117 D 9
San Pedro;river, 97 K 4
San Pietro;isl., 39 J 11
San Quintin, 106 B 1
San Rafael, 106 D 3
San Rafael;isl., 119 L 8
San Ramon, 114 G 5
San Ramon de la Nueva Oran, 116 D 5
San Ramon;river, 115 I 14
San Ricardo, 116 K 7
San Roque, 116 H 8
San Rosendo, 118 D 2
San Saba;river, 97 M 10
San Salvador, 108 J 4
San Salvador de Jujuy, 116 E 4
San Salvador;isl., 109 E 10
San Salvador;river, 116 K 8
San Sebastian, 106 I 6
San Simon;river, 115 I 13
San Sostenes, 106 C 6
San Souci, 108 K 3
San Telmo, 106 B 1
San Tiburcio, 106 G 8
San Timoteo, 112 C 7
San Vicente Tancuayalab, 107 I 10
San Vicente de Caguan, 112 I 5
San Vicente de Canete, 114 I 5
San Vicente de Castillos, 117 L 11
San el Hagar, 77 B 14
San'a, 70 M 4
San;river, 43 G 16
Sanafa, 77 C 12
Sanaga;river, 75 M 12
Sanana, 65 J 13
Sanandita, 114 L 4
Sanare, 112 C 8
Sanariapo, 113 F 9
Sanau;oasis, 70 L 6
Sanborn, 98 H 4
Sanchez, 112 I 6
Sancho Corral, 116 G 6
Sancti Spiritu, 116 K 6
Sand, 54 B 2
Sand;isl., 98 D 8
Sand;river, 102 D 11
Sanda;isl., 48 E 6
Sandai, 64 J 7
Sandakan, 65 G 10
Sandane, 54 F 3
Sanday;isl., 46 D 8
Sandbach, 50 A 7
Sandefjand, 55 J 5
Sanders, 97 J 5
Sanderson, 97 M 9
Sandersville, 105 I 5
Sandgate, 51 J 15
Sandhead, 48 G 8
Sandia, 115 I 9
Sandila, 69 I 9
Sanding;isl., 64 K 2
Sandnes, 55 I 2
Sandnessjoen, 52 G 2
Sando;isl., 54 B 2
Sandoval, 102 E 7
Sandover;river, 82 E 8
Sandoway, 67 K 11
Sandown, 51 K 10
Sandpoint, 96 A 2
Sandspit, 88 H 1
Sandstone, 82 H 3
Sandusky, 100 F 3
Sandusky;river, 100 I 2
Sandvik, 55 L 10
Sandvika, 54 D 7
Sandviken, 53 N 3
Sandvip, 69 K 15
Sandwich, 51 I 16
Sandy, 51 E 12
Sandy Hook, 101 I 12
Sandy Lake, 89 I 15
Sandy Point, 109 C 11
Sandy Pun, 95 A 4
Sandy Ridge, 104 E 8
Sandy;river, 91 E 10
Sandymount, 48 J 4
Sanford, 101 F 14
Sanga;river, 73 I 7
Sangaha, 77 C 13
Sangallan;isl., 114 I 5
Sangamon;river, 102 C 7
Sangan, 68 F 2
Sanganer, 68 I 6
Sangar, 57 F 12
Sangeang;isl., 65 M 10
Sangerfield, 101 G 10
Sanggau, 64 I 7
Sangha;river, 78 C 3
Sanghar, 68 I 2
Sangihe;isl., 65 H 13
Sangli, 67 L 3
Sangolqui, 112 J 3
Sangre de Cristo Range, 96 H 7
Sangrut, 68 F 6
Sangu;river, 69 L 16
Sangudo, 88 I 7
Sangue;river, 115 H 16
Sanhur el Medina, 77 B 10
Sanibel;isl., 105 N 9
Sanico, 118 E 4
Sanicolaul Mare, 40 C 5
Sanjo, 62 G 11
Sankisen, 62 E 5
Sankt Johann, 42 K 8
Sankt Veit, 43 L 9
Sankuru;river, 78 D 5
Sanni, 68 G 2
Sanniquellie, 75 L 3
Sannohe, 62 D 13
Sanogasta, 116 I 3
Sanok, 43 H 16
Sanquhar, 49 E 10
Sansanding, 75 I 6
Sansane Haoussa, 75 I 9
Sansanne Mango, 75 K 8
Sansapor, 65 A 13
Sant Antioco, 39 L 2
Santa, 114 F 3
Santa Agueda, 106 E 2
Santa Ana, 106 B 3
Santa Anita, 106 H 4
Santa Barbara, 113 C 12
Santa Barbara, 106 E 6
Santa Barbara do Sul, 117 H 11
Santa Barbara;river, 115 K 14
Santa Barbera, 112 E 7
Santa Catalina, 112 C 5
Santa Catarina, 106 C 1
Santa Catarina Juquila, 107 L 11
Santa Catarina;isl., 117 G 14
Santa Clara, 115 I 11
Santa Clara, 112 L 8
Santa Clara, 106 C 5
Santa Clara de Saguier, 116 J 6
Santa Clara;isl., 112 L 1
Santa Claus, 102 E 9
Santa Clotilde, 114 B 6
Santa Cruz, 65 B 10
Santa Cruz Islands;isl., 83 E 13
Santa Cruz Zenzontepec, 107 L 11
Santa Cruz de Juventino Rosas, 107 I 9
Santa Cruz de Tenerife, 37 M 11
Santa Cruz de la Sierra, 115 L 13
Santa Cruz del Sur, 109 G 9
Santa Cruz do Sul, 117 I 12
Santa Cruz;isl., 85 L 12
Santa Cruz;prov, 119 J 4
Santa Dorotea, 112 E 3
Santa Elena, 106 D 7
Santa Elena;river, 115 K 10
Santa Eufemia, 39 L 11
Santa Fe, 65 D 11
Santa Fe;isl., 112 G 2
Santa Fe;prov, 116 I 6
Santa Fe;river, 105 M 5
Santa Gertrudis, 106 D 6
Santa Ines;isl., 119 N 3
Santa Isabel, 106 D 2
Santa Isabel de Sihuas, 114 K 7
Santa Juana, 113 E 10
Santa Julia, 112 K 6
Santa Justina, 116 G 6
Santa Lucia, 112 K 2
Santa Lucia;isl., 85 L 13
Santa Lucia;river, 116 H 8
Santa Magdalena, 116 L 5
Santa Margarita, 116 H 6
Santa Maria, 117 F 9
Santa Maria, 106 C 1
Santa Maria Asuncion Tlaxiaco, 107 L 11
Santa Maria Petapa, 107 L 12
Santa Maria de Ipire, 113 D 11
Santa Maria de Nanay, 114 B 6
Santa Maria de Otaez, 106 G 6
Santa Maria del Oro, 106 I 6
Santa Marla, 116 G 4
Santa Marla Zacatepec, 107 L 10
Santa Marla de Erebato, 113 G 11
Santa Marta, 112 B 5
Santa Rita, 97 L 5
Santa Rita de Catuna, 116 J 4
Santa Rita do Weil, 115 B 9
Santa Rosa, 116 K 3
Santa Rosa, 97 J 7
Santa Rosa de Rio Primero, 116 J 6
Santa Rosa de Sucumbio, 112 J 3
Santa Rosa de Viterbo, 112 F 6
Santa Rosa de la Mina, 115 K 13
Santa Rosa de la Roca, 115 K 14
Santa Rosa de los Pastos-Grandes, 116 E 4
Santa Rosa del Palmar, 115 K 13
Santa Rosa;isl., 103 O 9
Santa Rosalia, 106 E 2
Santa Rosalla, 112 D 8
Santa Sylvina, 116 G 7
Santa Teresa, 113 G 13
Santa Teresa de la Una, 106 F 7
Santa Victoria, 116 D 6
Santa Vitoria do Palmar, 117 K 11
Santa;isl., 114 F 3
Santa;river, 114 F 3
SantaRoealla, 113 E 10
Santana, 113 J 10
Santana do Livramento, 117 I 10
Santander, 107 G 10
Santaquin, 96 G 3
Santarskije Ostrova;isl., 57 G 14
Santee;river, 85 J 9
Santiago, 106 G 4
Santiago Ixcuintla, 106 I 6
Santiago Jamiltepec, 107 L 11
Santiago Lachiguiri, 107 L 12
Santiago Papasquiaro, 106 F 6
Santiago Pinotepa Nacional, 107 L 10
Santiago Rodriguez, 109 G 11
Santiago Tlazoyaltepec, 107 L 11
Santiago de Cao, 114 E 3
Santiago de Chocorvos, 114 I 5
Santiago de Cuba, 109 G 9
Santiago de Huata, 115 K 9
Santiago de Machaca, 115 K 9
Santiago de Pacaguaras, 115 I 10
Santiago de Veraguas, 112 E 1
Santiago de los Caballeros, 109 G 11
Santiago del Estero;prov, 116 G 5
Santiago or San Salvador;isl., 112 G 1
Santiago;prov, 116 L 2
Santiago;river, 106 F 6
Santigi, 65 I 11
Santo Antao;isl., 75 O 1
Santo Antonio do Ica, 115 B 10
Santo Antonio do Zaire, 78 E 1
Santo Antonio;river, 113 K 14
Santo Corazon, 115 L 15
Santo Corazon;river, 115 L 15
Santo Domingo, 106 B 1
Santo Domingo Tehuantepec, 107 L 12
Santo Domingo Zanatepec, 107 L 13
Santo Domingo de los Colorados, 112 J 2
Santo Domingo;river, 107 K 11
Santo Tomas, 106 A 1
Santo Tomas-;river, 114 J 7
Santo Tome, 116 J 7
Santo Tome;isl., 32 H 7
Santos Reyes Nopala, 107 L 11
Sanza Pombo, 78 F 2
Sao Angelo, 117 H 11
Sao Antonio, 117 F 11
Sao Bento do Sul, 117 F 13
Sao Borja, 117 H 10
Sao Domingos, 75 J 2
Sao Francisco de Asis, 117 H 10
Sao Francisco de Paula, 117 I 13
Sao Francisco do Sul, 117 F 14
Sao Francisco;isl., 117 G 14
Sao Francisco;river, 117 E 11
Sao Gabriel, 117 I 11
Sao Hill, 78 F 10
Sao Jeronimo, 117 I 12
Sao Joaquim, 117 H 13
Sao Jose, 113 J 10
Sao Jose de Anava, 113 I 13
Sao Jose do Norte, 117 J 12
Sao Jose dos Pinhais, 117 F 13
Sao Leopoldo, 117 I 12
Sao Luis Gonzaga, 117 H 10
Sao Marcelino, 113 I 10
Sao Mateus do Sul, 117 F 13
Sao Miguel d'Oeste, 117 F 11
Sao Miguel;river, 115 H 13
Sao Nicolau;isl., 75 P 1
Sao Paulo de Olivenca, 115 B 9
Sao Paulo;isl., 110 F 12
Sao Pedro, 113 J 10
Sao Pedro do Sul, 117 I 11
Sao Salvadordo Congo, 78 E 1
Sao Sebastiao do Cai, 117 I 12
Sao Simao;river, 115 H 13
Sao Tiago;isl., 75 P 1
Sao Tome, 75 O 11
Sao Tome & Principe, 75 O 11
Sao Vicente;isl., 75 O 1
Sao;river, 111 I 9
Saona;isl., 109 H 12
Saone;river, 45 G 11
Saonek, 65 I 15
Saoner, 68 L 8
Saovinandriana, 79 N 12
Sapahaqui, 115 K 10
Sapai, 41 J 10
Sapallanga, 114 H 5
Saparua, 65 K 14
Saparua;isl., 65 K 14
Sapelo;isl., 104 K 7
Sapientza;isl., 41 O 6
Saposoa, 114 E 4
Sappho, 94 B 2
Sapporo, 62 D 3
Sapri, 39 J 10
Sapulpa, 97 I 12
Sapzurro, 112 D 3
Saquena, 114 C 6
Saquisili, 112 K 2
Sara, 75 J 6
Sara;river, 76 K 3
Saraguro, 112 L 2
Sarajevo, 40 F 2
Saranac, 99 I 13
Saranac;river, 101 E 11
Saranda, 41 K 4
Sarandi, 117 G 11
Sarandi Grande, 117 K 9
Sarandi del Yi, 117 K 10
Sarangani;isl., 65 G 12
Sarasota, 105 M 9
Saratoga, 96 F 6
Saratoga Springs, 101 G 11
Sarawak;state, 64 H 7
Saray, 41 I 12
Sarayacu, 112 K 3
Sarbaz, 71 I 10
Sarbisheh, 71 G 10
Sarbogard, 43 L 13
Sarco, 116 H 1
Sarda;river, 69 H 9
Sardalas, 74 E 12
Sardargarh, 68 G 5
Sardarshahr, 68 H 5
Sardegna;isl., 39 J 4
Sardhana, 68 G 7
Sardinata, 112 E 6
Sardis, 103 I 8
Sarepta, 103 L 2
Sargazos;sea, 85 J 11
Sargent, 96 F 10
Sargento Juan Bautista Cabral, 116 G 8
Sargento Lores, 114 B 6
Sargento Puno, 114 B 4
Sargoda, 68 E 5
Sarhari-Wakhan, 66 A 9
Saria;isl., 41 P 12
Saric, 106 B 3
Sarikei, 64 H 7
Sarila, 68 I 8
Sarina, 83 E 11
Sarine;river, 42 L 3
Sariyer, 41 I 13
Sark;isl., 47 P 9
Sarkad, 43 L 15
Sarkisla, 70 C 4
Sarlat, 107 K 14
Sarles, 98 A 1
Sarmi, 65 B 15
Sarmiento, 119 I 5
Sarn, 50 D 5
Sarna, 54 F 7
Sarnath, 69 J 10
Sarnau, 50 E 2
Sarnesfield, 50 E 5
Sarnia, 99 I 16
Sarobi, 71 F 13
Sarolang, 64 J 3
Saronikos Kolpos;gulf, 41 N 7
Saros korfez;gulf, 41 J 10
Sarospatak, 43 J 15
Sarpsborg, 55 I 6
Sarre, 51 H 15
Sarrebourg, 45 D 13
Sarreguemines, 45 C 13
Sartene, 45 M 15
Sarth, 76 J 3
Sarufutsu, 62 A 4
Sarvar, 43 K 11
Sarysu;river, 56 I 5
Sasa, 63 M 1
Sasabaneh, 77 K 12
Sasabe, 106 B 3
Sasd, 43 M 13
Sasebo, 63 M 1
Saskatchewan;prov, 86 J 6
Saskatchewan;river, 89 I 10
Saskatoon, 86 K 7
Sassafras;river, 95 G 4Sassandra, 75 M 5
Sassandra;river, 75 M 5Sassari
Sassari;prov, 39 I 3
Sasser, 105 K 3
Sastre, 116 J 6
Satadougou, 75 J 3
Satevo, 106 D 6
Satilla;river, 105 L 6
Satipo, 114 G 6
Satipo;river, 114 G 6
Sato, 63 L 12
Satoraljaujhely, 43 I 15
Satsuma, 103 O 8
Satsuma-Hanto;pen., 63 O 1
Satti, 68 D 7
Satu-Mare, 40 A 7
Satui, 65 K 9
Sauce, 114 D 5
Sauce Chico;river, 118 D 8
Sauce Grande;river, 118 D 8
Sauce de Luna, 116 J 8
Saucharkrokur, 52 A 3
Saucier, 103 O 7
Saucillo, 106 D 6
Sauda, 55 I 2
Saudi Arabia, 70 J 6
Saugatuck, 99 J 12
Saugerties, 101 H 11
Saujil, 116 H 4
Sauk Centre, 98 E 4
Sauk City, 99 I 9
Sauk Rapids, 98 F 5
Saukville, 99 I 10
Sauldre;river, 44 F 8
Saulieu, 45 F 10
Sault Sainte Marie, 99 D 13
Sault au Mouton, 91 I 10
Saumlaki, 65 M 15
Saumur, 44 F 7
Saurba, 52 A 1
Saurber, 62 A 3
Sautar, 78 G 3
Sautata, 112 E 3
Sauteur, 109 D 15
Sauzal, 116 M 1
Sava;river, 38 C 9
Savage, 103 J 6
Savalou, 75 L 9
Savan;isl., 109 C 15
Savanna, 98 K 8
Savanna-la-Mar, 108 H 8
Savannah, 98 M 4
Savannah Beach, 105 K 7
Savannah;river, 105 J 7
Savant Lake, 90 G 1
Save, 75 L 9
Save;river, 79 K 9
Savenay, 44 E 5
Saverne, 45 C 13
Savitaipale, 53 M 11
Savolina, 53 L 12
Savona, 88 K 4
Savona;prov, 38 E 2
Savonet, 112 B 8
Savoonga, 86 B 1
Savoy, 103 M 7
Savsjo, 55 L 8
Savu;isl., 80 H 1
Savu;sea, 80 H 1
Savukoski, 52 F 10
Sawahlunto, 64 I 2
Sawai, 65 K 14
Sawai Madhopur, 68 I 6
Sawbill, 91 F 11
Sawley, 49 J 15
Sawmill Bay, 88 A 6
Sawston, 51 E 13
Sawu;isl., 65 M 11
Sawyer, 98 D 6
Saxapahaw, 104 F 9
Saxilby, 51 A 11
Saxmundham, 51 E 16
Saxon, 98 E 8
Saxthorpe, 51 B 15
Say, 75 J 9
Sayabec, 91 I 10
Sayan, 114 G 4
Saylorsburg, 95 A 6
Sayola, 99 E 10
Sayre, 100 H 8
Sayreville, 95 C 8
Sayula, 106 C 3
Sayville, 95 B 11
Sazin, 66 C 8
Scammon, 102 F 1
Scammon Bay, 86 C 1
Scandia, 98 M 1
Scandinavia, 99 G 9
Scarba;isl., 48 A 5
Scarborough, 47 J 10
Scardale, 95 A 10
Scarlets Mill, 95 D 4
Scarva, 48 I 5
Scaur;river, 49 E 10
Scawton, 49 I 16
Sceale Bay, 82 I 7
Schaefferstown, 95 C 3
Scharding, 42 J 8
Schefferville, 91 E 10
Schell City, 102 F 2
Schenectady, 101 G 11
Schichinohe, 62 C 13
Schiltigheim, 45 D 13
Schleswig, 98 J 3
Schleswig-Holstein, 42 B 5
Schlitsselfd, 42 H 6
Schnecksville, 95 B 5
Schofield, 99 G 9
Scholcher, 109 M 15
Schore, 68 K 7
Schouten Islands, 83 A 13
Schouten;isl., 83 L 11
Schreiber, 99 A 11
Schroeder, 98 C 8
Schroon Lake, 101 F 11
Schruns, 42 L 6
Schuler, 88 L 8
Schumacher, 99 A 16
Schumansburg, 79 I 6
Schuyler, 98 K 2
Schuylerville, 101 G 11
Schuylkill Haven, 95 B 3
Schuylkill;river, 95 C 4
Schwarzelster;river, 42 E 8
Schwiebus, 43 D 10
Sciacca, 39 N 7
Scicli, 39 M 9
Scinawa, 43 F 11
Sciota, 95 A 6
Scioto;river, 100 L 2
Scobey, 96 A 7
Scole, 51 D 15
Scooba, 103 L 8
Scorff;river, 44 E 4
Scotch Corner, 49 H 15
Scotia Bay, 88 D 1
Scotland, 98 I 1
Scotland Neck, 104 E 11
Scotlandville, 103 O 4
Scotstown, 91 K 9
Scotsville, 104 C 10
Scott, 103 K 5
Scott City, 96 H 9
Scottdale, 100 K 5
Scottsbluff, 96 F 8
Scottsboro, 104 G 1
Scottsburg, 102 E 10
Scottsdale, 83 L 10
Scottsville, 104 E 1
Scottville, 99 H 12
Scranton, 99 K 4
Scraton, 103 I 2
Screven, 105 K 6
Scribner, 98 K 2
Scrub;isl., 109 B 11
Scunthorpe, 47 K 10
Scuscuiban, 77 J 15
Sea Isle City, 95 G 7
Sea Mill, 48 D 8
Seaboard, 104 E 11
Seaford, 51 K 13
Seaforth, 99 H 16
Seagraves, 97 L 8
Seagrove, 104 F 8
Seagrove Beach, 105 M 1
Seaham, 49 G 16
Seahouses, 49 D 15
Seal;river, 84 F 7
Seale, 105 J 2
Sealevel, 104 G 12
Seantonnach, 48 J 3
Seantrabh, 48 M 5
Searchmont, 99 D 14
Searcy, 103 I 4
Searsport, 101 E 15
Seascale, 49 I 11
Seaside, 94 D 3
Seaside Heights, 95 E 9
Seaside Park, 95 E 9
Seat Pleasant, 95 H 1
Seatoller, 49 H 12
Seaton, 50 K 5
Seaton Burn, 49 F 15
Seaton Delaval, 49 F 15
Seattle, 94 C 3
Seba, 65 M 11
Seba Beach, 88 I 7
Sebago Lake, 101 F 14
Sebakung, 65 J 9
Sebaldes;isl., 119 L 8
Sebanga, 64 I 3
Sebangka;isl., 64 I 4
Sebastian, 105 P 8
Sebba, 75 J 8
Sebderat, 77 H 10
Sebeka, 98 D 4
Seberi, 117 G 11
Sebes, 40 D 8
Sebewaing, 99 H 15
Seboeis;river, 101 B 15
Sebring, 105 P 6
Sebuhu, 65 G 9
Sebuku;isl., 65 K 9
Sebuku;river, 65 G 9
Sechelt, 88 L 3
Sechin;river, 114 F 3
Sechura, 114 D 2
Sechura;desert, 114 D 2
Seco;river, 116 D 5
Secovce, 43 I 15
Section, 104 G 1
Secure;river, 115 J 11
Security, 96 H 7
Sedalia, 88 K 8
Sedan, 97 I 12
Sedbergh, 49 I 13

Skjalfandafljot;river, 52 3
Skjem, 55 M 4
Skjoldungen, 87 D 15
Skjomen, 52 E 5
Sklingji, 68 J 5
Skogeroya;isl., 52 B 11
Skoki, 43 D 11
Skookumchuck, 88 L 6
Skopje, 41 I 6
Skorcz, 43 B 12
Skovde, 55 J 8
Skovorodino, 57 H 12
Skowhegan, 101 D 15
Skownam, 89 J 12
Skreia, 54 G 6
Skudeneshavn, 55 I 2
Skuna;river, 103 K 7
Skunk;river, 98 L 7
Skuo;isl., 54 B 2
Skye;isl., 46 G 5
Skykomish, 94 C 4
Slagelse, 55 N 6
Slaggyford, 49 G 13
Slagle, 103 N 2
Slaley, 49 G 14
Slana, 86 E 2
Slana;river, 43 I 14
Slate Islands;isl., 99 B 11
Slate Springs, 103 K 7
Slatedale, 95 B 5
Slater, 96 G 13
Slatina, 40 E 8
Slatington, 95 B 5
Slaton, 97 K 9
Slatvik, 52 4
Slaughter, 103 O 4
Slaughter Beach, 95 H 6
Slautnoje, 57 C 15
Slave Lake, 88 H 7
Slave;river, 88 D 8
Slavonska Pozega, 40 D 1
Slavonski Brod, 40 D 2
Slawno, 43 B 11
Slayton, 98 H 3
Sleaford, 51 B 11
Sledge, 103 J 6
Sleepy Eye, 98 G 4
Sleetmute, 86 D 1
Slidell, 103 O 6
Slinger, 99 I 10
Slip End, 51 F 12
Slippery Rock, 100 J 5
Slite, 55 K 11
Sliven, 40 H 10
Sloansville, 101 G 11
Slobozia, 40 E 11
Slocan, 88 L 5
Slocomb, 105 L 2
Slough, 51 H 11
Slovenske Rudohorie; range, 43 I 13
Slunj, 38 D 10
Slupca, 43 D 12
Slupsk, 43 A 11
Smackover, 103 K 3
Smallwood, 95 F 1
Smara, 74 D 4
Smeaton, 89 I 10
Smederevo, 40 E 5
Smederevska Palanka, 40 E 5
Smethport, 100 I 7
Smidta;isl., 57 B 9Smith, 88 H 7
Smith Center, 96 G 10
Smith Hiver, 88 D 3
Smith River, 86 G 4
Smith;isl., 90 A 6
Smith;river, 96 C 4
Smithers, 88 H 2
Smithfield, 49 F 12
Smithhorn, 102 E 7
Smithland, 98 J 3
Smiths Falls, 90 L 7
Smiths Grove, 104 D 1
Smithton, 102 E 3
Smithtown, 95 A 11
Smithville, 100 L 4
Smoke Creek Desert, 94 H 4
Smoky Bay, 82 I 7
Smoky Falls, 87 K 11
Smoky Hill;river, 96 H 10
Smoky Lake, 88 I 7
Smoky;river, 88 I 5
Smola;isl., 54 D 4
Smolensk-Moscu, 35 G 11
Smolyan, 41 I 9
Smooth Rock Falls, 90 H 5
Smyrna, 101 L 9
Smyrna Mills, 101 B 16
Smyrna;river, 95 G 5
Snake;river, 94 C 5
Snare River, 86 G 6
Snare;river, 88 B 7
Sneedville, 104 E 4
Sneeuberge;mts, 79 O 6
Snochomish, 94 C 4
Snonenjoki, 53 L 10
Snow Hill, 104 F 11
Snow Laka, 89 H 12
Snow Lake, 103 K 5
Snowdrift., 88 C 8
Snowflake, 97 J 4
Snowflake, 98 A 1
Snydertown, 95 A 2
Soa;isl., 48 A 5
Soacha, 112 G 5
Soalala, 79 M 11
Soar;river, 51 C 10
Soasiu, 65 I 14
Soata, 112 F 6
Sobat;river, 76 J 8
Sobger;river, 65 C 16
Sobo-Yama, 63 M 2
Sobradinho, 117 H 11
Sobrado;mount, 36 B 4
Sobrance, 43 I 16
Soca, 117 L 10
Socha, 112 F 6
Sochaczew, 43 D 14
Social Circle, 105 I 4
Sociedad;isl., 81 H 14
Society Hill, 104 H 8
Sococha, 114 M 3
Socompa, 116 E 2
Socopo;river, 112 D 7
Socorro, 97 K 6
Socorro;isl., 106 J 3
Socosani, 114 K 7
Socota, 112 F 6
Socotora;isl., 59 L 2
Soda Creek, 88 J 4
Soda Springs, 96 E 4
Sodankyla, 52 F 10
Soddy, 104 F 2
Soderhamn, 54 G 10
Sodermanland, 55 I 9
Sodiri, 76 H 7
Soe, 65 12
Soekmekaar, 79 L 8
Sofara, 75 I 6
Sofia;river, 79 M 12
Sofiya, 79
Sogamoso, 112 F 6
Sohagpur, 68 K 7
Soham, 51 D 13
Sohan;river, 66 F 6
Sohano, 83 B 16
Soini, 53 K 8
Soitue, 116 L 3
Sojat, 68 I 5
Sokode, 75 K 8
Sokolka, 43 B 16
Sokolo, 75 I 6
Sokolov, 42 G 8
Sokolow Podlaski, 43 D 15
Sokoto, 75 J 10
Sokoto;river, 75 J 10
Sol de Julio, 116 H 6
Sola de Vega, 107 L 11
Solano, 113 I 10
Solbad Hall, 42 K 7
Solden, 42 L 6
Soledad, 112 C 6
Soledad Diez Gutierrez, 107 H 9
Soledad de Doblado, 107 J 11
Soledad;isl., 119 L 9
Soledade, 115 D 9
Solihull, 47 M 9
Solleftee, 53 K 4
Solok, 64 J 2
Solola, 108 I 13
Solomon Sea;sea, 83 C 14
Solomon;river, 96 G 11
Solomons, 101 M 9
Solon, 98 K 7
Solon Springs, 98 E 7
Solor;isl., 65 M 12
Solvesborg, 55 M 7
Solway Firth;gulf, 49 G 11
Soma, 41 L 12
Somali;c., 78 B 13
Somalia;pen., 72 G 11
Sombe Dzong, 69 H 13
Sombor, 40 D 3
Sombrerete, 106 G 7
Sombrero;isl., 109 A 11
Somero, 53 N 8
Somers Point, 95 G 7
Somerset, 99 K 14
Somerset;isl., 87 D 7
Somersworth, 101 F 14
Somerville, 101 H 14
Somesul;river, 43 J 16
Somogyszob, 43 M 12
Somogyvar, 43 L 12
Somoto, 108 J 5
Somta, 115 J 9
Sona, 112 E 1
Sonar;river, 68 J 8
Sonaroa or Welle;isl., 83 C 15
Sondar, 66 F 12
Sonderborg, 55 O 5
Sondheimer, 103 M 4
Sondre Stromfjord, 87 D 14
Sondrio;prov, 38 B 4
Sonepat, 68 G 7
Song;no, 60 C 7
Songea, 78 G 10
Songhuajiang;river, 61 B 14
Songo, 78 E 1
Songo Songo;isl., 78 F 12
Soni, 106 B 2
Sonneberg, 42 G 7
Sonora, 106 C 3
Sonoyta, 106 A 2
Sonson, 112 F 5
Sonsonate, 108 J 4
Sonsorol Islands, 65 16
Soo Canals, 99 D 13
Sooke, 88 L 3
Sop's Arm, 91 H 15
Sopchoppy, 105 M 3
Soperton, 105 J 5
Sopley, 50 K 8
Sopor, 66 E 10
Sopot, 43 A 12
Soppero, 52 E 7
Sopur, 68 D 6
Sor, 68 G 1
Soraba, 113 F 13
Sorachi;river, 62 D 4
Sorbie, 49 G 9
Sord, 48 5
Sorel, 90 K 8
Sorell, 83 M 10
Sorfold, 52 E 4
Sorgona, 39 J 3
Soriano, 116 K 8
Soriano;prov, 117 K 9
Sorli, 53 I 2
Sorn, 49 D 9
Soro, 55 N 6
Soroda, 43 D 11
Sorong, 65 A 12
Soroya;isl., 52 A 7
Sorrento, 103 P 5
Sorris Sorris, 79 K 2
Sorsele, 54 A 10
Sorsogon, 65 C 12
Sortland, 52 D 4
Soruco, 116 J 1
Sorvattnet, 54 F 7
Sorvika, 54 E 6
Sotara;volc., 112 I 4
Soteapan, 107 K 12
Soterio;river, 115 H 12
Sotkamo, 53 J 11
Soto, 116 5
Soto la Marina, 107 G 10
Soto la Marina;river, 107 G 10
Sotuta, 107 I 16
Souanke, 78 B 1
Soubre, 75 L 5
Soudan, 82 E 8
Souderton, 95 C 6
Soufriere, 109 B 15
Soul, 61 E 14
Sound Beach, 95
Soures, 91 K 13
Souris, 89 L 12
Sourlake, 103 O 1
Sousse, 73 D 1
South Africa, 73 N 8
South Alligator;river, 82 B 6
South Amboy, 95 C 8
South Australia;state, 82 H 7
South Bank, 49 H 16
South Bay, 105 N 11
South Bend, 94 C 3
South Boston, 104 E 9
South Branch Patapsco; river, 95 F 1
South Branch Raritan; river, 95 B 7
South Carolina;state, 93 H 12
South Casco, 101 E 14
South Charleston, 100 K 2
South China, 101 E 15
South Coatesville, 95 D 4
South Cumberland;river, 104 E 3
South Dakota;state, 92 E 7
South Derfield, 101 H 12
South Edisto;river, 105 I 6
South Fabius;river, 102 C 4
South Fork Flathead; river, 96 3
South Fork;river, 94 F 6
South Grand;river, 102 E 1
South Haven, 99 J 12
South Hill, 104 E 10
South Holston;river, 104 E 6
South Indian Lake, 89 G 12
South Junction, 98 A 4
South Keeling;isl., 64 M 1
South Kentucky;river, 104 D 4
South Knife;river, 90 A 1
South Korea, 61 F 15
South Lano;river, 97 M 10
South Manitou;isl., 99 F 12
South Mills, 104 E 12
South Milwaukee, 99 I 11
South Molton, 50 J 3
South Nahanni, 88 D 4
South Nahanni;river, 88 C 3
South Newport, 105 K 7
South Otterington, 49 I 16
South Paris, 101 E 14
South Pekin, 99 M 9
South Petherton, 50 J 6
South Pittsburg, 104 G 2
South Platte;river, 96 G 8
South Ponte Vedra Beach, 105 M 7
South Poraipine, 99 A 16
South Porcupine, 90 I 5
South Portland, 101 F 14
South Queensferry, 49 B 11
South Range, 98 D 7
South Red;river, 102 G 9
South River, 90 K 5
South Ronaldsay;isl., 46 E 8
South Saint Paul, 98 G 6
South Sandwich;isl., 111 O 10
South Saskatchewan;river, 86 K 6
South Seal;river, 89 F 12
South Seaville, 95 G 7
South Shields, 49 F 15
South Sioux City, 98 J 2
South Sleeper Islands, 90 C 6
South Twin;isl., 90 F 6
South Tyne;river, 49 G 13
South Uist;isl., 46 G 5
South Whitley, 99 L 13
South Wilmington, 104 H 10
South;river, 104 G 10
Southam, 51 E 9
Southampton, 99 G 16
Southampton;isl., 89 A 15
Southbank, 88 H 3
Southbridge, 101 H 13
Southbury, 101 I 12
Southdean, 49 E 13
Southeast Depot, 91 J 11
Southend, 48 E 6
Southend on Sea, 51 H 14
Southern Alps;range, 83 L 14
Southern Cross, 82 I 3
Southern Pines, 104 G 9
Southminster, 51 G 14
Southmont, 104 F 8
Southold, 95 A 13
Southport, 49 L 12
Southwell, 51 A 10
Southwold, 51 D 16
Soven, 116 L 5
Sovetskaya Gavan, 57 H 15
Sowerby Birdge, 49 K 15
Soya, 62 A 3
Soyalo, 107 L 14
Soyopa, 106 D 4
Spalding, 98 K 1
Spanish, 99 E 15
Spanish Town, 109 H 9
Spanish;river, 99 D 16
Sparkford, 50 J 6
Sparkman, 103 K 3
Sparks, 105 L 4
Sparrows Point, 95 G 3
Sparta, 98 H 8
Spartanburg, 104 G 6
Sparti, 41 O 7
Spassk Dal'niy, 57 J 14
Spearhill, 89 J 13
Spearman, 97 I 9
Speartish, 96 D 8
Speculator, 101 F 10
Speightstown, 109 B 16
Spenard, 86 E 1
Spence Bay, 87 E 9
Spencer, 98 G 8
Spencer;gulf, 80 K 4
Spencerville, 101 J 1
Spences Bridge, 88 K 4
Spennymoor, 49 G 15
Sperryville, 100 M 7
Spesuitie;isl., 95 F 4
Spey;river, 46 F 8
Speyer, 42 H 4
Speyside, 109 E 16
Spezand, 68 F 1
Spezia;prov, 38 E 4
Spicer;isl., 84 D 8
Spiez, 42 L 4
Spillsby, 51 A 12
Spin Boldak, 71 H 12
Spindale, 104 F 6
Spirit Lake, 94 B 6
Spirit River, 88 H 5
Spisska Nova Ves, 43 I 14
Spisske Podhradie, 43 I 15
Spital, 43 L 9
Spitsbergen;isl., 58 B 5
Spittal, 49 C 14
Spitzberg;isl., 32 C 5
Spiwesk, 86 J 8
Split Lake, 89 G 13
Split Rock, 96 E 6
Spokane, 94 C 6
Spokane;river, 94 B 5
Spondin, 88 K 8
Spoon;river, 98 M 8
Spooner, 98 E 7
Sporting Hill, 95 D 3
Spoted Island, 91 E 15
Spotswood, 95 C 8
Spotsylvania, 104 C 11
Sprague, 98 A 4
Spray, 104 E 8
Spring City, 104 F 3
Spring Dale, 104 C 7
Spring Garden, 104 E 9
Spring Glen, 95 B 2
Spring Grove, 98 I 7
Spring Hill, 102 E 1
Spring Lake, 95 D 9
Spring Point, 109 F 10
Spring Valley, 98 G 6
Spring;river, 102 G 1
Springbrook, 98 E 7
Springburn, 88 H 6
Springdale, 91 H 15
Springer, 97 I 7
Springerville, 97 K 5
Springfield, 94 E 3
Springhill, 91 K 12
Springlands, 113 F 16
Springs, 96 G 13
Springs Creek, 104 D 4
Springsure, 83 F 11
Springvale, 101 F 14
Springview, 96 E 10
Springville, 100 H 6
Spruce Pine, 104 F 6
Spruce;river, 99 A 9
Spry, 95 E 2
Spurger, 103 O 1
Squamish, 88 K 3
Square Islands, 91 F 15
Squatteck, 91 J 10
Squaw Lake, 98 C 5
Squinzano, 39 J 13
Sraith an Domhain, 48 J 3
Srebrnica, 40 F 3
Sretensk, 57 I 12
Sri Lanka;c., 67 P 6
Sri Maddhopur, 68 H 6
Srimange, 69 J 15
Srinagar, 68 D 6
Staaten;river, 83 C 9
Stack Skerry;isl., 46 D 7
Stackpool, 99 B 15
Stacy, 104 D 5
Stadhampton, 51 G 10
Stadhastadhur, 52 B 1
Stadhur, 52 A 2
Stafford, 50 C 7
Stafford Springs, 101 H 12
Staffordsville, 104 C 6
Stagshaw Bank, 49 F 14
Staines, 51 H 11
Staines;pen., 119 L 3
Stainforth, 49 J 14
Stainz, 43 L 10
Stalham, 51 B 16
Stalingrad, 53 I 4
Stalling Busk, 49 I 14
Stalowa Wola, 43 F 16
Stalybridge, 49 L 14
Stamford, 51 C 11
Stamping Ground, 102 E 12
Stamps, 103 K 2
Stanardsville, 100 M 7
Stanberry, 98 M 4
Standish, 99 H 14
Stanford, 96 B 5
Stanford Bridge, 50 E 7
Stanhope, 49 G 14
Stanke Dimitrov, 40 H 7
Stanley, 49 G 15
Stanley Mission, 89 H 10
Stannington, 49 F 15
Stanovoy Khrebet;range, 57 H 12
Stans, 42 L 4
Stansted, 51 F 13
Stanton, 86 D 5
Stantonsburg, 104 F 11
Stanwood, 99 H 13
Stapleford, 50 I 8
Staplehurst, 51 I 14
Staples, 98 E 4
Stapleton, 96 F 9
Star, 103 M 6
Star City, 89 J 11
Star Lake, 101 E 10
Stara Lubovna, 43 H 14
Stara Pazova, 40 E 4
Stara Planina;range, 40 G 7
Stara Zagora, 40 H 10
Starachowice, 43 F 14
Starbuck;isl., 81 F 14
Stargard Szczecinski, 43 C 10
Starheim, 54 F 2
Starina, 43 H 16
Starke, 105 M 6
Starkville, 103 K 7
Starkweather, 98 A 1
Starlake, 99 E 9
Starmyri, 52 B 4
Starouck, 98 F 4
Starr, 104 H 5
Staszow, 43 G 15
State Center, 98 K 6
State College, 100 J 7
State Line, 103 N 8
Stateburg, 104 H 8
Staten;isl., 95 B 9
Statenville, 105 L 5
Statesboro, 105 J 6
Statesville, 104 F 7
Statham, 104 H 4
Statsbuoyen, 54 F 5
Staunton, 50 F 7
Stavanger, 55 I 2
Staveley, 49 M 16
Stavroupolis, 41 J 9
Stawiski, 43 C 15
Steam Boat, 88 E 4
Steamboat Springs, 96 G 6
Stearns, 104 E 3
Steebenville, 100 K 4
Steele, 96 C 9
Steeleville, 102 F 6
Steelpoort, 79 L 8
Steelton, 95 C 2
Steelville, 102 F 4
Steen, 98 H 3
Steen River, 88 E 6
Steenkool, 65 B 13
Steenwijk, 42 C 3
Steep Falls, 101 F 14
Steep Rock Lake, 98 A 7
Steepholm;isl., 50 I 5
Steiermark, 43 K 9
Steinach, 42 L 7
Steinbach, 89 L 14
Steinhansen, 79 K 3
Stella, 104 G 11
Stellarton, 91 K 13
Stephen, 98 B 3
Stephens, 103 K 2
Stephens City, 100 L 7
Stephenson, 99 F 11
Stephenville, 91 I 14
Sterling, 96 C 9
Sterlington, 103 L 3
Stetler, 88 J 7
Steuben, 99 E 12
Stevenage, 51 F 12
Stevens, 95 D 3
Stevens Point, 99 G 9
Stevens Pottery, 105 I 4
Stevenson, 94 D 4
Stevenston, 48 D 8
Stevensville, 95 H 3
Steventon, 51 G 9
Stewardson, 102 D 7
Stewart, 88 G 1
Stewart Valley, 89 K 9
Stewart;isl., 83 M 14
Stewart;river, 86 E 3
Stewarton, 49 C 9
Stewartstown, 95 E 2
Stewartsville, 102 C 1
Stewartville, 98 H 6
Stewiacke, 91 K 12
Steyning, 51 J 12
Steytlerville, 79 O 6
Stibb Cross, 50 J 2
Stickney, 51 A 12
Stikine;river, 88 E 1
Stilesville, 102 D 10
Still Pond, 95 G 4
Stillington, 49 J 16
Stillmore, 105 J 6
Stillwater, 95 A 7
Stillwater, 97 I 11
Stillwell, 99 K 12
Stilwell, 102 E 1
Stimson, 99 A 16
Stinchar;river, 48 F 8
Stinnett, 97 J 9
Stip, 41 I 6
Stirling, 49 B 10
Stites, 96 C 2
Stjordal, 54 D 6
Stobo, 49 D 11
Stockbridge, 51 J 9
Stockertown, 95 B 6
Stockholm, 55 I 11
Stockholm;land., 55 I 10
Stockport, 49 M 14
Stocksbridge, 49 L 15
Stockton, 50 E 6
Stockton Springs, 101 E 15
Stockton on Tees, 49 H 16
Stockton;isl., 98 D 8
Stockville, 96 G 9
Stod, 42 H 8
Stode, 53 L 4
Stoke Ferry, 51 C 13
Stoke Fleming, 50 M 4
Stoke Mandeville, 51 G 10
Stokenchurch, 51 G 10
Stokes Bay, 99 F 16
Stokes;isl., 118 H 2
Stokesdale, 104 E 8
Stokesley, 49 H 16
Stoksund, 54 C 5
Stolac, 40 G 2
Stolbovoj;isl., 57 D 11
Stomion, 41 K 7
Ston Easton, 50 I 6
Stone, 50 B 7
Stone Harbor, 95 H 7
Stone Mountain, 104 H 3
Stoneboro, 104 H 7
Stonehaven, 46 G 9
Stonehenge, 83 F 10
Stonehouse, 49 C 10
Stoneville, 104 E 8
Stonewal, 103 M 7
Stonewall, 89 K 14
Stoneykirk, 48 G 8
Stonington, 99 F 11
Stony Brook, 95 A 11
Stony Creek, 104 D 11
Stony Plain, 88 I 7
Stony Point, 104 F 7
Stony Rapids, 89 E 10
Stony Run, 95 B 4
Stony Stratford, 51 F 10
Stony;isl., 101 F 9
Store Belt;detroit, 55 N 5
Store Dimon;isl., 54 B 2
Store Lule alv;river, 52 G 6
Store Sotra;isl., 54 H 1
Storen, 54 D 5
Storjuktan, 52 H 4
Storlien, 54 D 6
Storm Lake, 98 J 4
Stornoway, 46 E 7
Storsjo, 54 E 7
Story, 102 D 10
Story City, 98 J '5
Stosch;isl., 119 K 2
Stouchsburg, 95 C 3
Stoughton, 89 L 11
Stour;river, 50 J 7
Stourbridge, 50 D 7
Stourpaine, 50 J 7
Stourport, 50 D 7
Stover, 102 E 3
Stow, 49 C 12
Stowall, 104 E 10
Stowan, 86 K 8
Stowe, 101 E 12
Stowell, 103 P 1
Stowmarket, 51 E 15
Stoystown, 100 K 6
Strabane, 64 G 2
Strachur, 48 A 7
Stradbroke, 51 D 15
Stradsett, 51 C 13
Strahan, 83 L 10
Stranda, 54 E 3
Strangford, 48 I 6
Strangford Lough;loch, 48 H 6
Strangnas, 53 P 4
Stranraer, 48 G 8
Strasbourg, 45 D 13
Strasburg, 96 C 9
Strasswalchen, 42 J 8
Stratford, 37 H 8
Stratford on Avon, 50 E 8
Strathaven, 49 D 10
Strathmere, 95 G 7
Strathmore, 88 K 7
Strathnaver, 88 I 4
Strathroy, 90 M 4
Straton, 50 K 2
Stratton, 101 D 14
Strausstown, 95 C 3
Strawberry, 98 J 7
Strawberry;river, 102 H 4
Strawn, 99 M 10
Streaky Bay, 82 I 7
Streatley, 51 H 10
Streator, 99 L 10
Stredocesky;prov, 43 H 9
Stredoslovensky;prov, 43 I 13
Streiul;river, 40 D 7
Strelka-Cuna, 57 G 10
Strensall, 49 J 16
Stretford, 49 M 14
Streton, 51 A 9
Stribro, 42 H 8
Strickland;river, 83 B 12
Strimon;river, 41 J 8
Strimonikos Kolpos;gulf, 41 J 8
Stringer, 103 M 7
Stroeder, 118 E 8
Stromboli;isl., 39 L 10
Stromness, 46 D 8
Stromo;isl., 54 B 1
Strong, 103 K 8
Strong;river, 103 M 6
Stronghurst, 98 L 8
Strongsville, 100 I 3
Stronsay;isl., 46 D 8
Strood, 51 H 13
Strother, 104 H 7
Stroud, 50 G 7
Stroudsburg, 95 A 6
Struma;river, 41 I 7
Strumica, 41 I 7
Stryker, 99 K 14
Strzegom, 43 F 11
Strzyzow, 43 G-15
Stuart, 96 E 10
Stuart;isl., 86 B 1
Stuart;river, 88 H 3
Stuartburn, 98 A 3
Stuarts Draft, 104 C 9

Stuarts Draft,

Studland, 50 K 8
Studley, 50 E 8
Stumpy Point, 104 F 13
Sturgeon Bay, 99 G 11
Sturgeon Falls, 90 K 5
Sturgeon Landing, 89 I 11
Sturgeon; river, 99 C 16
Sturgis, 89 J 11
Sturminster Newton, 50 J 7
Sturt; river, 82 D 5
Stuttgart, 103 J 4
Stykkisholmur, 52 B 1
Su-kkur, 68 H 2
Suaita, 112 F 6
Suakin, 77 F 10
Suam; river, 78 B 10
Suapure; river, 113 E 10
Suaqui Grande, 106 D 4
Suata; river, 113 D 11
Subanxirihe; river, 69 G 16
Subcuti, 112 D 3
Subei, 60 F 7
Subi; isl., 64 H 6
Subk, 77 E 11
Sublette, 97 I 9
Subotica, 40 C 4
Subterraneo, 106 C 5
Sucarnoochee; river, 103 L 8
Succasunna, 95 A 7
Suceava, 40 A 10
Suchana, 61 B 9
Sucila, 107 I 16
Sucio; river, 112 E 3
Sucre, 112 K 1
Sucre, 112 D 5
Sucre; state, 113 C 12
Sucuaro, 113 G 9
Sucunduri; river, 115 D 15
Sudan, 97 K 8
Sudan; c., 76 H 5
Sudbury, 50 B 8
Suddie, 113 E 15
Sudlersville, 95 G 4
Sue; river, 76 L 7
Suedberg, 95 B 3
Sueyoshi, 63 O 2
Suffern, 101 I 11
Suffield, 88 L 8
Suffolk, 104 E 12
Sufton, 51 H 12
Sugar Creek; river, 99 M 12
Sugar Tree, 102 H 8
Sugar Valley, 104 G 2
Sugar; river, 99 J 9
Sugartown, 103 O 2
Sugluk, 87 G 12
Sugut; river, 65 F 9
Suhait, 64 I 7
Suhar, 71 J 9
Sui, 68 G 3
Suibara, 62 G 11
Suiding, 60 D 4
Suigam, 68 J 4
Suileng, 61 C 14
Suipacha, 114 L 3
Suiteh, 61 G 10
Suitland, 95 H 1
Suiyang, 61 C 15
Sujangarh, 68 H 5
Sujanpur, 68 E 7
Sujawal, 68 J 2
Sukadana, 64 J 6
Sukagawa, 62 H 12
Sukamara, 64 J 7
Sukaradja, 64 J 7
Sukau, 65 G 10
Sukhna, 76 B 2
Sukhona; river, 58 D 3
Sukkertoppen, 87 E 14
Sukki, 62 E 1
Sula Sgeir; isl., 46 D 6
Sula; isl., 54 G 1
Sula; isl., 80 F 2
Sulabesi; isl., 65 J 13
Sulanheer, 61 E 10
Sulat, 65 D 13
Sulawesi Selatan; prov, 65 J 10
Sulawesi Utara; prov, 65 I 11
Sulawesi; isl., 65 J 10
Sulayyil As-, 70 K 5
Sulby; river, 49 I 9
Sule Skerry; isl., 46 D 7
Sulecin, 43 D 10
Suleskar, 55 I 3
Suli, 60 E 4
Sulina, 40 D 13
Sulitelma, 52 F 4
Sullana, 114 C 2
Sulligent, 103 K 8
Sullivan, 100 I 3
Sulphur, 94 G 5
Sulphur; river, 103 L 1
Sultan, 99 C 15
Sultanpur, 66 H 13
Sulu, 83 B 15
Sulu Archipelago, 65 G 11
Sulu Sea, 65 E 10
Suluq, 76 A 3
Sulzbach, 42 H 7
Sumaca, 112 F 6
Sumaeima, 115 E 15
Sumampa, 116 H 6
Sumatera Selatan; prov, 64 K 4
Sumatera; prov, 64 I 2
Sumatra, 105 M 2
Sumatra; isl., 59 M 10
Sumba; isl., 65 M 16
Sumbat, 77 C 12
Sumbawa, 65 M 9
Sumbawa; isl., 65 M 9
Sumbay, 114 K 8
Sumburi, 100 J 8
Sumeg, 43 L 12
Sumen, 40 G 11
Suminit, 98 F 2
Sumiton, 103 K 9
Summer, 105 N 5
Summer Shade, 104 D 2
Summer; isl., 99 F 12
Summerland, 88 L 4
Summerside, 91 K 12
Summersville, 100 M 4
Summerton, 105 I 8
Summertown, 103 I 9
Summerville, 104 H 2
Summit, 86 D 2
Summit Hill, 95 B 4
Summit Lake, 88 E 4
Summit Station, 95 B 3
Summitville, 99 M 13
Sumner, 98 J 7
Sumpter, 94 E 5
Sumrall, 103 N 6
Sumter, 104 H 8
Sun, 103 O 6
Sun City, 97 K 3
Sun Kosi; river, 69 H 12Sun Prairie, 99 I 9
Sun; river, 96 B 4
Sunaer, 60 F 3
Sunagawa, 62 C 3
Sunam, 68 F 6
Sunamganj, 69 J 16
Sunba, 68 C 8
Sunbright, 104 E 3
Sunburg, 98 F 4
Sunbury, 99 M 15
Sunchales, 116 I 7
Sunciya; river, 112 I 5
Suncook, 101 G 13
Sundblad, 116 M 6
Sundbyberg, 55 I 10
Sunderland, 49 F 16
Sundown, 98 A 3
Sundre, 88 K 7
Sundsvall, 54 E 10
Sunflower, 103 K 5
Sunflower; river, 103 L 5
Sungaibatu, 64 I 7
Sungailiat, 64 J 5
Sungaipenuh, 64 J 3
Sungaipinang, 64 I 8
Sungaiselan, 64 J 5
Sungaitiram, 65 I 9
Sunman, 102 D 11
Sunndalsora, 54 E 4
Sunniland, 105 N 11
Sunnynook, 88 K 8
Sunnyside, 91 I 16
Suno; river, 112 J 3
Sunray, 97 J 9
Suntar, 57 G 11
Suntsar, 71 J 11
Sunwe, 61 B 14
Suoche, 60 F 2
Suomi, 99 B 9
Suomussalmi, 53 I 11
Supamo; river, 113 E 13
Supe, 114 B 8
Superb, 89 K 9
Superior, 94 C 7
Suphka, 66 H 11
Supiori; isl., 65 A 14
Supu, 65 H 14
Surab, 68 G 1
Surad, 77 B 11
Surada, 69 M 11
Suramana, 65 J 10
Suratgarh, 66 H 3
Surbiton, 51 H 12
Surcabamba, 114 H 6
Surendranagar, 68 K 4
Surf City, 95 F 9
Surgoinsville, 104 E 5
Surgut, 56 F 7
Surigao, 65 E 13
Surimena, 112 G 6
Suriname; c., 113 G 16
Surkhet, 69 G 9
Surnadal, 54 D 4
Surotgarh, 68 G 5
Surprise, 88 D 1
Surprise Valley, 94 G 4
Surrency, 105 K 6
Surry, 104 D 12
Surt, 76 B 2
Surtanahu, 68 I 3
Suru; river, 66 D 12
Surumu; river, 113 G 14
Surup, 65 F 13
Susacon, 112 F 6
Susitna; river, 86 D 2
Susquehanna, 101 H 9
Susquehanna; river, 95 E 3
Susques, 116 D 3
Sussex, 91 K 11
Susuman, 57 E 14
Suterton, 51 B 12
Sutherlin, 94 F 3
Sutlej; river, 68 F 5
Sutsu, 62 D 2
Sutton, 98 L 1
Sutton, 49 I 16
Sutton Coldfield, 51 D 9
Sutton Scotney, 51 I 9
Sutton in Ashfield, 51 A 9
Sutton; river, 90 E 4
Suttor; river, 83 E 11
Suvorov Islands, 83 E 16
Suwalki, 43 B 15
Suwannee, 105 N 4
Suwannee; river, 105 N 5
Suyo, 114 C 2
Suzu, 62 G 9
Svaerholthalveya; pen., 52 A 9
Svappavaara, 52 E 7
Svarta, 55 J 8
Svartarkot, 52 B 3
Svartenhuk Halvo; pen., 87 C 12
Sveg, 53 L 2
Svelgen, 54 F 2
Svelo, 55 I 2
Svelvik, 55 I 5
Svendborg, 55 O 5
Sverdrup; isl., 84 C 7
Svetozarevo, 40 F 5
Svihov, 42 H 8
Svilengrad, 41 I 10
Svina; isl., 54 A 2
Svobodny, 57 H 13
Svolvaer, 52 D 3
Svullrya, 53 N 1
Swa Tenda, 78 E 3
Swadlincote, 51 C 9
Swaffham, 51 C 14
Swains; isl., 83 E 16
Swainsboro, 105 J 5
Swainsthorpe, 51 C 15
Swakop; river, 79 K 2
Swakopmund, 79 K 2
Swale; river, 49 I 5
Swan Hills, 88 I 6
Swan Islands; isl., 108 H 6
Swan Quarter, 104 F 12
Swan River, 89 J 12
Swan; river, 82 I 2
Swanage, 50 L 8
Swannanoa, 104 F 5
Swans; isl., 101 E 16
Swansboro, 104 G 11
Swansea, 50 G 3
Swanton, 99 K 14
Swarzedz, 43 D 11
Swat; river, 68 C 5
Swaziland; c., 79 M 8
Sweden, 55 I 6
Swedesboro, 95 E 6
Swedru, 75 M 8
Sweetwater, 104 F 3
Swepsonville, 104 F 9
Swell; pen., 119 J 2
Swidnica, 43 F 11
Swidwin, 43 B 10
Swiecie, 43 C 12
Swift Current, 89 L 9
Swift River, 88 D 2
Swifton, 102 H 5
Swindon, 50 H 8
Swineshead, 51 B 12
Swinton, 49 C 13
Switzerland; c., 76 L 5
Sycamore, 99 L 15
Sycow, 43 F 12
Sydero; isl., 54 B 2
Sydney, 91 K 14
Sydney Mines, 91 J 13
Syiva, 104 F 5
Sykesville, 95 F 1
Sykesville, 100 J 6
Sylne, 49 J 13
Sylvan Lake, 88 J 7
Sylvania, 105 J 6
Sylvester, 105 K 4
Symon, 106 G 8
Synders, 95 4
Syracuse, 96 H 9
Syria; c., 70 D 5
Siyriam, 67 L 12
Syrian; desert, 58 H 1
Syston, 51 C 10
Sytle, 54 E 3
Szamotury, 43 D 11
Szany, 43 K 12
Szarvos, 43 L 15
Szczebrceszyn, 43 F 16
Szczecin, 43 C 9
Szczecinek, 43 B 11
Szczekoeiny, 43 G 14
Szeged, 43 L 14
Szeghalom, 43 K 15
Szekesfehervar, 43 K 13
Szekszard, 43 L 13
Szendro, 43 J 15
Szentes, 43 L 14
Szentgotthard, 43 L 11
Szentlorinc, 43 M 12
Szerencs, 43 J 15
Szigetvar, 43 M 12
Szikszo, 43 J 15
Szombathely, 43 K 11
Szprotawa, 43 E 10
Sztum, 43 B 13
Szubin, 43 C 12
Szydlowiec, 43 F 14
Tab, 43 L 12
Tabacal, 116 D 5
Tabaco, 65 C 12
Tabaconas, 114 D 3
Tabacundo, 112 J 3
Tabajara, 115 F 13
Tabalosos, 114 D 4
Tabankort, 74 H 8
Tabaquite, 108 L 2
Tabar Islands, 83 A 15
Tabas, 71 F 9
Tabasca, 113 D 13
Tabasco; state, 107 K 13
Taber, 88 L 8
Tabili, 78 C 7
Tabir; river, 64 J 3
Tabitenea; isl., 81 E 9
Tablas Las-, 112 E 1
Tablas; isl., 65 D 11
Tabor, 98 L 3
Tabor City, 104 H 9
Tabu, 75 M 5
Tacheng, 60 C 4
Tachichilte; isl., 106 F 4
Tachira; state, 112 E 6
Tachourou, 75 K 9
Tachov, 42 H 8
Tacloban, 65 D 13
Tacna, 114 L 8
Taco Pzo, 116 F 6
Tacoma, 94 C 3
Tacopaya, 115 L 11
Tacotalpa, 107 K 14
Tacuaras, 116 F 8
Tacuarembo, 117 J 10
Tacuarembo; prov, 117 J 10
Tacuarembo; river, 117 J 10
Tacuati, 117 D 9
Tacupeto, 106 D 4
Tacutu; river, 113 H 14
Tadami, 62 G 11
Tadami; river, 62 G 12
Tadcaster, 49 K 16
Tadjakant, 74 G 2
Tadjemout, 74 E 10
Tadjerhi, 76 D 1
Tadjoura, 77 I 12
Tadley, 51 I 10
Tadmarton, 51 F 9
Tado, 112 G 4
Tadoussac, 91 J 9
Taf; river, 50 G 2
Tafahi; isl., 81 H 11
Taff; river, 50 G 4
Tafi Viejo, 116 G 4
Taft, 94 K 5
Taga Dzong, 69 H 14
Tagbilaran, 65 E 12
Taggart; isl., 119 K 2
Tagna, 112 L 7
Tagna; river, 112 K 7
Tagua, 115 M 10
Tagua La-, 112 J 5
Taguke, 60 H 4
Tagula; isl., 83 D 15
Tahiti; isl., 81 I 14
Tahleguah, 97 I 13
Tahltan, 88 E 1
Tahoka, 97 K 9
Tahoua, 75 I 10
Tahquamenon; river, 99 D 13
Tahrud, 71 H 9
Tahuamanu, 115 H 9
Tahuamanu; river, 115 G 9
Tahuere; isl., 81 I 16
Tahulandang; isl., 65 H 13
Tahuna, 65 H 13
Tai, 75 L 5
Taibahat, 68 J 8
Taileleo, 64 J 2
Taili, 69 G 16
Taim, 117 K 11
Taio, 117 G 13
Taiping, 64 G 2
Tais, 64 K 3
Taitao; pen., 119 I 2
Taiwan; c., 61 K 14
Taiwan; isl., 59 J 13
Taiwara, 71 G 12
Taiyuan, 61 F 11
Taizhaozong, 60 I 6
Tajibo, 115 I 11
Tajima, 62 H 12
Tajjal, 68 H 2
Tajset, 57 H 9
Takada, 62 G 12
Takagushi, 63 L 1
Takahama, 63 N 1
Takaharu, 63 N 2
Takama, 113 F 15
Takanabe, 63 N 2
Takanosu, 62 D 12
Takaoka, 63 N 2
Takarabe, 63 O 2
Takasaki, 63 N 2
Takatu; river, 113 H 14
Takaw, 67 J 12
Take; isl., 63 P 1
Takela, 60 F 2
Takelama Ganshamo; desert, 60 E 3
Takenake, 60 G 3
Takev, 63 M 1
Takeze; river, 77 H 10
Taki, 83 B 16
Takieta, 75 J 11
Takikawa, 62 C 3
Takingeun, 64 G 1
Takinoue, 62 C 4
Takipy, 89 H 12
Takko, 62 D 13
Takla Landing, 88 G 3
Takla Makan; desert, 59 I 7
Takoma Park, 95 H 1
Taku, 63 L 1
Takum, 75 L 12
Takutea; isl., 81 I 13
Takutu, 113 E 14
Tal, 66 C 7
Tala, 77 D 11
Talafa, 77 D 15
Talagang, 66 G 7
Talagante, 116 L 2
Talamanca; range, 108 L 6
Talamuyuna, 116 I 4
Talan, 64 I 6
Talar Grande, 116 E 3
Talara, 114 C 2
Talasea, 83 B 14
Talata Mafara, 75 J 10
Talaud; isl., 80 E 2
Talawdi, 76 J 7
Talayan, 65 F 12
Talbot; isl., 105 M 7
Talbotton, 105 J 3
Talca, 116 M 2
Talcahuano, 118 C 2
Talcher, 69 M 12
Talgatalga, 82 E 3
Taliabu; isl., 65 J 12
Talina, 114 L 3
Talisayan, 65 E 12
Taliwang, 65 M 9
Talkeetna, 86 D 2
Talkha, 77 B 12
Talkot, 69 G 9
Tall Uwainat, 70 D 5
Tallahassee, 105 M 3
Tallapoosa, 104 H 2
Tallapoosa; river, 105 J 1
Tallasee, 105 J 1
Talleyville, 95 E 5
Tallington, 51 C 11
Tallulah, 103 M 4
Talmage, 98 L 3
Talmin, 38 C 8
Taloda, 68 L 5
Talode, 68 M 6
Taloga, 97 I 10
Talok, 65 I 10
Talorza, 74 G 3
Talpa de Allende, 106 I 6
Talparo, 108 L 2
Taltal, 116 F 1
Taltal; river, 116 F 1
Taltson; river, 89 D 9
Talu, 64 I 2
Taludaa, 65 I 12
Taluk, 64 I 3
Taluqan, 71 E 13
Talvik, 52 B 7
Taly-Kurgan, 56 J 6
Talybont, 50 D 3
Tama, 116 I 4
Tama; river, 63 I 11
Tamagawa, 62 D 12
Tamagua, 101 J 9
Tamala, 82 G 1
Tamalai, 77 D 11
Tamalameque, 112 D 5
Tamale, 75 K 8
Tamana, 63 M 1
Tamana; isl., 81 E 10
Tamanaco; river, 113 D 10
Tamangueyu, 118 D 10
Tamaniqua, 115 B 11
Tamanrasset, 74 F 10
Tamaqua, 95 B 4
Tamar; river, 50 L 2
Tamara, 112 F 7
Tamarack, 98 D 6
Tamarike, 65 E 16
Tamaroa, 102 E 7
Tamarome, 65 B 13
Tamasi, 43 L 13
Tamassee, 104 G 5
Tamatave, 79 N 13
Tamaulipas; state, 107 G 10
Tamaya; river, 114 F 6
Tamazula, 106 F 4
Tamazula de Gordiano, 106 J 7
Tamazulapan, 107 K 11
Tamazunchale, 107 I 10
Tamba; river, 69 H 12
Tambacounda, 75 I 3
Tambalan, 65 H 9
Tambelan; isl., 64 I 5
Tamberias, 116 J 2
Tambilahan, 64 I 3
Tambisan, 65 G 10
Tambo, 83 F 11
Tambo Gmnde, 114 C 2
Tambo de Mora, 114 I 5
Tambo; river, 114 G 6
Tambobamba, 114 I 7
Tambohorano, 79 I 13
Tambolongeng; isl., 65 L 10
Tambopata; river, 115 I 9
Tambores, 117 J 9
Tamboryacu; river, 114 A 6
Tambura, 76 L 6
Tamchakett, 74 H 4
Tame, 112 F 7
Tamesi; river, 107 H 10
Tamhlacht, 48 M 4
Tamiahua, 107 I 11
Tamiami; canal, 105 O 12
Tamil Nadu; state, 67 N 5
Tampa, 105 P 5
Tampere, 53 M 8
Tampico, 107 H 10
Tampin, 64 H 3
Tamshiyacu, 114 C 7
Tamsweg, 43 L 9
Tamu, 67 I 10
Tamuin, 107 H 10
Tamur; river, 69 H 13
Tamworth, 50 C 8
Tana, 52 B 10
Tana; river, 52 B 10
Tanacross, 86 E 3
Tanaga; isl., 84 C 1
Tanagra, 42 D 9
Tanahgrogot, 65 J 9
Tanahmerah, 65 G 9
Tanakeke; isl., 65 L 10
Tanami, 82 E 6
Tanami Desert, 82 D 6
Tanan, 77 E 12
Tanana, 86 C 2
Tanana; river, 86 D 2
Tananarive, 79 N 12
Tanda, 75 L 7
Tandag, 65 E 13
Tandara, 66 C 11
Tanderagee, 48 I 4
Tandia, 74 A 7
Tandiala, 68 E 6
Tandil, 118 C 10
Tandjun, 64 J 8
Tandjung; river, 64 K 8
Tandjungbalai, 64 H 2
Tandjungbatu, 65 H 10
Tandjungkarang, 64 K 4
Tandjungpinang, 64 I 4
Tandl-ungpandan, 64 J 5
Tando Adam, 68 I 2
Tando Bago, 68 J 2
Tando Muhammad Khan, 68 J 2
Tanega; isl., 63 P 2
Taneichi, 62 D 13
Tanel Aike, 119 J 4
Taneytown, 95 E 1
Tanfield, 49 I 15
Tanga, 78 E 12
Tanga Islands, 83 A 15
Tangail, 69 J 14
Tangarana o Pucacuro; river, 114 A 5
Tangba, 60 I 3
Tanggulajiaka, 60 H 6
Tangi, 66 E 6
Tangier, 91 L 13
Tangier; isl., 101 M 9
Tangipahoa; river, 103 O 5
Tangue, 75 J 3
Tangyuan, 61 C 14
Tanimbar; isl., 80 G 3
Taninthari; river, 64 C 2
Tanivama, 63 O 1
Tanjay, 65 E 12
Tankhala, 68 L 5
Tankse, 68 D 7
Tanna-s, 54 E 7
Tannin, 90 H 1
Tano; river, 75 M 7
Tanque, 106 D 7
Tanque Verde, 97 L 4
Tansing, 69 H 11
Tanta, 77 C 11
Tantan, 74 D 4
Tantoyuca, 107 I 10
Tanunak, 86 C 1
Tanzania; c., 78 E 8
Taoerhe; river, 61 C 13
Taormina, 39 M 10
Taos, 97 I 7
Tapacari, 115 L 11
Tapachula, 107 M 14
Tapaktuan, 64 H 1
Tapalquen, 116 M 7
Tapan, 64 J 3
Tapanatepec, 107 L 13
Tapaua, 115 D 12
Tapaua; river, 115 D 12
Tapebicua, 117 H 9
Tapenaga; river, 116 G 8
Taperas, 115 L 14
Tapes, 117 I 12
Tapia, 115 L 16
Tapiche; river, 114 D 6
Tapijulapa, 107 K 14
Tapira, 113 I 9
Tapirapua, 115 J 16
Tapolca, 43 L 12
Tappahannock, 104 C 11
Tappita, 75 L 4
Tapti; river, 59 K 5
Tapul Group; isl., 65 G 11
Tapung Kiri; river, 64 I 2
Tapuruquara, 113 J 11
Taquara, 117 I 12
Taquari, 117 I 12
Tar; river, 104 E 10
Taraba; river, 75 L 12
Tarabuco, 115 M 11
Taraco, 115 J 9
Taragi, 63 N 2
Tarahouahout, 74 F 10
Tarairi, 114 L 4
Tarakan, 65 H 9
Taraken; isl., 65 H 9
Taranaquis, 115 L 15
Taranto, 39 J 12
Taranto; prov, 39 J 11
Tarapaca, 112 L 8
Tarapaca; prov, 115 L 9
Tarapoto, 114 D 5
Taraqua, 113 J 9
Tarariras, 117 L 9
Tarascon-sur-Ariege, 44 L 8
Tarata, 115 L 9
Tarauaca, 114 E 8
Tarauaca; river, 114 E 8
Tarbert, 46 F 5
Tarbolton, 49 D 9
Tarboro, 104 F 11
Tarcoola, 82 H 7
Tardara; river, 112 K 8
Tarendo, 52 F 7
Tarfaia, 74 D 4
Tarheel, 104 G 10
Tarhuna, 76 A 1
Tari, 83 B 12
Tariana, 113 J 9
Tarija, 114 L 3
Tariku; river, 65 B 15
Tarim, 70 M 6
Tarim; river, 58 H 7
Tarin Kot, 109 G 12
Tarina, 77 J 11
Taringuiti, 114 L 4
Tariquia, 114 M 3
Taritatu; river, 65 C 16
Tarja, 57 D 9
Tarkio, 98 M 3
Tarko-Sale, 56 F 8
Tarkwa, 75 M 7
Tarlac, 65 B 11
Tarleton, 49 L 12
Tarma, 114 B 7
Tarnaby, 52 H 3
Tarpon Springs, 105 P 5
Tarporley, 50 A 6
Tarrafal, 75 P 1
Tarragona; prov, 37 D 12
Tarrant, 103 K 10
Tarrytown, 105 J 5
Tartagal, 116 D 5
Tartarskiy Proliv; detroit, 58 E 12
Tarumizu, 63 O 1
Tarut, 77 D 13
Tarutung, 64 H 2
Tarver, 105 L 5
Tarvisio, 38 B 8
Tarvita, 115 M 12
Tas; river, 51 D 15
Tasauz, 56 I 3
Taschereau, 90 K 6
Tash Kurghan, 60 F 2
Tashi Gang Dzong, 69 H 15
Tashiro, 62 D 13
Tashota, 90 G 2
Tasinge; isl., 55 O 5
Tasitel, 83 A 14
Tasjo, 53 J 4
Taskul, 83 A 14
Tasman Sea, 80 L 7
Tasmania; state, 83 L 10
Tassagh, 48 I 4
Tastiota; state, 106 D 3
Tatamagouche, 91 K 12
Tatamy, 95 B 6
Tatau; isl., 83 A 15
Tate, 104 H 3
Tatelang, 60 F 4
Tatlayoko Lake, 88 J 3
Tatta, 68 J 2
Tattershall, 51 A 12
Taudenni, 74 F 7
Taumaturgo, 114 F 7
Taung-Gyi, 67 J 12
Taunsa, 68 F 3
Taunton, 50 J 5
Taura, 112 L 2
Tauramena, 112 G 6
Tauranga Taurianova,

Valkeakoski, 53 M 9
Valkeala, 53 N 10
Valladares, 106 B 1
Valladolid, 107 I 16
Valle, 112 M 2
Valle Daza, 118 C 6
Valle El-, 112 F 3
Valle Grande, 115 L 12
Valle Hermoso, 107 F 10
Valle Nacional, 107 K 11
Valle de Bravo, 107 J 9
Valle de Olivos, 106 E 6
Valle de Pascua, 113 D 10
Valle de Topia, 106 F 6
Valle de Zaragoza, 106 E 6
Valle del Rosario, 106 E 6
Valle;river, 116 E 5
Valledupar, 112 C 6
Vallee La-, 98 B 5
Vallenar, 116 H 2
Valles Mines, 102 F 5
Valletta, 39 P 9
Valley, 49 M 9
Valley City, 98 D 2
Valley Forge, 95 D 5
Valley Head, 100 M 5
Valley Station, 102 E 10
Valley View, 104 C 3
Valleyfield, 90 L 7
Valleyview, 88 H 6
Vallnas, 52 H 4
Vallo della Lucania, 39 J 10
Valora, 90 G 1
Valparaiso, 99 K 11
Valsetz, 94 E 3
Valtimo, 53 J 11
Valverde, 109 G 11
Van Alstyne, 97 K 12
Van Buren, 97 J 13
Van Diemen Gulf, 82 B 5
Van Etten, 100 H 8
Van Horn, 97 M 7
Van Lear, 104 C 5
Van Wert, 98 L 5
Van Wyck, 104 G 7
Vanaja, 53 N 9
Vanavana;isl., 81 J 16
Vanavara, 57 G 10
Vanceboro, 104 F 11
Vanceburg, 100 M 2
Vancleave, 103 O 7
Vancouver, 88 L 3
Vancouver;isl., 88 K 1
Vandalia, 100 K 1
Vandemere, 104 G 12
Vanderbilt, 99 F 13
Vandergrift, 100 J 6
Vanderhoof, 88 H 3
Vanderlin;isl., 82 C 8
Vandervoort, 103 J 1
Vandiola, 115 K 11
Vanduser, 102 G 6
Vanga, 78 D 12
Vangaindra, 79 O 12
Vangunu;isl., 80 G 7
Vanier;isl., 86 C 8
Vanikoro Islands, 83 E 13
Vanikoro;isl., 80 G 8
Vanimo, 83 A 12
Vankaner, 68 K 3
Vanlaiphai, 69 K 16
Vanleer, 102 H 9
Vanna;isl., 52 B 6
Vannes, 44 E 5
Vansant, 104 D 6
Vansbro, 54 H 8
Vansittart;isl., 87 F 10
Vanylen, 54 E 2
Var;river, 45 J 13
Vara, 55 K 7
Varangal, 67 L 5
Varanger;pen., 35 A 9
Varas Las-, 106 I 6
Varberg, 55 L 6
Vardaman, 103 K 7
Vardar;river, 41 I 6
Varde, 55 N 3
Vardo, 52 B 11
Varese;prov, 38 C 3
Varillas, 116 E 1
Varillas Las-, 116 J 6
Varina, 104 F 9
Varjistrask, 52 G 6
Varkaus, 53 L 11
Varmland, 54 H 7
Varna, 40 G 12
Varnado, 103 O 6
Varnamo, 55 L 8
Varnsdorf, 43 F 9
Varnville, 105 J 7
Varoslod, 43 K 12
Varpalota, 43 K 12
Varrag el Arab, 77 F 12
Varzea;river, 117 G 11
Vashti, 98 C 1
Vaslui, 40 C 11
Vassar, 99 I 14
Vassijaure, 52 D 5
Vasugan;river, 56 G 7
Vasvar, 43 L 11
Vatican City, 38 H 6
Vatomandry, 79 N 13
Vatta Dornei, 40 B 9
Vattern, 55 K 8
Vattholma, 53 O 4
Vauclin Le-, 109 M 16
Vaupes;river, 112 I 8
Vauxhall, 88 L 8
Vavau;isl., 81 H 11
Vaygach;isl., 58 C 5
Vealwater, 79 L 7
Veblen, 98 E 2
Vechta, 42 D 5
Veddige, 55 L 7
Vedea;river, 40 F 10
Vedia, 116 L 7
Veedersburg, 99 M 11
Vega, 97 J 8
Vega La-, 109 H 11
Vega;isl., 54 A 6
Vegas;Las-, 94 J 7
Vegreville, 88 I 8
Veinticinco de Mayo, 116 L
Veintiocho de Noviembre, 119 L
Vejle, 55 N 4
Veki Kapusany, 43 I 16
Vela La-, 112 B 8
Vela Luka, 40 G 1
Velardena, 106 F 7
Velazquez, 112 F 5
Veleka;river, 40 H 12
Velez, 112 F 6
Velikaja;river, 57 C 15
Veliko Bonjince, 40 G 6
Veliko Turnovo, 40 G 10
Vella Lavella;isl., 80 F 5
Velva, 96 B 9
Venadillo, 112 G 5
Venado, 107 H 9
Venado Tuerto, 116 K 6
Venancio Aires, 117 I 12
Vendee;river, 44 G 6
Venetie, 86 D 3
Venezia;prov, 38 C 6
Venezuela;c., 112 D 8
Venezuela;gulf, 112 B 7
Vengreen, 50 J 2
Venice, 103 P 7
Venice, 38 C 6
Venison Islands, 91 F 15
Venlo, 98 D 2
Ventnor, 51 L 9
Ventosa La-, 107 L 12
Ventura La-, 107 G 9
Venturai;river, 113 G 10
Venturosa La-, 113 F 9
Venus, 105 M 11
Venustiano Carranza, 107 L 14
Vera, 116 H 7
Veracruz;state, 107 I 10
Veranopolis, 117 H 12
Veranos, 106 G 6
Veraval, 68 L 3
Verbena, 105 J 1
Vercelli;prov, 38 C 2
Verchoyansk, 57 E 12
Verde La-, 116 M 5
Verde;bay, 91 H 15
Verde;isl., 112 B 5
Verde;river, 97 K 3
Verdigre, 98 J 1
Verdura, 106 F 4
Vergara, 117 K 11
Vergas, 98 D 3
Vergennes, 101 E 11
Vermilion, 88 J 8
Vermilion Chutes, 88 F 7
Vermilion;river, 100 I 3
Vermillion, 98 I 2
Vermillion;river, 98 I 2
Vermont, 98 M 8
Vermont;state, 93 E 13
Vernal, 96 G 5
Verne Lu-, 98 I 5
Vernigerode, 42 E 6
Vernon, 88 K 5
Vero Beach, 105 P 8
Veroia, 41 J 6
Verona, 90 L 6
Verona;prov, 38 C 5
Veronica, 117 L 9
Versailes, 96 H 13
Versalles, 115 H 13
Versalles;river, 114 H 7
Vertou, 44 F 6
Verwood, 50 J 8
Vesique, 114 F 3
Vesper, 99 G 9
Vesta, 104 E 8
Vesteralen;isl., 52 D 3
Vesterhavet;sea, 55 L 1
Vestmannaeyjar, 52 C 2
Vestmannaeyjar;archp., 52 C
Vestregoy;isl., 52 D 3
Vesuvius, 104 C 9
Veszprem, 43 K 12
Vetasjarvi, 52 F 7
Veteran, 88 J 8
Vetlanda, 55 L 9
Vetovo, 40 F 10
Vevay, 102 E 11
Veveno;river, 76 L 8
Viacha, 115 K 10
Viale, 116 J 7
Viamonte, 116 K 6
Vibank, 89 K 11
Vibo Valentia, 39 L 11
Vibora La-, 106 E 7
Viborg, 55 M 4
Vicco, 104 D 5
Vichadero, 117 J 10
Vichigasta, 116 I 3
Vichina;river, 116 H 3
Vichoca, 114 L 3
Vichuquen, 116 M 1
Vichy, 45 G 10
Vicksburg, 99 J 13
Victor, 96 E 4
Victoria, 64 G 8
Victoria Downs, 83 E 11
Victoria La-, 112 E 8
Victoria Mine, 99 D 16
Victoria Nile;river, 78 B 9
Victoria Point, 67 N 13
Victoria River Downs, 82 C
Victoria de Durango, 106 G 6
Victoria de las Tunas, 109 G 9
Victoria;isl., 119 I 2
Victoria;river, 82 C 6
Victoria;state, 83 J 9
Victoriaville, 91 K 9
Victorica, 116 M 5
Victorino, 113 H 9
Victorino de la Plaza, 116 M 5
Vicuna, 116 I 2
Vicuna Mackenna, 116 K 5
Vida, 40 F 10
Vida Dela-;isl., 114 F 3
Vidal Gormaz;isl., 119 M 2
Vidalia, 103 N 4
Videira, 117 G 12
Videla, 116 I 7
Videre;isl., 54 A 2
Viderejde, 54 A 2
Vidhidalsa;river, 52 B 2
Vidhidalur, 52 A 4
Vidin, 40 F 7
Vidor, 103 O 1
Vidoure;river, 45 J 10
Vidrios Los-, 106 A 2
Viedma, 118 F 7
Vieja Carmen, 119 N 5
Viejo, 106 D 4
Vienna, 98 G 2
Vienna, 43 J 11
Vienne;river, 44 F 7
Viento;range, 118 C 4
Vientos Los-, 116 E 2
Vieques;isl., 109 H 14
Vierema, 53 J 10
Vierzon, 45 F 9
Viesca, 106 F 8
Vietnam, 64 D 6
Vieux Comptoir, 90 F 6
Vieux Fort, 109 M 11
Vieux Habitants, 109 M 11
Viga La-, 107 H 9
Vigan, 65 A 11
Vigia Chico, 107 J 16
Vigia El-, 112 D 7
Vigia de Curvarado, 112 E 4
Vigia;isl., 119 L 9
Vihanti, 53 I 9
Vihelmina, 53 I 4
Vihowa, 68 F 3
Vihti, 53 O 9
Viitasaari, 53 K 10
Vik, 52 C 3
Viking, 88 J 8
Vikna, 54 6
Viksmanshvttan, 54 H 9
Vila Arriaga, 78 H 1
Vila Cabral, 78 H 10
Vila Conecicao, 113 J 12
Vila Coutinho, 78 H 10
Vila Fontes, 79 I 10
Vila General Machado, 78 G
Vila Junqueiro, 79 I 11
Vila Marechal Carmona, 78 F 2
Vila Mariano Machado, 78 H 2
Vila Murtinho, 115 G 11
Vila Norton de Matos, 78 G 2
Vila Nova do Seles, 78 G 2
Vila Paiva Couceiro, 78 H 2
Vila Pereira d'Eca, 79 I 2
Vila Rica, 115 B 14
Vila Teixeira da Silva, 78 G 2
Vila Verissimo Sarmento, 78 F 4
Vila da Ribeira Brava, 75 P 1
Vila de Manatuto, 65 M 13
Vila de Mocuba, 79 I 11
Vila de Sena, 79 I 10
Vila do Chinde, 79 J 10
Vila do Dondo, 79 J 10
Vilacaya, 115 M 11
Vilaine;river, 44 E 5
Vilancuios, 79 K 10
Vilavila, 114 L 2
Vilcabamba;range, 114 H 6
Vilcanota;river, 114 I 8
Vileany, 43 M 13
Vilelas, 116 G 6
Vilhena, 115 H 14
Vilini;river, 58 D 10
Villa Abecia, 114 L 3
Villa Aberastain, 116 J 3
Villa Alberdi, 116 G 4
Villa Allende, 107 F 9
Villa Ana, 116 H 8
Villa Angela, 116 G 7
Villa Atamisqui, 116 H 5
Villa Atuel, 116 L 3
Villa Azueta, 107 K 12
Villa Bella, 115 G 11
Villa Brana, 116 G 6
Villa Brugua, 112 D 8
Villa Bustos, 116 H 4
Villa Canas, 116 K 6
Villa Carios Paz, 116 J 5
Villa Castelli, 116 H 3
Villa Cisneros, 74 E 2
Villa Colon, 116 J 3
Villa Constitucion, 106 F 2
Villa Corona, 106 G 6
Villa Coronado, 106 E 6
Villa Dolores, 116 J 5
Villa Escalante, 106 J 8
Villa Escobedo, 106 E 6
Villa Federal, 116 I 8
Villa Flores, 107 L 13
Villa Florida, 117 F 9
Villa Frontera, 106 E 8
Villa General Roca, 116 K 4
Villa Gonzalez Ortega, 106 H 8
Villa Guillermina, 116 G 8
Villa Hayes, 117 E 9
Villa Hernandarias, 116 J 7
Villa Hidalgo, 106 E 6
Villa Huidobro, 116 L 5
Villa Independencia, 116 J 3
Villa Ingavi, 114 M 4
Villa Iris, 118 D 7
Villa Jose Cardel, 107 J 11
Villa Juarez, 107 H 9
Villa Klause, 116 J 3
Villa La Angostura, 118 F 3
Villa Lopez, 106 E 6
Villa Mainero, 107 G 9
Villa Maria, 116 J 6
Villa Maria Grande, 116 J 7
Villa Martin, 114 L 2
Villa Matamoros, 106 E 6
Villa Minetti, 116 H 6
Villa Montes, 114 L 4
Villa Nueva, 116 K 6
Villa Ocampo, 106 E 6
Villa Oliva, 117 F 9
Villa Portales, 118 D 3
Villa Ramirez, 116 J 7
Villa Regina, 118 D 5
Villa Rey, 117 E 9
Villa Rica, 115 G 10
Villa Rio Hondo, 116 G 5
Villa San Giovanni, 39 M 10
Villa San Jose, 116 J 8
Villa San Martin, 116 H 5
Villa Sauze, 116 L 6
Villa Sena, 116 M 6
Villa Serrano, 115 M 12
Villa Traful, 118 F 3
Villa Tulumba, 116 I 5
Villa Tunari, 115 K 11
Villa Union, 106 G 7
Villa Vaca Guzman, 115 M 12
Villa Valeria, 116 L 5
Villa Victoria, 107 I 9
Villa Zorraquin, 117 J 9
Villa de Allende, 107 J 9
Villa de Comaltitlan, 107 M 14
Villa de Coss, 106 H 8
Villa de Cura, 113 C 10
Villa de Guadalupe, 107 H 9
Villa de Mendez, 107 F 10
Villa de Praga, 116 K 4
Villa de Ramos, 106 H 8
Villa de Reyes, 107 I 9
Villa del Pueblito, 107 I 9
Villa del Rosario, 116 J 6
Villach, 43 L 9
Village, 103 O 1
Village The-, 97 J 11
Villagran, 107 G 9
Villaguay, 116 J 8
Villahermosa, 107 K 14
Villaldama, 107 E 9
Villalonga, 118 E 8
Villano;isl., 37 A 9
Villano;river, 112 K 3
Villanueva, 106 H 7
Villapiana, 39 K 11
Villar, 115 M 12
Villarica, 117 F 9
Villarrica, 118 E 3
Villas, 95 H 7
Villas Las-;prov, 108 F 8
Villavicencio, 112 G 6
Villavieja, 112 H 5
Villazon, 114 M 3
Ville Marie, 90 J 5
Ville Platte, 103 O 3
Villeparisis, 45 C 9
Villeta, 117 F 9
Villisca, 98 L 4
Vilna, 88 I 8
Vilos Los-, 116 K 1
Vilppula, 53 L 9
Vilshofen, 42 I 8
Viluse, 40 G 3
Vilyujsk, 57 F 11
Vilyuy;river, 57 F 12
Viola, 95 H 5
Violet Grove, 88 J 6
Vipos, 116 F 4
Viqueque, 65 M 13
Virden, 89 L 12
Vire, 44 C 6
Vire;river, 44 C 6
Virgil, 98 G 1
Virgilina, 104 E 9
Virgin Gorda;isl., 109 A 9
Virgin Islands, 109 B 9
Virgin;river, 97 I 2
Virginia, 98 C 6
Virginia City, 96 D 4
Virginia;state, 93 G 12
Virginiatown, 90 I 5
Viroque, 98 I 8
Virrat, 53 L 8
Virtaniemi, 52 D 10
Viru, 114 F 3
Viru;river, 114 F 3
Visavadar, 68 L 3
Visayan Sea, 65 D 12
Viscount, 89 J 10
Visegrad, 40 F 3
Vishakhapatnam, 67 L 7
Visnaga, 106 D 3
Vista, 102 F 2
Vista Alegre, 113 I 13
Vit;river, 40 G 8
Viterbo;prov, 38 G 6
Viti Levu;, 81 H 10
Vitichi, 114 K 3
Vitor, 114 K 7
Vitor;river, 114 K 7
Vitre, 44 D 6
Vitry-le-Francois, 45 D 11
Vittangi, 52 E 7
Vitteaux, 45 E 11
Vittoria, 39 N 9
Vitu Islands, 83 B 14
Vitvattnet, 52 G 8
Vivian, 103 L 1
Vivorata, 118 D 10
Vivoritica, 40 D 2
Vizcaino;desert, 106 D 1
Vizcaya;gulf, 37 A 9
Vizcaya;prov, 37 B 9
Vizianagram, 67 L 7
Vjosa;river, 41 J 4
Vladivostok, 57 J 14
Vlasenica, 40 F 3
Vlore, 41 J 3
Vohemar, 79 L 13
Voi, 78 D 11
Voinjama, 75 L 3
Voitsberg, 43 L 10
Volcan, 116 E 4
Volcan El-, 116 L 2
Volcanica;range, 108 K 5
Volda, 54 E 3
Volta;sea, 75 L 8
Vordingborg, 55 O 6
Voriai Sporadhes;isl., 41 L 8
Vrtoce, 38 D 10
Vukovar, 40 D 3
Vulcan, 40 D 7
Vulcano;isl., 39 L 9
Vuoggatjalme, 52 G 4
Vuolijoki, 53 J 10
Vuotta, 52 G 7
Vuotzo, 52 E 10
Vyborg, 56 C 4
Vyrnwy;river, 50 C 4
Wa, 75 K 7
Wabag, 83 12
Wabana, 91 I 16
Wabasca, 88 H 7
Wabash, 99 L 13
Wabash;river, 102 E 8
Wabasha, 98 G 7
Wabasso, 105 P 8
Wabeno, 99 F 10
Wabi Shebele;river, 77 K 11
Wabiskaw;river, 88 G 7
Wabowden, 89 H 13
Wabron, 88 J 5
Wabrzezno, 43 C 13
Waccamaw, 104 H 10
Waccamaw;river, 104 H 10
Waccasassa;river, 105 N 5
Wach El-, 77 M 11
Wachapreague, 104 C 13
Wacissa, 105 M 3
Wad, 68 H 1
Wad Banda, 76 I 6
Wad Hamid, 76 G 8
Wad Medani, 76 H 8
Wada, 62 D 12
Waddan, 76 B 2
Wadden Eilanden;archp., 42 C 2
Waddesdon, 51 F 10
Waddington, 49 K 13
Waddy, 102 E 11
Wade, 104 G 10
Wadebridge, 50 L 1
Wadena, 98 D 4
Wadesboro, 104 G 8
Wadeville, 104 G 8
Wadi Halfa, 76 E 7
Wadi Shebele;river, 77 L 12
Wading River, 95 A 12
Wading;river, 95 F 8
Wadley, 105 I 2
Wagasaki, 62 F 13
Wagener, 105 I 7
Wager;isl., 119 J 2
Wagin, 82 J 3
Wagner, 98 I 1
Wagoner, 97 I 12
Wagontown, 95 D 4
Wagram, 104 G 9
Wagrowiec, 43 D 11
Wagu, 83 B 12
Wahai, 65 J 14
Wahego;river, 83 H 10
Wahoo, 98 K 2
Wahpeton, 98 E 3
Waianae;river, 65 K 10
Waibeem, 65 A 13
Waigama, 65 J 15
Waigeo;isl., 80 F 3
Waikabubak, 65 M 10
Waikato;river, 83 J 15
Wainfleet, 51 A 13
Waingapu, 65 M 11
Waini;river, 113 E 15
Wainwright, 86 B 3
Waitaki;river, 83 L 14
Waiteville, 104 D 8
Wajam;isl., 65 I 15
Wajiahe;river, 61 E 10
Wajima, 62 H 9
Wajir, 78 B 12
Wak El-, 78 B 12
Waka, 77 B 10
Wakarusa, 99 K 12
Wakaw, 89 J 10
Wakayanagi, 62 F 13
Wake Forest, 104 F 10
Wake;isl., 80 B 8
Wakeeney, 96 H 10
Wakefied, 104 D 11
Wakefield, 49 L 16
Wakeham Bav, 87 G 12
Wakhan Daria;river, 66 A 8
Wakinosawa, 62 C 13
Wakkanai, 62 A 3
Wakre, 65 I 15
Wakulla, 105 M 3
Wakulla Beach, 105 M 3
Wakusimi;river, 99 A 15
Wakuya, 62 F 13
Walbridge, 100 I 2
Walbrzych, 43 F 11
Walcott, 88 H 2
Walcz, 43 C 11
Walden, 96 G 6
Waldkirchen, 43 I 9
Waldo, 100 J 2
Waldron, 103 I 1
Waleabahi;isl., 65 I 11
Wales, 98 D 7
Wales;isl., 87 F 10
Waleska, 104 H 3
Walewale, 75 K 8
Walhalla, 98 A 2
Walhallow, 82 D 8
Walikale, 78 C 7
Walkaway, 82 H 2
Walker, 98 D 5
Walkerton, 90 L 4
Walkertown, 104 E 8
Walkerville, 99 H 12
Wall, 96 E 8
Wall Lake, 98 J 4
Walla Walla, 94 D 5
Wallace, 90 K 6
Wallaceburg, 90 M 4
Wallaceburv, 99 J 16
Wallal, 82 E 3
Wallasey, 49 M 12
Walley Stream, 95 B 10
Walleyview, 86 I 5
Wallingford, 51 H 10
Wallis;isl., 81 G 10
Wallkill;river, 101 I 11
Wallndse, 49 F 15
Wallowa, 94 D 5
Walls, 103 J 6
Walney;isl., 49 J 11
Walnut, 99 K 9
Walnut Cove, 104 E 8
Walnut Grove, 102 F 2
Walnut Hill, 103 O 9
Walnut Ridge, 102 H 5
Walnutport, 95 B 5
Walsall, 50 C 8
Walsenburg, 97 I 7
Walsingham, 51 B 14
Walterboro, 105 J 7
Walters, 104 E 12
Walthall, 103 K 7
Walthill, 98 J 2
Walton, 91 K 12
Walton Junction, 99 G 13
Walton on the Naze, 51 F 15
Waltonville, 102 E 7
Walworth, 99 J 10
Wamba, 78 B 7
Wamba;river, 78 E 3
Wamego, 96 G 12
Wami;river, 78 E 11
Wamma;river, 65 B 14
Wampsville, 101 G 9
Wan, 65 E 15
Wana, 68 E 3
Wanamakers, 95 B 4
Wanaque, 95 A 8
Wanblee, 96 E 9
Wancheese, 104 F 13
Wanda, 117 F 11
Wando;river, 105 J 8
Wandoan, 83 G 12
Wanggaimiao, 61 D 12
Wangi Wangi;isl., 65 L 12
Wangqing, 61 B 15
Wanlockhead, 49 E 10
Wanopo, 83 B 15
Wansbeck;river, 49 F 14
Wantage, 51 H 9
Wapakoneta, 100 J 1
Wapato, 94 D 4
Wapella, 89 K 11
Wapello, 98 L 7
Wapiti;river, 88 H 5
Wapoga;river, 65 B 14
Wapsipinicon;river, 98 J 7
Wapwallopen, 95 A 4
Waqid, 77 C 10
War, 104 D 6
War Eagle, 104 D 6
Warah, 68 H 2
Warai Post, 66 D 7
Waratah, 83 L 10
Warba, 98 D 6
Warboys, 51 D 12
Warburg, 88 J 7
Warcop, 49 H 13
Wardensville, 100 L 6
Wardington, 51 E 9
Wardlow, 88 K 8
Ware, 51 G 12
Ware Shoals, 104 H 6
Wareham, 50 K 7
Waren, 65 B 14
Warenford, 49 D 14
Waresboro, 105 L 5
Waretown, 95 E 9
Waringstown, 48 I 5
Waris, 65 B 16
Wark, 49 D 13
Warka, 43 E 14
Warkopi, 65 A 13
Warkworth, 49 E 15
Warlingham, 51 I 12
Waroona, 82 I 2
Warra, 83 G-12
Warrawagine, 82 E 3
Warren, 99 J 9
Warren, 98 B 3
Warren, 90 K 5
Warren Glen, 95 B 5
Warren Grove, 95 E 8
Warren Landing, 89 I 13
Warrenpoint, 48 J 5
Warrens, 98 H 8
Warrensburg, 96 G 13
Warrenton, 94 D 3
Warri, 75 M 10
Warrington, 49 M 13
Warrior, 103 K 9
Warroad, 98 A 4
Warsa, 65 A 14
Warsaw, 96 H 13
Warsaw, 43 D 14
Warsop, 51 A 10
Warta;river, 43 D 10
Wartburg, 104 E 3
Warton, 49 K 12
Wartrace, 104 F 1
Waru, 65 K 15
Warwick, 49 G 12
Wasam, 66 A 8
Wasatch Range, 96 G 3
Wasdale, 49 I 11
Wase, 75 K 12